ENCYCLOPEDIA OF PHYSICS

EDITED BY
S. FLÜGGE

VOLUME LIV
ASTROPHYSICS V: MISCELLANEOUS

WITH 181 FIGURES

SPRINGER-VERLAG BERLIN HEIDELBERG GMBH

HANDBUCH DER PHYSIK

HERAUSGEGEBEN VON

S. FLÜGGE

BAND LIV

ASTROPHYSIK V: VERSCHIEDENES

MIT 181 FIGUREN

SPRINGER-VERLAG BERLIN HEIDELBERG GMBH

© by Springer-Verlag Berlin Heidelberg 1962

Ursprünglich erschienen bei Springer-Verlag OHG. Berlin • Gottingen • Heidelberg 1962

Softcover reprint of the hardcover 1st edition 1962

Library of Congress Catalog Card Number A 56 — 2942

ISBN 978-3-642-45986-3   ISBN 978-3-642-45984-9 (eBook)
DOI 10.1007/ 978-3-642-45984-9

Inhaltsverzeichnis.

Verlag und Herausgeber bedauern, daß sich das Erscheinen dieses Bandes durch unvorhersehbare Umstände beträchtlich verzögert hat. Die drei ersten Artikel (Solar Instruments, Radio Astronomy Techniques, Photographic Photometry), welche bereits im Herbst 1957 im Manuskript vorlagen, wurden im Herbst 1960 ausgedruckt, so daß seitdem keine Änderungen mehr an ihnen vorgenommen werden konnten.

Um das Erscheinen des Bandes nicht noch länger aufzuhalten, mußte der ursprünglich geplante umfangreiche Artikel „Telescopes and Spectrographs" wegfallen. Verlag und Herausgeber hoffen, einen derartigen Beitrag noch in Band XXIX (optische Instrumente) veröffentlichen zu können.

Solar Instruments.

By

R. R. McMATH and O. C. MOHLER.

With 27 Figures.

1. Introduction. Nearly every device developed for measurement of terrestrial radiation or for the measurement of small angles has been applied, at some time or other, to the study of the Sun. Only a few of these instruments have met with acceptance and use by astronomers, but it is important to review as many as possible of them, for they may suggest new approaches to some of the problems encountered in the design and construction of instruments for solar observation. We shall try to describe the major types of instruments that have played roles in solar research, and we shall try to indicate the part that each has played in the accumulation of the almost numberless volumes of solar observations. Where we find it possible, we shall attempt to suggest additional uses of the instruments.

Some of the oldest remnants of prehistoric civilizations are structures that were built to assist the observation of solstitial and equinoctial risings and settings of the Sun, but, without question, 1611, the date of the first use of the telescope in the observation of the Sun, marks the start of an epoch in the design and construction of solar instruments. We shall treat first of all those instruments that make no use of image-forming lenses and mirrors, and proceed from these to descriptions and discussions of devices that are designed to try to produce the perfect images now required for significant solar observation.

I. Instruments for the measurement of the total solar radiation.

Instruments for measurement of the total solar radiation received at the surface of Earth are called *pyrheliometers*. Such measurements are fundamental for the determination of the total radiation from the Sun.

2. Pyrheliometers. The distinctive features of pyrheliometers are: 1. The readings of the instruments can be transformed directly into units of absolute radiant flux without reference to laboratory standard sources of radiation; (2) The field of view of the instruments is limited by diaphragms, but without the use of image-forming elements, to the solar disk (practically, a small area of the sky surrounding the Sun will always be included).

Two types of pyrheliometers are in common use. One of these was developed at the Astrophysical Observatory of the Smithsonian Institution, using principles first published by POUILLET[1]; the other was developed from the designs of ÅNGSTRÖM[2], and KURLBAUM[3].

There are two variants of the Smithsonian design, both based on the same principle; the measurement of the change in temperature produced by the Sun

---

[1] C. S. POUILLET: C. R. Acad. Sci., Paris **7**, 24 (1828). — Pogg. Ann. **45**, 25 (1836); **45**, 481 (1838).

[2] K. ÅNGSTRÖM: Nova Acta Upsala **13**, 16 (1893). — Phys. Rev. **1**, 365 (1893). — Wied. Ann. **67**, 633 (1899). — Astrophys. Journ. **9**, 339 (1899).

[3] F. KURLBAUM: Z. Instrumentenkde. **13**, 122 (1893).

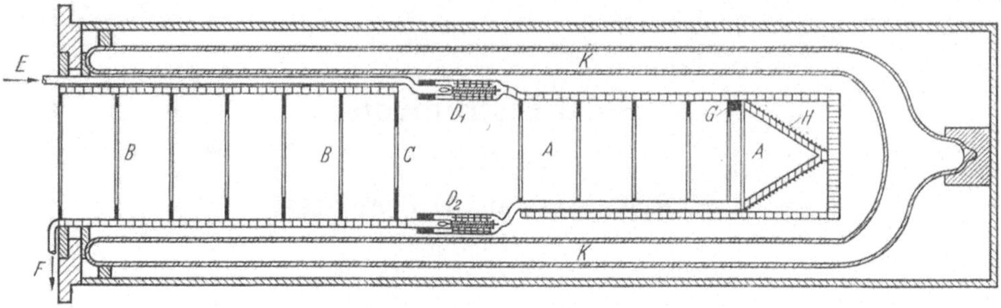

Fig. 1. Water-flow Pyrheliometer. *AA* Black body cavity for absorbing sunlight; *BB* Diaphragms for limiting field of view. *C* Final limiting diaphragm. D_1, D_2 Differential platinum resistance thermometer. *E* Entrance tube for water stream. *F* Exit tube for water stream. *G* Electrical heating coil for instrument calibration. *H* Cone shaped blackened absorber. *KK* Dewar tube.

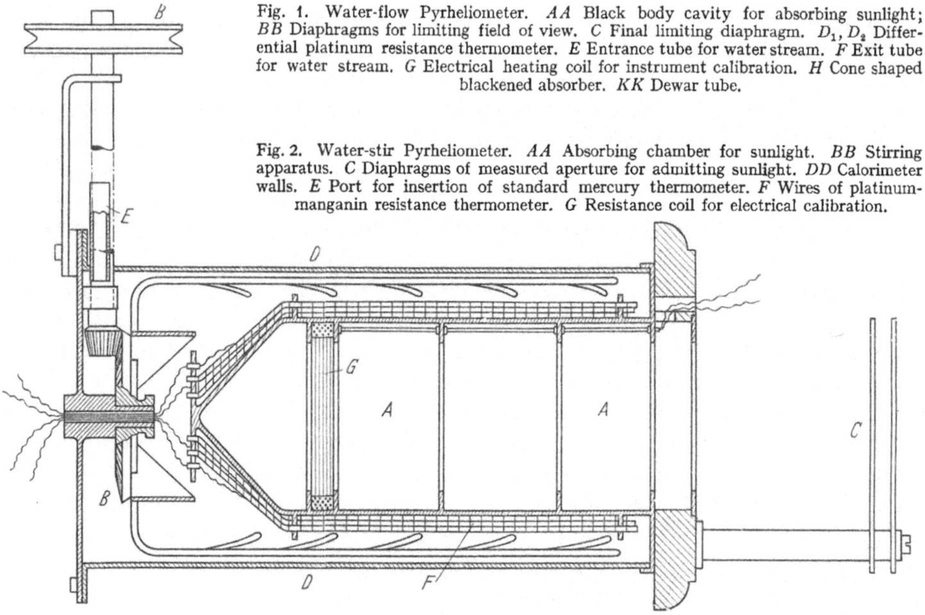

Fig. 2. Water-stir Pyrheliometer. *AA* Absorbing chamber for sunlight. *BB* Stirring apparatus. *C* Diaphragms of measured aperture for admitting sunlight. *DD* Calorimeter walls. *E* Port for insertion of standard mercury thermometer. *F* Wires of platinum-manganin resistance thermometer. *G* Resistance coil for electrical calibration.

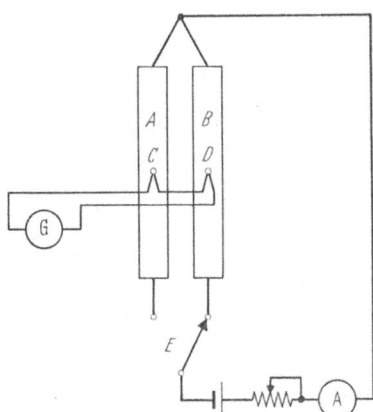

Fig. 3. Diagram of the essential elements of the Angstrom Compensation Pyrheliometer. *A, B* Identical thin resistance strips that are alternately exposed to, and shielded from, sunlight. *C, D* Identical thermocouples connected in opposition so that the galvanometer reads zero when *A* and *B* are the same temperature. *E* Electrical switch for switching heating current to the unilluminated strip.

in a known mass of water. The Smithsonian instruments are called the *water-flow* and the *water-stir* pyrheliometers[1]. In both of these instruments (Figs. 1 and 2) solar radiation is absorbed within blackened conical cavity. The cavity is surrounded with either flowing or rapidly stirred water. In the water-flow type the temperature difference of the water flowing into and out of the instrument depends on the constants of the instrument and the absolute value of the Sun's radiation. In the water-stir type, the temperature rise in the stirred water becomes the measure of the Sun's radiation.

The essential elements of the Ångström pyrheliometer are two blackened strips (*A* and *B*, Fig. 3) either of which may be exposed to the Sun's radiation, and either of which may be electrically heated. In use, one of the strips is

[1] C. G. ABBOT: Ann. Astrophys. Obs Smithson. Inst. **2**, 39 (1908); **3**, 52, 64 (1913).

heated by the Sun, the other is shielded from the Sun's radiation, but is heated electrically until there is no temperature difference, as measured by thermocouples, between the two strips. The electrical power supplied to the shielded strip must then be equal to the power supplied by the Sun to the exposed strip. The measurement of the electrical power is generally effected by a measurement of the heating current and reference to a laboratory calibration that does not require comparison with a radiation standard.

The operation of these pyrheliometers, and others of somewhat modified design[1,2], have been described in detail in various publications[3-6], and the results of observations made with the several types have been discussed at great length[7-11].

If only relative measurements of solar radiation are required, the silver-disk pyrheliometer, developed by the Smithsonian observers, is the most nearly reproducible and accurate of the present designs (Fig. 4)[12]. A tube with suitable

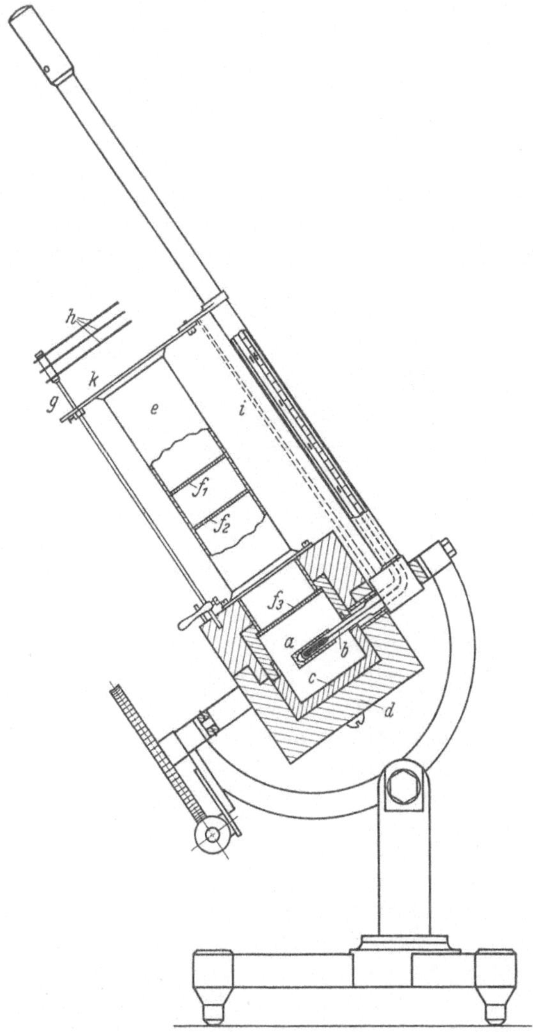

Fig. 4. The silver disk pyrheliometer. a Silver disk, painted dead black with lampblack, with radial hole to admit thermometer. b Thermometer with right-angle bend in tube. c Cylindrical copper box, blackened on inside. d Wooden box for thermal insulation. e Cylindrical tube for supporting diaphragms. f_1, f_2, f_3 Diaphragms with circular apertures smaller than diameter of silver disk. g Rotatable shutter made of three nickeled parallel metal plates. h, h, h Nickeled parallel metal shutter plates. i Small beam of sunlight for guiding. k Screen to shade wooden box, d.

[1] C. Tingwaldt: Z. Instrumentenkde. **51**, 593 (1931).

[2] G. Falckenberg: Meteor. Z. **53**, 303 (1936).

[3] C. G. Abbot: Ann. Astrophys. Obs. Smithson. Inst. **3** (1913).

[4] K. Ångström: Astrophys. Journ. **40**, 274 (1914).

[5] F. Albrecht: Meteorologisches Taschenbuch, ed. by F. Linke, Vol. 2, p. 46. Leipzig: Akademische Verlagsgesellschaft 1933. — Handbuch der Meteorologischen Instrumente, ed. by V. Kleinschmidt, p. 109. Berlin: Springer 1935.

[6] B. Strömgren: Handbuch der Experimentalphysik, ed. by W. Wien and F. Harms, Vol. 26, p. 797. Leipzig: Akademische Verlagsgesellschaft 1937.

[7] B. Strömgren: Handbuch der Experimentalphysik, ed. by W. Wien and F. Harms, Vol. 26, p. 818. Leipzig: Akademische Verlagsgesellschaft 1937.

[8] J. Guild: Proc. Roy. Soc. Lond., Ser. A **161**, 1 (1937).

[9] M. Nicolet: K. R. Met. Inst. Belg. No. 21, 1948.

[10] A. Unsöld: Physik der Sternatmosphären, 2. Aufl., S. 27. Berlin: Springer 1955.

[11] C. W. Allen: Gerlands Beitr. Geophys. **46**, 32 (1935). — The Solar System, Vol. 1: The Sun, ed. by G. P. Kuiper, p. 592. Chicago: Chicago University Press 1953.

[12] C. G. Abbot: Ann. Astrophys. Obs. Smithson. Inst. **2**, 36, 72 (1908); **3**, 47 (1913).

diaphragms limits the field of view of this instrument to six degrees. The Sun's radiation is absorbed on the blackened surface of a silver disk. The temperature of the silver disk is measured by a sensitive mercury-in-glass thermometer, and the radiation of the Sun is measured by observing the rate of rise and fall of the temperature of the silver disk as it is alternatively exposed to, and shielded from the Sun. The instrument is calibrated by comparison with the primary instruments, the *water flow*, or *water stir* pyrheliometers, or other absolute instruments. Excellently reproducible results are obtained with this instrument, but its usefulness depends entirely on precise calibration by comparison with the fundamental instruments.

3. Pyranometers. Pyrheliometers are intended for measurement of radiation coming from the direction of the Sun only, but a part of the Sun's radiation is scattered by Earth's atmosphere, and special instruments, called pyranometers have been devised to measure radiation from the entire sky. Generally, the sky radiation is a negligible fraction of the total, but on overcast days, nearly all sunlight is scattered, and pyranometers are very useful for measurement of variations in cloudiness. In connection with pyrheliometers and radiometers, these instruments have markedly improved the precision of solar constant determinations[1]. Reasonably complete descriptions of pyranometers and their uses, and references to the original literature can be found in BERNHEIMER's paper[2].

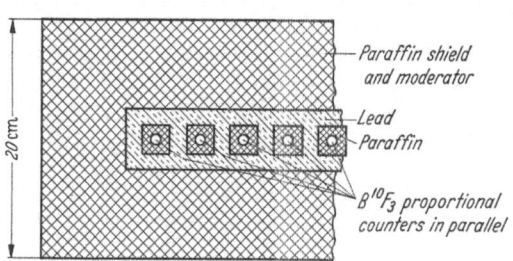

Fig. 5. Schematic cross section of cosmic ray pyranometer for indirect measurement of primary cosmic ray intensity.

Cosmic ray pyranometers have found some application to the detection of low energy cosmic rays that apparently originate in the Sun. It has been found that small intensity changes in the low energy components of the primary cosmic rays can be observed on Earth's surface with neutron detectors. The secondary protons and neutrons that are produced by collision of primary cosmic rays with atmospheric nuclei are reduced in energy by additional repeated collisions to a level at which they may be absorbed efficiently in blocks of lead at Earth's surface. Absorption in lead results in the release of neutrons, and the rate of production of neutrons in the lead is a measure of the primary cosmic ray intensity. A special detecting device for measuring the local neutron production is shown schematically in Fig. 5, and the detecting methods are described in detail in Vol. XLVI of this Encyclopedia.

The multiple scattering of the primary cosmic rays makes it necessary for neutron detectors to record radiation from the entire sky, and these instruments are properly classed as a kind of pyranometer.

II. Radio telescopes.

4. If observations of details of solar structure are to be interpreted satisfactorily, the size of the telescope aperture must be at least one hundred thousand times the wavelength of the radiation employed. Telescopes for observation

[1] C. G. Abbot: Ann. Astrophys. Obs. Smithson. Inst. **5**, 110 (1932).

[2] W. E. Bernheimer: Handbuch der Astrophysik, ed. by G. Eberhard, A. Kohlschütter, and H. Ludendorff, Vol. 1, p. 443. Berlin: Springer 1933.

of the Sun with radiation in the approximate range one millimeter to thirty-five meters wavelength (radio radiation) are not often of these dimensions, or if of sufficient size, are too imperfect in figure to realize their theoretical resolving power. They thus stand in intermediate position between pyrheliometers and pyranometers, in which no image is formed, and those solar instruments able to resolve extremely small angular detail. The difficulties introduced into solar observation by these circumstances, and the attempts to circumvent them are discussed in BRACEWELL's article of this volume. Radio telescopes are similar in general performance to instruments with two or three millimeters aperture operating in the 0.7 to 0.3 micron region of the spectrum. Yet, despite the feeble ability to define detail on the Sun, the results from radio telescopes are of the greatest importance.

III. The image-forming instruments of solar astronomy.

5. The equatorial refractor. The telescopes and spectrographs in general astronomical use are described in a companion article in this volume. Therefore, our description of telescopes and spectrographs will treat in detail only the instruments that have found rather wide use in solar astronomy. The simple refractor was used by SCHEINER[1] for systematic observation of the Sun almost as soon as this instrument began its astronomical career. Solar astronomy profited from the many improvements in refracting telescopes made throughout the nineteenth century, and indeed, some of the most minute structures ever observed on the Sun were recognized during this time. The refracting telescopes used for solar observation were, almost without exception, equatorially mounted stellar instruments that, through some chance, were turned on the Sun.

The first important modifications of the refracting telescope that finally resulted in the evolution of an exclusively solar instrument were embodied in the coronagraph developed during the 1930's by B. LYOT. The coronagraph originated from attempts to solve one of the major problems of solar research; the observation of the solar corona when the Sun is not eclipsed. The changes made in the refracting telescope to solve this problem were: (1) all optical surfaces had to be cleaned scrupulously; (2) instrumental sources of stray light had to be eliminated as far as possible.

The principle sources of stray light in the ordinary refracting telescope are in the object glass. Complete removal of these sources requires, above all else, the production of a perfect blank of optical glass, free from striations, bubbles, inclusions, and other blemishes. Unfortunately, perfection has not yet been attained, but small disks (about 100 millimeters in diameter) can be produced that have, at most, one or two detectable sources of stray light.

Given a perfect disk, it must be figured and polished absolutely without a scratch or irregularity. The surface must be shaped so that the images formed by reflected light can be caught and absorbed on screens. The perfect polish and the exact figure must be achieved simultaneously, and this can be accomplished by specially skilled opticians. The diffraction maxima of the patterns formed by the circular aperture of the objective must also be imaged and absorbed on screens. The lens must be free from dust at all times. The detailed procedures that accomplish these ends are described in an extensive group of publications listed in the Refs.[2,3].

---

[1] C. SCHEINER: De Maculis Solaribus. Augsburg 1612.

[2] B. LYOT: Ann. Astrophys. **7**, 31 (1944).

[3] J. W. EVANS: The Solar System: The Sun, ed. by G. P. KUIPER, Vol. 1, p. 635. Chicago: Chicago University Press 1953.

The principles of the re-imaging system of the Lyot coronagraph (Fig. 6) are useful in other instruments, since an important source of stray light common to many optical systems, the bright ring around the image of the objective lens, is absorbed on the annular diaphragm. It is wise to pay some attention to the quality of the field lens and the image lens, for these, too, can be sources of stray light[1].

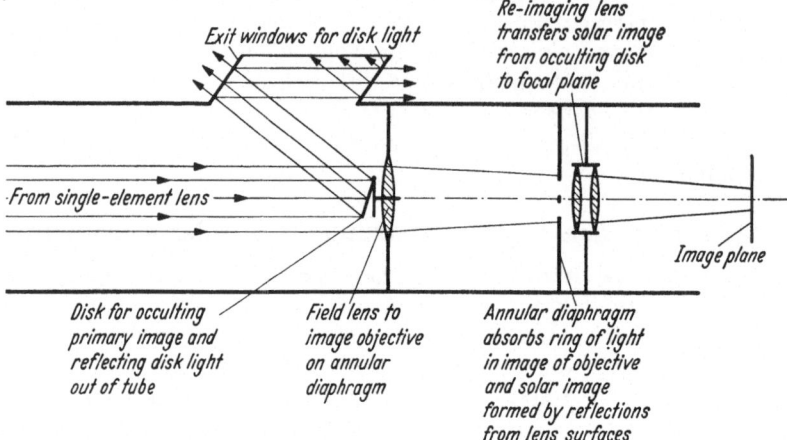

Fig. 6. Re-imaging system of the Lyot coronagraph.

One of the defects of the original coronagraph design is the poor degree of achromatism that results from the use of a simple lens for its objective. The occulting disk (Fig. 6) can be in focus for only a single wavelength, although the image lens can be constructed to produce an achromatic image in the final focal plane of the instrument. Many suggestions of procedures for eliminating the

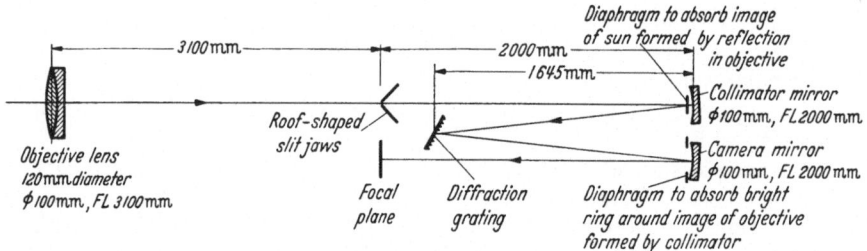

Fig. 7. Optical system of the achromatic coronagraph-spectrograph of the Pulkovo Observatory.

chromatic defects have been made, but only in recent years has it become feasible to replace the single lens by a two-component objective with improved achromatic qualities, and, at the same time hold the stray instrumental light at the three to five millionths of the brightness of the Sun's disk required for success in coronal observation.

An achromatic coronagraph-spectrograph, using a two-component objective, has been constructed and is now in operation at the Pulkovo Observatory near Leningrad, USSR. Fig. 7 is a schematic drawing of the optical system of the Pulkovo coronagraph-spectrograph. The spectrograph is an essential part of this instrument, for its collimator and imaging mirror serve the same purpose as the field-lens and image-transfer-lens in the original Lyot design. The clear

[1] I. A. Prokofieva: Solnechnye Dannye Bull. 2, 119 (1956).

aperture of the objective of the Soviet instrument is 100 millimeters, the focal length is 3.1 meters. The mating surfaces of the two components of the lens are in optical contact, and the edges of the disks have been polished and coated with a special lacquer to diminish internal reflections. It is reported that with this instrument the principal coronal lines, from the visual through the infrared spectral regions, can be photographed on ordinary days at Pulkovo at an altitude of 75 meters above sea level[1].

The infrared lines are photographed with the aid of an image transformer[2] (Fig. 8). The infrared spectral image is first formed on a Cs—O—Ag photo-emissive surface from which photo-electrons are released. The electrons are accelerated and focused by electrical fields to impinge on a fluorescent surface where they reproduce the infrared image incident on the first surface of the tube. The fluorescent light is blue, with peak intensity well with in the spectral region of good sensitivity of ordinary photographic plates. The ratio of the exposure time using the image transformer to that required for photography directly on infrared sensitive plates is about 1 to 100.

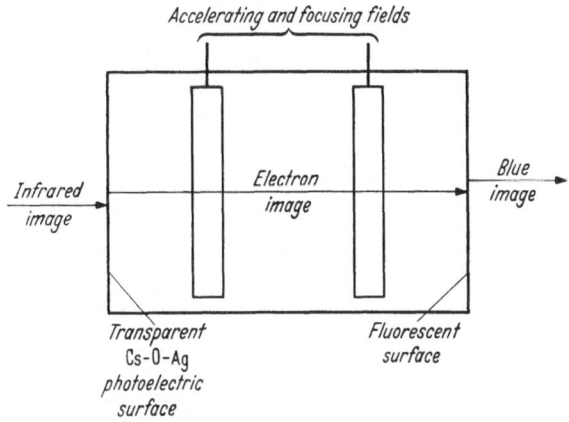

Fig. 8. Infrared image transformer tube.

It might be thought that an achromatic coronagraph could be constructed with a concave mirror for the objective. All attempts along these lines have failed because the large amount of light scattered at the mirror surface masks the corona.

Numerous experimenters have tried to make the corona visible in ordinary telescopes with the aid of detecting instruments sensitive to characteristic features of coronal radiation. Increase of contrast by observing within a dark absorption line of the Sun's spectrum[3], the use of infrared radiation[4], the isolation of one of the bright coronal emission lines[5], and measurement of polarization[6] have repeatedly failed to detect the corona. Mechanical scanning of the region surrounding an occulted image of the Sun, with photoelectric detection at the scanning frequency also has proved unsuccessful in enhancing the faint contrasts of coronal structure[7, 8].

[1] I. A. Prokofieva: Solnechnye Dannye Bull. **1**, 110 (1956); **2**, 119 (1956).

[2] T. H. Pratt: J. Sci. Instrum. **24**, 312 (1947). — A. A. Kalinyak, V. B. Krasouski and V. B. Nikonov: Izvest. Krim Astrophys. Obs. **6**, 119 (1951). — G. S. Ivanov-Kholodny: Izvest. Krim. Astrophys. Obs. **8**, 115 (1952). — D. E. Dean: Amer. J. Phys. **24**, 66 (1956).

[3] G. E. Hale: Astron. and Astrophys. **13**, 658 (1900).

[4] H. Deslandres: C. R. Acad. Sci., Paris **138**, 1378 (1904).

[5] G. Millochau and M. Stefanik: C. R. Acad. Sci., Paris **142**, 945 (1906). — B. Lyot: Ann. Astrophys. **7**, 3 (1944).

[6] R. W. Wood: Astrophys. Journ. **12**, 281 (1900).

[7] A. M. Skellett: Proc. Nat. Acad. Sci. U.S.A. **20**, 461 (1934). — Bell Syst. Techn. J. **19**, 249 (1940). -- Bell Labor. Rec. **18**, 62 (1940).

[8] H. W. Babcock: Astrophys. Journ. **96**, 242 (1942).

However, Fig. 9 is a diagram of a successful system used with an ordinary refracting telescope[1]. In this design, a small scanning aperture, 0.7 millimeter in diameter in an opaque screen can be moved to any position angle in a circle about the Sun's image. The aperture is adjusted to move at a fixed radial distance corresponding to one minute of arc from the edge of the Sun. Light passing through the aperture is transmitted by a polarizing monochromatic filter in three bands: one centered at 5302.85 Å, the wavelength of the brightest coronal emission line; and two flanking side bands at 5300.85 and 5304.85 Å. The plane of polarization of the central band is at right angles to the plane of polarization of the flanking bands. The intensity of the central band is equal to the combined intensity in the two side bands. A rotating modulator alternates the plane of polarization of the two systems, and a Wollaston prism deviates the beams to two photoelectric cells connected to produce a signal only when their illumination differs.

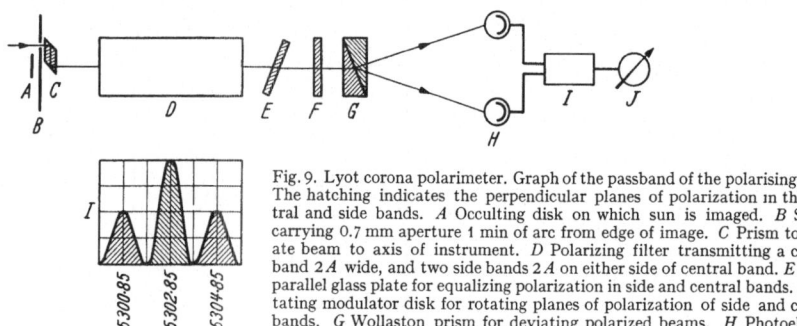

Fig. 9. Lyot corona polarimeter. Graph of the passband of the polarising filter. The hatching indicates the perpendicular planes of polarization in the central and side bands. A Occulting disk on which sun is imaged. B Screen carrying 0.7 mm aperture 1 min of arc from edge of image. C Prism to deviate beam to axis of instrument. D Polarizing filter transmitting a central band 2 Å wide, and two side bands 2 Å on either side of central band. E Plane parallel glass plate for equalizing polarization in side and central bands. F Rotating modulator disk for rotating planes of polarization of side and central bands. G Wollaston prism for deviating polarized beams. H Photoelectric cells. I Difference amplifier tuned to frequency of rotation of F. J Meter.

The photocell current is amplified by an amplifier tuned sharply to the modulation frequency. The degree of discrimination against scattered light that results from the use of the monochromatic filter and the beam-switching technique makes measurement of the solar corona easily possible. The polarization of the continuous radiation from the corona can be measured with a similar instrumental arrangement[2].

Two of the largest aperture refracting telescopes built especially for solar observation are the duplicate coronagraphs of the Sacramento Peak Observatory of the U.S. Air Force Cambridge Research Center (Sunspot, New Mexico, U.S.A.), and the High Altitude Observatory of the University of Colorade (Boulder, Colorado, U.S.A.) (Fig. 10)[3]. The most distinctive feature of the mounting of these telescopes is a rectangular parallelepiped, called a spar, $810 \times 1020 \times 7920$ millimeters in size, which is in the position of the tube of the usual telescope. The spar is supported on a system of oil pads designed to maintain flexural distortion within negligible limits for all positions. The telescope is limited to declinations between $-25°$ and $+25°$, but it has unobstructed motion from the east to the west horizon. The rate of motion and the position of the telescope in hour angle and declination are controlled by a 100 mm aperture, 7620 mm focal length, lens attached to the spar near its west upper end. This telescope forms an image of the Sun on an occulting disk that is small enough to permit light from the Sun's edge to enter four photoelectric cells located at the north, east,

[1] B. Lyot: C. R. Acad. Sci., Paris **226**, 137 (1948); **231**, 461 (1950).
[2] Sky and Telescope **16**, 122 (1957) (Abstract). — G. Wlérick and J. Axtell: Astrophys. Journ. **126**, 253 (1957). — C. R. Acad. Sci., Paris **244**, 1143 (1957). — Dollfus, A.: C. R. Acad. Sci., Paris **241**, 1717 (1955); **246**, 2345 (1958).
[3] J. W. Evans: Sky and Telescope **15**, 436 (1956).

south, and west positions. Any deviation from exact centering is corrected automatically by a signal from the pair of activated cells.

A camera for the chromosphere consists of a 380 mm single element lens,
carried on a bracket on the east side of the pier. There is no tube enclosing the light
path from the lens to its focal plane. When set up for photography of the chromosphere, the Sun's image is projected on an occulting disk and only a fringe of
chromospheric light is transmitted beyond. The occulting disk is followed by a
beam splitting device that produces four images which are sent through a monochromatic filter. The chromosphere finally is imaged on 35 mm motion pictur film.
The scale of the final images corresponds to a solar diameter of about 200 mm.

On the north side, or top, of the spar, near its upper end, a 410 mm 7620 mm
focal length single element coronagraph lens is mounted. This telescope, like the

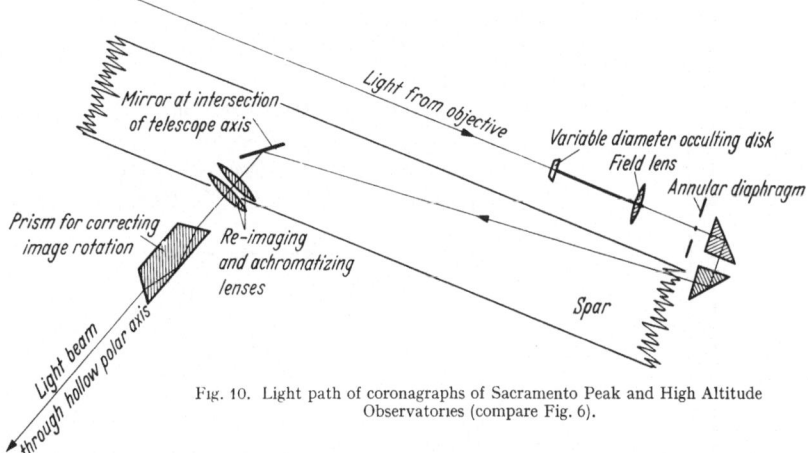

Fig. 10. Light path of coronagraphs of Sacramento Peak and High Altitude
Observatories (compare Fig. 6).

guiding telescope and the chromospheric camera, is tubeless. An image of the
Sun is formed on an occulting disk of variable diameter located near the lower
end of the spar (Fig. 10). A field lens is about 60 cm behind the occulting disk.
The disk, following Lyot's design, is supported on a shaft through a hole in the
lens. This construction eliminates shadows from the support for the occulting
disk. The telescope can be used for the study of the surface of the Sun and the
region simmediately surrounding the solar disk, and in conjunction with laboratory
instruments.

For the study of the solar surface the special field lens carrying the support
for the occulting disk is removed, along with the occulting disk, and an ordinary
field lens is installed in their place. Closely behind the field lens a pair of 150 mm
totally reflecting prisms reflect the beam to the inside of the spar and toward a
fused silica mirror at the intersection of the hour-angle and declination axes. The
mirror turns the beam downward through the hollow hour angle axis. Auxiliary
transfer lenses and prisms for achromatizing and counteracting the rotation of
the image are installed in this part of the light path. At the lower end of the hour-
angle axis additional lenses and mirrors make possible a wide choice of image
sizes and direct the light beam to any one of a number of fixed laboratory instruments.

6. Equatorial reflectors. Equatorially mounted reflecting telescopes have not
been used to any great extent for the observation of the Sun. Most of the reflectors for stellar observation have such a small ratio of focal length to aperture
that they produce very hot solar images which are difficult to use with safety.

Where the focal ratios are larger and the heat of the image more controllable, the instruments have generally been designed to operate with very small fields of view, and this situation is not especially suitable for solar observation.

Several equatorial Cassegrainian reflecting telescopes are under construction for use in solar studies. The schematic design of one such telescope is shown in Fig. 11. The dimensions of the Cassegrainian secondary and the diagonal flat

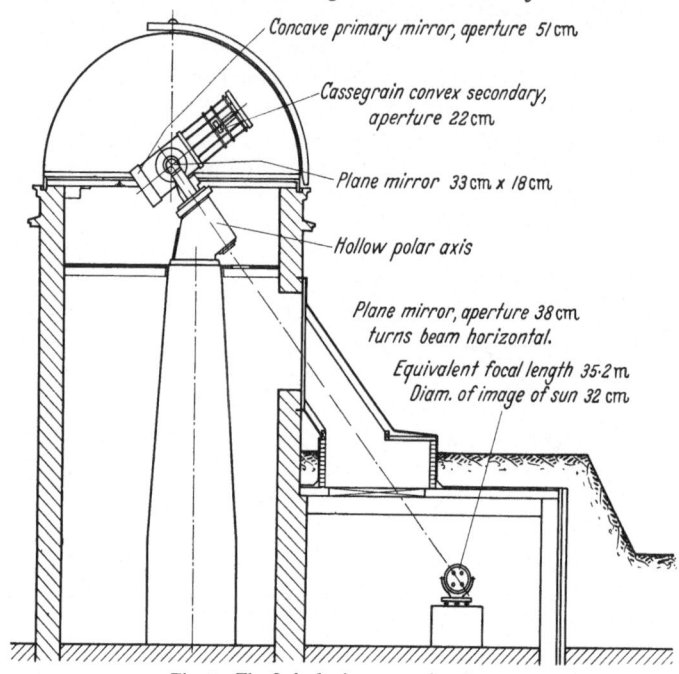

Fig. 11. The Oxford solar cassegrain reflector.

must be carefully chosen large enough to provide as wide a field as possible. These sizes are limited, because the central obstruction may seriously reduce the resolving power of the telescope. A telescope of this type has been installed at the University Observatory, Oxford, England[1], and others are under construction.

7. Equatorial telescopes using combinations of mirrors and lenses. Telescopes using both mirrors and lenses have been enjoying a great vogue in stellar astronomy. The Schmidt system (see the article on Telescopes and Spectrographs, in this volume) leads to very long telescope tubes if focal lengths reasonable for solar research are used. The Schuppman[2] telescope places a correcting lens and mirror system very close to the focal plane of the telescope where the heating effect of the Sun's image is most likely to be troublesome. The Maksutov meniscus telescope can be easily adapted to the Cassegrainian heliostat system of Fig. 11. Excellent images can be obtained with this system, but it is vulnerable to solar heating and is probably not the best design for day long observation. Maksutov telescopes in the Cassegrainian form shown in Fig. 12 can be used as a very compact photo-

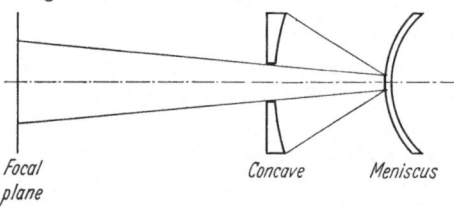

Fig. 12. Maksutov meniscus telescope suitable for use as a photoheliograph.

[1] H. H. Plaskett: Monthly Notices Roy. Astronom. Soc. London **115**, 542 (1955).
[2] L. Schupmann: Z. Instrumentenkde. **33**, 308 (1913).

heliograph for direct solar photography. Telescopes of this sort are in use in the Soviet Union as solar patrol cameras recording the Sun with an image diameter of 50 mm.

8. Fixed telescope systems. The possibility of directing the radiation from an astronomical object in a fixed direction by means of a moving plane mirror, or system of mirrors, and observing the object through a telescope fixed relative to Earth was realized very soon after the introduction of telescopes into astronomy. The first system that was correct in principle was described by BOUFFART in 1682. The advantages of having the long and cumbersome solar telescopes and their equally awkward accessories stationary are obvious, and most solar research for the past century has been carried out with the aid of such instruments. (See HARTMANN[1] for a summary of the history of plane mirror systems and a detailed treatment of the geometry of their construction and motion.)

There are two general classes of mirror systems in common use. The first is the *uranostat*. The uranostat consists of a single plane mirror provided with two perpendicular axes of rotation. The mirror is, in general, in continuous motion about the two axes, and usually not at a uniform rate. A special type of uranostat is an equatorially mounted plane mirror. A plane mirror thus mounted can be adjusted to reflect any point of the sky in any direction at a given instant, and if it is moved at the correct variable rate about both axes, it will maintain the direction of the reflected ray, but the image will rotate about a single fixed point in its plane. This arrangement has been used frequently to reflect a beam of sunlight southward into a long focal length telescope mounted horizontally in a fixed position. Uranostat-horizontal-telescope combinations of this kind are the photoheliographs used at many observatories for daily photographic recording of sunspot positions. Exposure times are so short that only very crude approximations to guiding motions are provided for the mirror and the rotation of the image is not important.

The second type of plane mirror system is the *heliostat*. The essential feature of the heliostat system is the provision of continuous guiding motion for the mirror about a single axis only. The plane mirror in a heliostat system may be mounted equatorially, just as in the case of the uranostat, but a continuous driving motion is supplied only in hour angle. The reflected light beam is directed either north or south along the axis of hour angle rotation in the case of the equatorial mounting. This arrangement provides a solar image with one point of the image perfectly stationary. The stationary point is that point of the image that coincides exactly with the direction of the hour angle axis. All other points in the image rotate slowly around the single fixed point. A telescope receiving light from an equatorially mounted heliostat mirror must have its optical axis exactly in the direction of the polar axis of the heliostat mirror.

At, or near, 45° latitude, this requirement results in rather awkward observing positions, and perhaps for this reason the heliostat mounting has not been used very often for solar observation. (Note that the solar telescope shown in Fig. 11, and the "coudé" equatorials for stellar observation, are essentially heliostats in which the telescope precedes the plane mirror.)

A special uranostat mounting is called the *siderostat*[2]. In a common form of this instrument the mirror is rotatable about altitude and azimuth axes. The motions of the mirror on the two axes are transferred by a linkage from the hour angle and declination axes of an ordinary equatorial mounting. The linkage

---

[1] W. HARTMANN: Astronom. Abh. Hamburg Obs. Bergedorf **4**, 1 (1928).
[2] E. PETTIT: Astrophys. Journ. **91**, 161 (1940).

moves the mirror on the alt-azimuth mounting so that it reflects a light beam from an object (toward which the equatorial is directed) in a fixed direction, generally south in the northern hemisphere. In common with all forms of the uranostat and the equatorial heliostat, the field of view of the siderostat rotates.

The most commonly used plane mirror mounting for solar astronomy is a special form of heliostat, the *coelostat*. The plane mirror of the coelostat mounting is attached rigidly to a polar axis so that a line element of its plane surface coincides with the direction of the polar axis. If the polar axis is rotated at half the normal rate, so that it makes one complete rotation in forty-eight hours, a fixed point in the sky will be reflected from the mirror surface in a fixed direction with respect to the Earth. The image formed by a telescope that receives the reflected beam will be completely fixed and will not rotate in its own plane. Objects at different declinations in the sky will be reflected in different fixed directions relative to Earth, and since the Sun changes its declination, the simple coelostat is not a very convenient instrument for daily observation. Any telescope used with it would have to be shifted in position for each day's observations. This difficulty can be eliminated by adding a second mirror to receive the reflected beam from the coelostat and direct it to the telescope[1]. The relative positions of the second mirror and the coelostat can then be adjusted once each day to compensate for the slowly changing declination of the Sun, and the telescope can be installed in an absolutely fixed position. The coelostat mirror alone is often used for eclipse observations, and it is perfectly satisfactory for observations that extend over a small part of one day out of the year, but for daily use the two-mirror system is almost essential.

The following are some of the general features of mirror systems for use with fixed telescopes that must be considered in choosing an instrument for a particular application.

If observations are to run from sunrise to sunset, then in order to illuminate fully a telescope of given aperture at a given location, the polar heliostat will require the smallest mirror and the coelostat the largest. The uranostat, heliostat, and siderostat are used as single-mirror systems, although auxiliary mirrors can be, and often are, used to direct the light beam into convenient observing positions. The coelostat requires a second mirror for satisfactory continuous daily operation. The uranostat is about equally well-suited to all latitudes; the polar heliostat is best for low latitudes, where the siderostat is completely unsuitable; and the coelostat is best adapted to intermediate latitudes.

The uranostat, the polar heliostat, and the siderostat each introduce a large and annoying rotation of the field of view. There are some solar observations, such as direct photography with very short exposure times, or the measurement of the Sun's integrated radiation, in which field rotation is unimportant, and there are relatively simple mechanical ways of rotating a plate holder, a spectrograph, or a reversing prism to counteract the image rotation. Nevertheless, the coelostat and second mirror is the most satisfactory scheme for producing a fixed image in a fixed telescope, even at the cost of the additional mirror.

Sunlight reflected from a mirror is always polarized to a greater or lesser degree, and the direction of the axes of polarization change with the angle of incidence. The polar heliostat maintains a fixed angle of incidence throughout a day's observations and in this respect it is the most satisfactory of the mirror systems. The coelostat and second mirror system are very unsatisfactory in this regard and the angle of incidence runs through a wide range of values from sunrise to sunset.

---

[1] C. G. Abbot: Ann. Astrophys. Obs. Smithson. Inst. **2**, 22 (1908).

IV. Construction and housing of solar telescopes.

In years past, solar telescopes seem to have been designed with two major guiding principles in mind. First, since the Sun is very luminous, no particular efforts need be made to get a bright image; second, the image formed by the telescope should be as large as possible. These two criteria have worked together to produce a large number of instruments with ratios of focal length to aperture in the range 30 to 200. The largest of the solar installations, whether judged on the basis of aperture or focal length, all consist of a coelostat and second mirror directing sunlight to a fixed telescope. A consideration of two of the early tele- scopes of this type should help make clear some of their characteristics.

9. Horizontal telescopes. The horizontal telescope of the Astrophysical Section of the Observatory of Paris, at Meudon (1908), was among the first of the exclusively solar telescopes[1]. The coelostat is housed in a small wooden shelter with a roof that is moved on rails to the south to admit sunlight to the mirrors. The coelostat mirror is 50 cm in diameter. It is driven electrically to keep the sunlight on the second mirror. The second mirror is 40 cm in dia- meter and it reflects the light beam horizontally northward to the telescope objective. The telescope objective has a diameter of 25 cm and a focal length of 400 cm. It is mounted in a cloth bellows that forms an air tight seal to the wall of the observing room. The objective thus forms the window and the air seal by which light enters the observing room and unwanted air currents are suppressed. When the centers of the coelostat and second mirror are in the meri- dian plane, the north-south line of the projected solar image is vertical. When the Sun is at low declinations, the supporting members for the second mirror cast shadows on the coelostat and it must be shifted along an east-west line to avoid the shadow. In the east or west positions of the coelostat the north- south line of the solar image is no longer vertical. The second mirror is provided with a set of ways along which it can be shifted in the north-south direction, and a set of axes so that the mirror can be tilted about a diameter. On beginning observation in the early morning, positions for the coelostat and second mirror are selected that will make possible a long period of observation before the second flat shadow intervenes. The focal setting of the objective is checked visually and with the aid of a table of differential settings, the objective can be placed in position for observation at any wavelength. The same visual checking method is used to correct for small changes in focal setting that arise from the heating effect of the solar radiation on the mirrors and the lens.

Another early example of the horizontal solar telescope is the Snow Telescope, first erected at the Yerkes Observatory. It is the direct predecessor of the solar tower telescope, and it, therefore, is of great importance for an understanding of the growth of present day ideas of an ideal solar instrument. The Rumford Spectroheliograph[2], a rather elaborate instrument, had made use of the 40-inch refractor of the Yerkes Observatory as a solar telescope, but the combination was restricted in performance by the instability of the telescope mounting. Attempts to remove these restrictions led to the adoption of a horizontal telescope design, following the general lines of coelostat telescopes used on eclipse expedi- tions. Before the instrument had been adequately tested at its original site, it was moved to Mount Wilson, California, where conditions seemed to promise more successful observation. The coelostat and second mirror arrangement is very similar to that of the Meudon instrument. Sunlight is incident on the coelo-

[1] H. DESLANDRES: Ann. Obs. Meudon **4**, 1 (1910).
[2] G. E. HALE and F. ELLERMAN: Publ. Yerkes Obs. **3**, 3 (1903).

stat mirror, a plane 76 cm in diameter. The second plane mirror is southward from the coelostat. It is 61 cm in diameter and it reflects the beam from the coelostat in a horizontal direction to the north, to a concave mirror. The concave mirror has the dimensions: diameter 61 cm; thickness 10 cm; focal length 18.3 m. A second concave with the same dimensions, but with focal length 45.7 m was also provided. The scheme for making adjustments to suit the constantly changing declination of the Sun is the same as that of the Meudon telescope: the second mirror can be translated in a north-south direction, and inclined to the horizontal as required. The coelostat may be moved east or west to avoid the noontime shadow of the second mirror and its supports.

The Snow telescope did not perform as well as had been expected on Mount Wilson, and it was soon discovered that some of the difficulty was inherent in the horizontal mounting of the telescope near Earth's surface. The chief source of trouble was found in convection currents that result from heating of the ground and the mirror mountings by the Sun's radiation. Since the Snow telescope is mounted at a small height above ground level, observing conditions were investigated, using small telescopes, from ground level to heights of more than 20 meters. These tests showed that conditions were least favorable at ground level, and improved with height above ground. The improvement is rapid within a few feet of the ground and much slower as the heights reach a few tens of meters. A result of the tests was the construction of a second instrument on Mount Wilson during the years 1906 to 1907. This was the first solar tower telescope[1-4].

10. Solar tower telescopes. In the original construction, a structural steel tower carried the coelostat and second mirror on a platform unprotected from weather or solar radiation. The necessity for such protection soon became apparent and the tower eventually was completely rebuilt into the form it now has— the 60-foot solar tower telescope. At present, the coelostat is equipped with the usual east-west ways so that the noonday shadow of the secondary mirror can be avoided, and an additional set of ways parallel to the direction of the Earth's axis is provided to make possible adjustment for changes in the Sun's declination. The second mirror is carried on a fixed pedestal, and is provided with two perpendicular axes about which it may be rotated. One of the axes is parallel to the Earth's axis; the other is a diameter of the mirror. Rotation of the mirror by means of these axes directs the light beam vertically downward, and the same axes provide for small guiding corrections.

A second, larger tower telescope, known as the 150-foot, incorporating in its design many features derived from the experience with the first, was constructed on Mount Wilson in 1912[1].

The 60-foot and 150-foot towers on Mount Wilson remained for some years the only instruments of their kind, but the number of similar structures has gradually increased until there now exist nearly two dozen   All of the tower telescopes constructed recently such as those at the McMath-Hulbert Observatory, at the Crimean Astrophysical Observatory and at the Institute for Theoretical Astrophysics in Oslo, have been built with cylindrical outer shells of steel, masonry, or concrete. The outer cylindrical shells carry a dome that can be closed for protection of the coelostat, second mirror, and telescope from the weather, and

---

[1] G. E. Hale and S. B. Nicholson: Magnetic Observations of Sunspots 1917—1924, Part 1, 1. Washington: Carnegie Institution of Washington 1938.
[2] G. E. Hale and F. Ellerman: Astrophys. Journ. **23**, 54 (1906).
[3] G. E. Hale: Astrophys. Journ. **25**, 78 (1907).
[4] G. E. Hale: Publ. Astronom. Soc. Pacific **20**, 35 (1908).

when open provides a measure of shielding from wind and sunlight. The coelostat, second mirror, and telescope are supported on an inner structure that may be either cylindrical, or an open framework of steel, wood, or concrete. Every effort is made to achieve an inner tower structure that is thermally and mechanically stable, and special care is taken to isolate it from vibrations induced in the outer tower by wind or dome motions.

The telescopes in the original solar towers were ordinary refractors, but many of the new tower telescopes are reflectors. The majority of the reflecting telescopes are off-axis designs that provide a number of different focal lengths. Many of the older solar tower telescopes have added mirrors as alternatives for the previously existing refracting systems.

The refracting telescopes used with coelostats in the first solar instruments, whether of the horizontal or vertical construction, were distinguished by very large ratios of focal length to aperture. Such telescopes provided faint solar images of large size. In most instances, the color curves of the objectives were not known, and this made the use of the instruments difficult in any but visual wavelength regions. It is perhaps the difficulty of focusing a lens system in the infrared or ultraviolet that has been a major factor in the replacement of lenses by mirrors in the recent solar telescopes.

The transition from the outward form of the original solar towers on Mount Wilson to the cylindrical structures of the present, and the increasing use of all-reflecting image-forming systems can be traced in three later observatories. The first solar tower of German construction, the Einstein tower at Potsdam, built during the years 1920 to 1924, adopted a massive, essentially cylindrical, concrete structure for the outer tower that protects the telescope proper. The coelostat and telescope are mounted on a completely independent inner tower made of wood in an attempt to reduce mechanical vibration[1]. In 1926, the Hale Solar Laboratory (Pasadena, California, U.S.A.) was completed with all-mirror optical systems in the telescope, spectrohelioscope, and spectroheliograph, but an autocollimating lens system was retained for the 23 m focal length spectroscope[2]. The final stage in the transition can be marked by the completion of the 15 m (1936) and 21 m (1940) towers of the McMath-Hulbert Observatory of the University of Michigan (Pontiac, Michigan U.S.A.)[3]. Both of these towers are built with concentric steel cylinders, the outer shell forming the protecting cover for the inner, which serves to support the coelostat and optical parts of the telescope. The 15 m tower was designed with the guidance of results obtained during the development of astronomical motion picture techniques at this observatory, especially to produce time-lapse photographs of solar activity[4]. The double steel-cylinder construction and the recording of solar activity by the motion picture method have been so widely adopted in solar astronomy that they are now commonplace.

The solar tower telescope of the Crimean Astrophysical Observatory (near Partizanskoye, U.S.S.R.) is an example of a recently completed instrument that combines many features of the Hale Solar Laboratory and the solar tower tele-

---

[1] E. Freundlich: Das Turmteleskop der Einstein-Stiftung. Berlin: Springer 1927. — E. v. d. Pahlen: Z. Instrumentenkde. **46**, 49 (1926).

[2] G. E. Hale: Year Book (Washington: Carnegie Institution of Washington) **23**, 110 (1924); **25**, 135 (1926). — G. E. Hale: Astrophys. Journ. **70**, 306 (1929).

[3] R. R. McMath: Publ. Obs. Univ. Michigan **7**, 1 (1937). — H. D. Curtis: Pop. Astron. **48**, 348 (1940).

[4] F. C. McMath, H. S. Hulbert and R. R. McMath: Publ. Obs. Univ. Michigan **4**, 53 (1931).

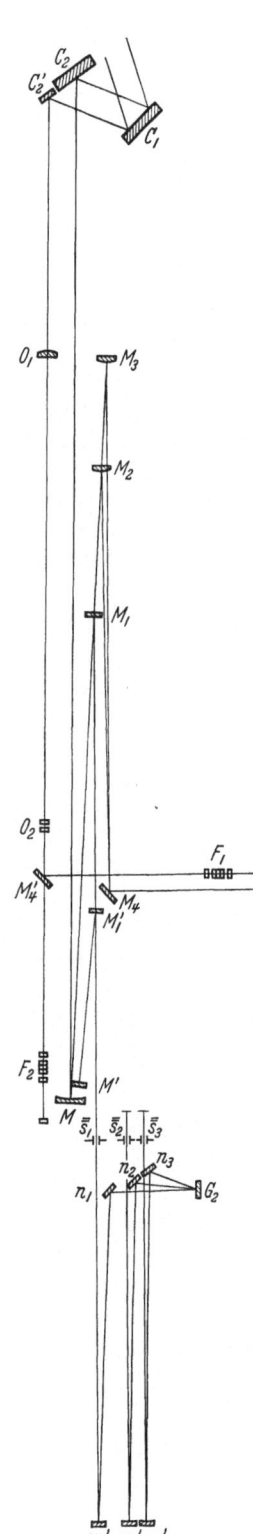

Fig. 13. Light paths in the solar tower telescope and spectroscopes of the Crimean Astrophysical Observatory. C_1 Coelostat plane mirror ⌀ 650 mm. C_2 Second plane mirror-⌀ 500 mm. C_2' Second plane mirror for guiding ⌀ 214 mm. F_1 Polarizing monochromatic filter for guiding. F_2 Polarizing monochromatic filter for guiding. G_1 Diffraction grating, ruled surface 150×150 mm, 600 g/mm. G_2 Diffraction grating, ruled surface 150×150 mm, 600 g/mm. M Concave spherical mirror ⌀ 400 mm, FL 12 m. M_1 Plane mirror ⌀ 220 mm (used with M). M_2 Convex cassegrainian secondary ⌀ 180 mm (used with M, EFL 21 m). M_3 Convex cassegrainian secondary ⌀ 140 mm (used with M, EFL 35 m). M_4 Plane mirror ⌀ 300 mm. M' Concave spherical mirror ⌀ 150 mm, FL 5 m. M_1' Plane mirror ⌀ 135 mm (used with M'). M_4' Plane mirror shifted to divert light from F_2 to F_1. m_1 Concave spherical collimator ⌀ 230 mm, FL 10 m. m_2 Concave spherical collimator ⌀ 300 mm, FL 10 m, or ⌀ 300 mm, FL 20 m. m_1' Concave spherical collimator ⌀ 170 mm, FL 5 m. m_2' Concave spherical camera ⌀ 170 mm, FL 5 m, for λ_1. m_3' Concave spherical camera ⌀ 170 mm, FL 5 m, for λ_2. n_1 Plane mirror-collimator to grating G_2. n_2 Plane mirror-grating to camera mirror m_2' for λ_1. n_3 Plane mirror-grating to camera mirror m_3' for λ_2. O_1 Guiding lens ⌀ 100 mm. O_2 Collimator lens for guiding beam. O_3 Camera lens for 50 mm Hα guiding image. P Plateholder 13×18 cm, or plane mirror for use with m_2 FL 20 m. S_1 Entrance slit horizontal spectroscope, 30 mm long. S_1' Entrance slit pit type spectroheliograph, 55 mm long. S_2'' Exit slit of spectroheliograph for λ_1. S_3'' Exit slit of spectroheliograph for λ_2.

scopes of the McMath-Hulbert Observatory[1]. The general optical arrangement is shown in Fig. 13.

A collection of data, general descriptions, and extensive lists of references useful for the study of solar tower telescopes will be found in Ref.[2]. Additional data for new solar tower telescopes are given in Table 1 at the end of this article (p. 38).

V. Auxiliary instruments for solar telescopes.

Telescopes equipped only for visual observation and direct photography can play but a relatively small part in solar research. Nonetheless, visual observations and direct photography are still too important to be ignored completely, particularly since modest modification of a simple telescope may greatly enhance the value of its work.

11. Light filters. Visual observation of the sun always requires that the light intensity be decreased in some way, and this is equally true for all radiation detectors except the very most insensitive. The eye can look without much discomfort on an image of the noonday Sun projected on a white card, if the equivalent focal ratio of the projecting system is 150, or greater, but this focal ratio is too small if the image is intended for examination directly with an eyepiece. An equivalent focal ratio of 500 to 1000 is much better for direct viewing. The most common method of attenuating the beam makes use of the light reflected from polished planes of optical glass placed near the focal plane. (The aperture of the telescope should not be reduced, except when conditions of seeing are bad.) Each reflection from a polished glass surface results in the achromatic loss of about 95% of the incident energy. If a non-reflecting coating is placed on the glass, the loss can be increased to 99.5%.

[1] A. B. Severny: Izvest. Krim Astrophys. Obs. 15, 31 (1955).
[2] H. Gollnow: Naturwiss. 36, 213, 175 (1949).

One or two reflections from such surfaces make direct observation of the Sun possible with the ordinary stellar telescopes. The same methods can be used to lengthen photographic exposures, to convenient intervals of time.

Exposure times with a reflecting telescope at a focal ratio of 50, and the least sensitive emulsions, are about one-thousandth of a second. Exposure times as short as this are almost completely uncontrollable under the conditions of solar observation. Exposure times can easily be lengthened by restricting the range of spectrum wavelengths transmitted by the telescope and selected filters. Simultaneously, the contrast between various solar features may be increased. Thus, a set of high-grade colored optical-glass filters is an indispensable auxiliary for any solar telescope.

Good quality colored glass filters are used chiefly in isolating fairly broad ranges of wavelength. These are particularly useful for separating the shorter from the longer wavelength radiation, since a large number of different glasses are available with broad transmission bands which are sharply limited on the short wavelength side.

An interesting filter for separating the Sun's radiation into two parts, one containing all wavelengths longer than a pre-selected limit, and the other the shorter wavelengths, can be constructed using the internal reflections in prisms[1]. This filter is especially useful for rejecting the intense solar radiation in the blue and green spectral regions during observations with ultraviolet sunlight. Unfortunately, internal totally reflecting filters, and colored glass filters cannot be combined in simple fashion to produce very narrow pass bands. Band widths of two or three hundred angstroms are the narrowest that can be achieved in this way. To obtain higher spectral purity, one must turn to other devices.

It is now possible, by using controlled evaporation, to produce systems of thin films with narrow pass-band characteristics. The thin-film filters can be obtained with almost every imaginable combination of transmission, reflection, and absorption characteristics. These, the interference filters, can be constructed to transmit band widths only a few, in some cases not more than five, angstroms wide. With no other aid than such an interference filter and a clean telescope solar prominences can be observed. The new and very active field of optics dealing with the production of interference filters is reviewed in detail in Vol. XXIV, p. 461 of this Encyclopedia. The interference filter has become an instrument of enormous importance in all fields of astronomy for isolating narrow spectrum regions.

The polarizing monochromatic filter[2] is still another means for limiting the band of wavelengths transmitted from a solar image This device has met with such wide acceptance that there are many detailed descriptions of the theory of its operation and of methods for its construction and use (see Françon, Vol. XXIV, p. 449 of this Encyclopedia). The polarizing monochromatic filter, like the thin-film interference filter can be constructed with a wide range of transmission characteristics. In particular, it can be built to transmit a wavelength band about one-half angstrom wide. This is the present practical lower limit for the width of the pass band of a polarizing filter because of two difficulties: (1) adequately perfect birefringent materials are extremely rare in the required sizes; (2) the demands on the thermostatic control of the filter become excessive as the pass-band is made narrower.

---

[1] E. M. Brumberg: C. R. Acad. Sci. USSR. **2**, 467 (1935). — V. H. Regener: Quart. J. Roy. Meteorol. Soc. (Suppl.) **62**, 9 (1936).

[2] B. Lyot: C. R. Acad. Sci., Paris, **197**, 1593 (1933).

Polarizing monochromatic filters have found their greatest use with the transmitted pass-band centered at Hα (6563 Å) because the solar prominences, the flares, the filaments, and the plages are most easily recorded at this wavelength. A filter can be placed in the converging beam of radiation near the focal plane of the usual solar telescope of large focal ratio, and the monochromatic image produced in this way may be viewed directly with an eyepiece, or photographed.

In order to capitalize on the excellent quality of the monochromatic images obtained with a simple lens and this type of filter, monochromatic (Hα) photo-

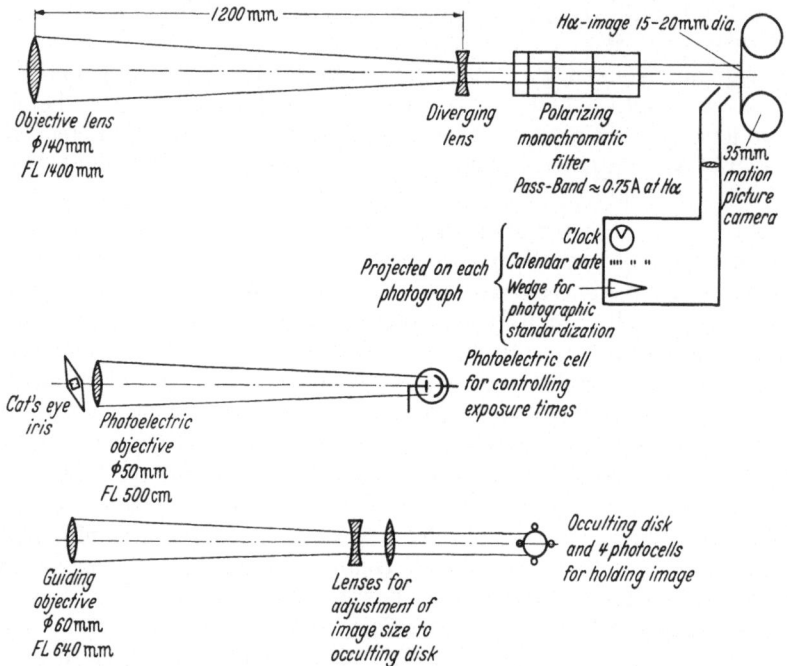

Fig. 14. Arrangement of optical paths and control telescopes in the IAU Hα Photoheliograph.

heliographs following a standardized design (recommended by the International Astronomical Union[1]) have been constructed for installation at observing stations distributed around the world. It is expected that the photographic records from these instruments can be combined to provide day by day, twenty four hours a day, observations of the disk of the Sun in Hα light. Fig. 14 shows the dimensions and the schematic arrangement of the optical paths of the single-lens telescope of the Hα photoheliograph, and also the three auxiliary telescopes: the photoelectric guiding device; the photoelectric exposure control; the time recording device, and the photographic standardization equipment. The telescope and its controls are mounted on equatorial axes. Once pointed to the Sun, the instrument will continue taking motion pictures at a uniform rate (generally two pictures per minute) as long as the sunlight is bright enough to operate the photoelectric exposure control.

Another arrangement, lacking the elaborate controls of the IAU Hα photoheliograph is diagrammed in Fig. 15. Instruments of this type attached to the

---

[1] B. Lyot: Trans. Internat. Astronom Union ed. by P. T. Oosterhoff, Vol. 8, p. 146. Cambridge: University Press 1954.

tubes of stellar equatorial telescopes have been operated with great success for daily solar photography. Since they do not have automatic controlling mechanisms, these instruments require the attention of observers at approximately hourly intervals.

If a pass-band still narrower than the one-half angstrom provided by the polarizing monochromatic filter is required, solar astronomers make use of the spectroscope in one of its many forms. Several solar spectroscopes now in use provide resolving powers from 350000 at 11000 Å to 1000000 at 3000 Å, or a nearly constant resolving limit, measured in frequency units, of 10^9 Hz. Thus, it is apparent that a spectroscope can provide a pass-band of less than one-tenth the width of the narrowest pass-band attainable with the polarizing monochromatic filter.

12. Spectroscopes. A small prism spectroscope, attached to the eyepiece end of a refracting telescope was used extensively by early solar observers, essentially as a monochromatic filter. The discovery of the yellow line of helium, the green

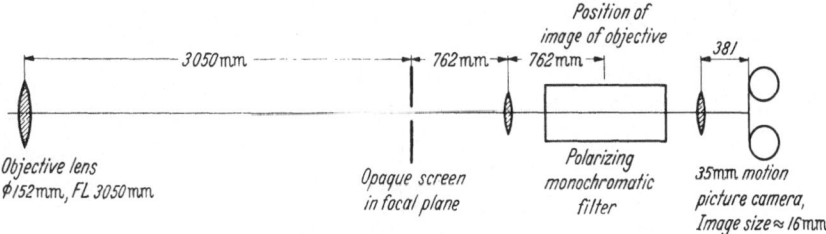

Fig. 15. Simplified Hα Photoheliograph.

coronal line, and the observation of prominences, the chromosphere, and occasional flares on the solar disk were accomplished in the last third of the 19th century, with spectroscopes of very modest size and resolving power.

During the course of the one hundred years that have passed since its introduction as an auxiliary instrument for the study of the Sun, the solar spectroscope has grown into a structure as large as the telescope with which it is used. Its most recent evolutionary changes have been brought about by the improvement of the techniques for the ruling of reflection diffraction gratings. Diffraction gratings of modern manufacture diffract as much as 70% of the incident light in a given direction: a much larger amount of the incident light than can be dispersed on the same angular scale in a given direction by a prism or a train of prisms. The fact that gratings can be produced with larger apertures than optically precise prisms is an additional advantage for their use in solar spectroscopy.

Two of the most important of the advance of recent years in the techniques of grating production are: (1) the use of evaporated metallic films for the stratum in which the grooves are formed; (2) the use of glass, or fused silica, optically figured with precision of a small fraction of a wavelength of light, for the base on which the grating films are evaporated and supported. Item (1) implying the adoption of soft, easily formed, non-abrasive metals, lubricated during the ruling process, has made possible gratings with large ruled areas containing grooves of precisely pre-determined shape. Item (2) has made possible gratings with accurately figured optical surfaces suitable for use at large angles of incidence. These, and other improvements in diffraction gratings have produced a series of changes in the art of construction of spectroscopes. The elimination of faulty and infelicitous designs has gradually transformed the solar instrument until only a few distinct types survive in common use.

2*

Perhaps the simplest of solar spectroscopes is that using a *concave grating* as the collimating, dispersing, and image-forming element. Variants of this type of instrument are called the Rowland mounting, the Paschen-Runge mounting, the Abney mounting, the Radius mounting, and the Eagle mounting of the concave grating (see Vol. XXIX of this Encyclopedia). In all of these, the concave grating, the entrance slit, and the diffracted spectrum lie on a circle (the Rowland circle) which has as its diameter the radius of curvature of the grating blank, and to which the grating is tangent. Some of the advantages of this principle of construction for solar instruments are: (1) the number of optical elements is the absolute minimum, one; (2) the system may be used for all wavelength regions;

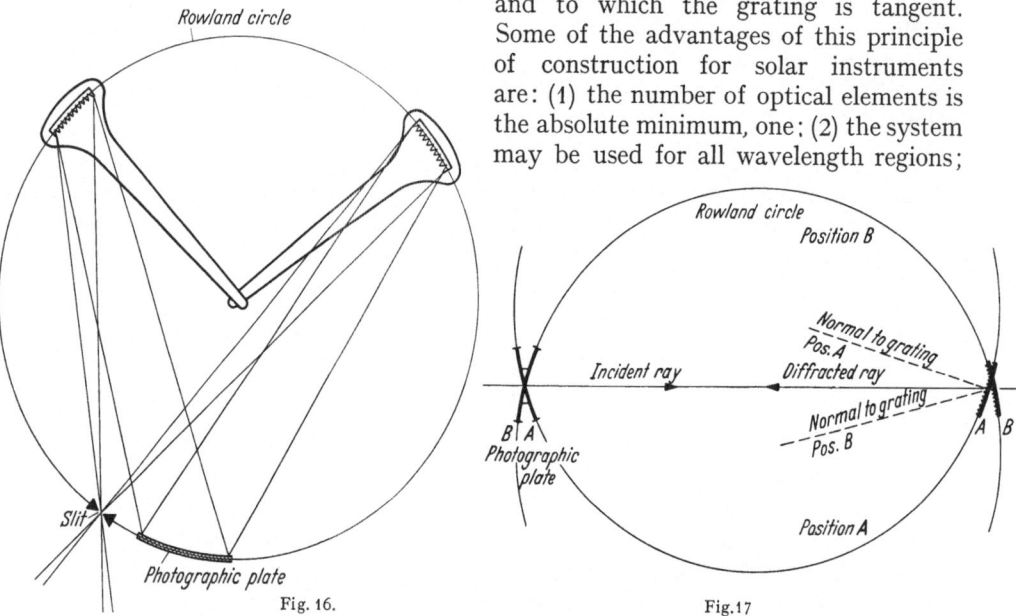

Fig. 16. The radius mounting of the concave grating. The grating is shown in two different positions to illustrate the large floor space required for concave grating mountings using the Rowland circle principle.

Fig. 17. The Eagle mounting of the concave grating. The grating is shown in two positions. The rotation of the photographic plateholder for the two gratings positions is indicated.

(3) spectra formed in the direction of the normal to the grating have constant and linear dispersion; (4) large ranges in wavelength and several spectral orders can be photographed simultaneously; (5) multiple slits and multiple gratings may be used to attain high efficiency in various regions. The principal disadvantage of each of the five types of spectroscope that have been mentioned is the presence of a large degree of astigmatism in the image. Additionally, there are several difficult problems connected with the installation of these instruments. All but the Eagle mounting occupy a very large space (Fig. 16) because of the location of the slit, spectral image, and grating on the circumference of a circle of large diameter. (A large diameter is required to obtain a plate scale adequate for solar research.) This condition nearly prohibits the use of these types as vertical spectroscopes in relatively small pits beneath solar tower telescopes, and they are cumbersome in any position if it is required (as is frequently the case) that they rotate about the line connecting the center of the entrance slit with the center of the grating[1]. However, the concave grating spectroscope can be given a convenient form as a fixed horizontal instrument.

The Eagle mounting (Fig. 17) leads to a compact instrument in which the slit and focal surface are in coincidence at the same point on the Rowland circle.

---

[1] A. S. King: Astrophys. Journ. **40**, 205 (1914).

The dispersion of an Eagle spectroscope varies with the wavelength, but the astigmatism is near the minimum value obtainable with a concave gratings alone. Solar installations of Eagle mountings have been used for investigation of the ultraviolet spectrum, and for general spectroscopy[1].

It is a comparatively simple matter to introduce small cylindrical lenses to correct the astigmatism of concave-grating spectroscopes for short ranges of wavelengths. However, another instrument commonly used for solar study, the Wadsworth mounting, provides many advantages over types using the Rowland circle. In the original Wadsworth mounting, a lens was used to collimate the radiation passing through the slit and the concave grating was placed in the collimated beam. Modern practice employs a concave mirror in place of the lens, thus achieving compactness and perfect achromatism. The slit is placed near the grating and the collimator and focal plane are near together at the other end of the instrument (Fig. 18). Astigmatism, coma, and spherical aberration are zero on the grating

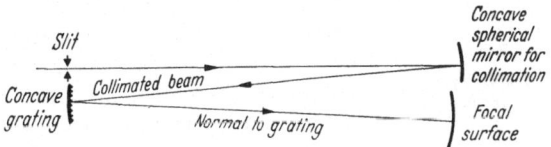

Fig. 18. The all-reflecting Wadsworth mounting of the concave grating.

normal, and do not increase rapidly with distance from the normal. The perfection of the spectrum image makes it possible to correlate line widths and intensities in the spectrum point by point with a solar image on the slit. The spectra of small solar structures can be recorded and identified with this instrument.

If it is necessary to change to a new region of wavelengths, the procedure is a bit complicated. The grating must be rotated about an axis parallel to its grooves and passing through its center. The focal surface will change its shape and also its distance from the grating. For photographic observation of the spectrum this requires that the plateholder be shifted so that its center is on the normal to the grating, and the plate must be curved to fit the focal surface as well as possible. Changes to new ranges of wavelengths are usually made infrequently, since the Wadsworth mounting will produce good spectral images over long spectral regions at any given adjustment.

Although concave-grating spectroscopes are attractive because of their simplicity and excellent achromatic qualities, and despite their generally satisfactory use in solar spectroscopy at solar eclipses and for the measurement of wavelengths of solar spectrum lines, only a few instruments of this type are now in use or under construction because of the slow improvement in the availability of concave gratings of high quality.

However, the number and quality of *plane gratings* has increased rapidly since their introduction into solar spectroscopy. The autocollimating spectroscopes of the Mount Wilson Observatory were among the first large solar instruments to exploit the relatively easy availability and high degree of perfection of the plane grating. There are four spectroscopes of similar size at the Mount Wilson Observatory and among these the spectroscope of the 150-foot Tower Telescope will be described as typical. The image of the Sun, 4250 mm

[1] E. PETTIT: Astrophys. Journ. **91**, 161 (1940). — H. A. BRÜCK and T. MARY: Vistas in Astronomy, Vol. 1, p. 433. London and New York: Pergamon Press 1955.

in diameter, is formed by the tower telescope ($\varnothing = 305$ mm, focal length 45.7 m) on the entrance slit of the spectroscope about at ground level (Fig. 19). The radiation passing through the slits proceeds vertically downward along the axis of a cylindrical pit to a two-element collimator lens ($\varnothing = 200$ mm, focal length 22.8 m). A plane grating (150×200 mm² ruled area, 600 grooves per mm) is mounted just below the collimator. Light diffracted by the grating in an upward direction will pass again through the collimator lens which will focus it at the top of the pit near the entrance slit A window is provided at this position so that the spectrum may be studied. The spectrum may be observed visually, a photographic plate may be placed at the focal surface, or a slit and photocell may be driven along the spectrum to record the variations in intensity electrically. A spectroscope of this kind (sometimes called a Littrow, or auto-collimating spectroscope) has many properties in common with the Eagle mounting of a concave grating, but it gives better definition over the usable spectral range (150 Å or less, at a single setting). To change from one spectral region to another, the collimator-image lens is re-focused, the grating is rotated, the plate-holder is adjusted to the focal surface produced by the color curve of the lens, and the spectrum is re-centered in the observing window. In addition to these minor inconveniences, serious troubles may arise because of stray light. The incident and diffracted beams travel along almost identical paths from slit to lens and from lens to focal surface. Therefore, it is almost impossible to keep stray light from the incident beam out of the image. The incident light reflected at the collimator lens surfaces is the most troublesome source of stray light.

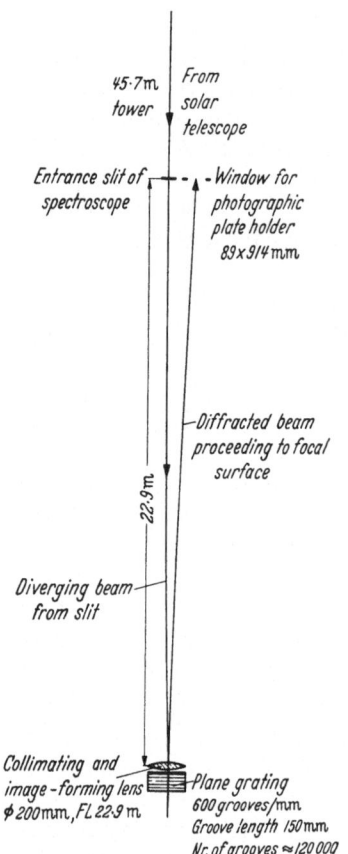

Fig. 19. Diagram of the optical path in the spectroscopy of the 150-foot Tower Telescope of the Mount Wilson Observatory.

With re-arrangement of the light path and the use of an all-reflecting system an instrument can be devised that is free from the nuisance of refocusing for different wavelengths and in which diaphragms can be introduced to control stray light (Fig. 20). The two mirrors shown in the diagram must form images at considerable angles with their optical axes, and it might be expected that serious aberrations would appear, but since the light path is perfectly symmetrical, except for dispersion at the grating, the aberrations introduced by the first mirror are almost perfectly corrected by the second[1,2]. Furthermore, the light paths

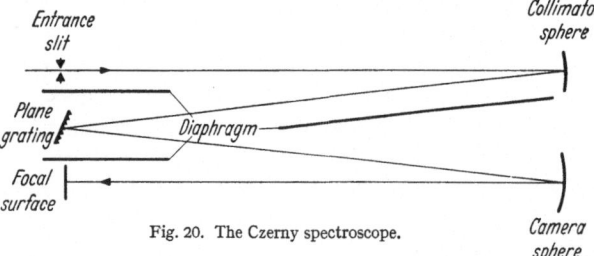

Fig. 20. The Czerny spectroscope.

[1] M. Czerny and A. F. Turner: Z. Physik 61, 792 (1930).
[2] M. Czerny and V. Plettig: Z. Physik 63, 580 (1930).

make possible good control of stray light with absorbing diaphragms. The range of wavelengths imaged with high quality at a given grating setting is small, 100 Å or less on the scale of the usual solar instrument. The aberrations increase at about the same rate as in the Wadsworth spectroscope as distance from the line in the focal plane that is exactly symmetrical to the slit increases. However, only rotation of the grating is required to shift to a new spectral region. This design, the Czerny spectroscope, has been adopted in many recent solar installations (Figs. 13 and 21).

In 1955 a very large spectroscope of the Czerny type, completely enclosed in a vacuum tank, was installed at the McMath-Hulbert Observatory (Fig. 21)[1]. The spectroscope is enclosed in a vacuum tank for the following reasons: (1) all air currents within the instrument are eliminated; (2) the instrument is insensitive to temperature and pressure changes in the surrounding air; (3) the mirrors and grating are protected from deterioration in air; (4) the spectroscopic observing conditions are controlled

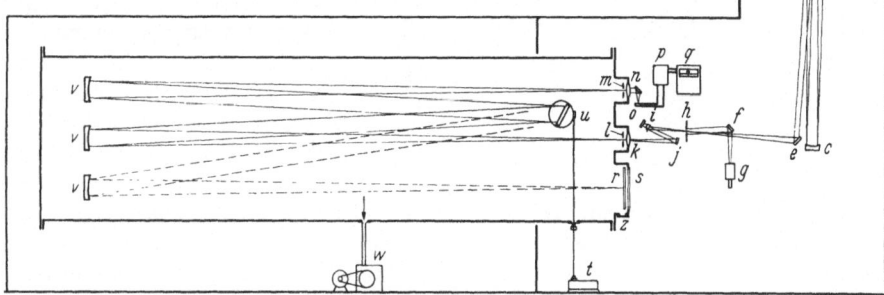

Fig. 21. The McGregor tower telescope and vacuum sepctroscope of the McMath-Hulbert Observatory of the University of Michigan. *a* Coelostat mirror. *b* Second plane mirror. *c* Concave spherical mirror ⌀ 457 mm, FL 30.8 m. *d* Plane mirror. *e* Plane mirror. *f* Plane mirror. *g* Lyot polarizing monochromatic filter. *h* Reflecting slits. *i* CaFe prism and mirror. *j* Concave mirror. *k* Fused silica field lens. *l* Entrance slit of grating spectrosope. *m* Exit slit. *n* Field lens and mirror. *o* Photocell. *p* Amplifier. *q* Recorder. *r* Photographic plate. *s* Window for visual observation. *t* Motor for turning grating. *u* Plane diffraction grating, groove length 134 mm, width of ruled area 204 mm, 600 grooves per mm. *v* Spherical concave mirror ⌀ 305 mm, FL 15.2 m. *w* Vacuum pump.

and reproducible within close limits from day to day. The vacuum spectrograph of the McMath-Hulbert Observatory may be used in either of two slightly different arrangements: one for photoelectric recording; the other for photography of the spectrum.

The photoelectric recording system deviates from the original Czerny design in that the axis of rotation of the grating is displaced 690 mm toward the mirrors from the line joining the centers of the entrance and exit slits. This is of little importance in decreasing the excellence of the image since the mirrors are used only 24 minutes of arc off-axis.

The photographic image-forming mirror is operated 1° 12′ off-axis, and its aberrations add to those of the collimating mirror because of the position of the grating, but since the focal ratio of the instrument is greater than fifty, the image quality is still well-matched to the resolving power of fast photographic emulsions (see Vol. LII, this Encyclopedia, p. 163, for reproductions of spectra photographed with the McMath-Hulbert Czerny-type vacuum spectrograph). The

---

[1] R. R. McMath: Astrophys. Journ. **123**, 1 (1956).

small prism spectroscope that precedes the main instrument acts as a mono-chromator whose exit slit is simultaneously the entrance slit of the vacuum instrument. It serves to separate the overlapping spectra formed by the grating. Colored glass and interference filters are used in addition to the prism monochroma-tor to limit the band of wavelengths admitted to the main instrument, and an elaborate system of absorbing screens of black felt is installed to diminish stray light.

During photoelectric recording, the photographic image mirror is covered by a blackened screen. The grating rotates slowly about an axis parallel to its grooves and passing through its center. The motion of the grating moves the spectrum across the photoelectric exit slit and the emergent radiation is focused by a quartz field lens to form an image of the image-forming mirror on the detector.

The change to photography of the spectrum is made by first shifting the ab-sorbing screen from the photographic image mirror to the photoelectric mirror by means of a remote controlled electric motor. The grating is then rotated so that it diffracts the desired range of wavelengths to the photographic plate. The

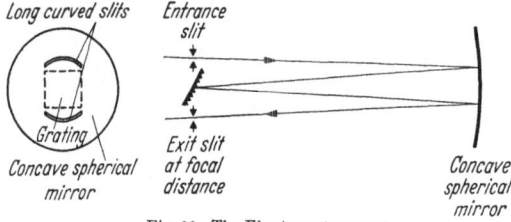

Fig. 22. The Ebert spectroscope.

monochromator must next be adjusted to transmit the appro-priate region of the spectrum by means of a small tangent screw which the observer adjusts manu-ally. Photographic plates are placed in, and removed from the instrument through an air-lock so that the vacuum is not disturbed.

The Czerny spectroscope (Fig. 20) forms extremely good images over a small field for f-ratios greater than 10, but if the spectroscope must concentrate more energy in the focus for use with insensitive detectors such as the thermocouples or bolometers employed in infrared spectroscopy, other designs permitting the use of long slits are preferable. One of the best of the alternate designs is the Ebert system (Fig. 22)[1] in which a single large concave spherical mirror replaces the two concave sphericals of the Czerny spectroscope. (The Czerny spectroscope may be considered a special case of the Ebert spectroscope for large focal ratios.) The slits in an Ebert spectroscope may be very long, provided they are curved about the axis of symmetry of the system (Fig. 22), with an enormous gain in the quantity of radiation ad-mitted to the spectroscope. The advantages of the Ebert design have been realized only recently and are just now being applied in infrared solar spectroscopy.

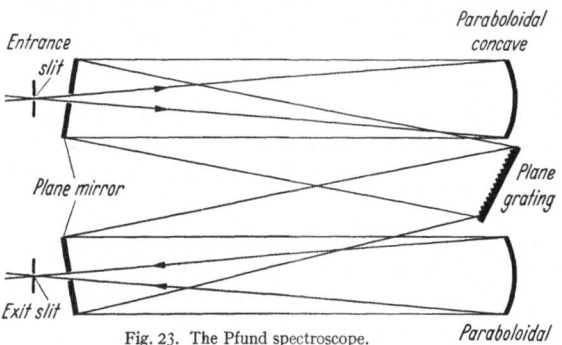

Fig. 23. The Pfund spectroscope.

The Pfund spectroscope[2] (Fig. 23) has been widely used in solar spectroscopy, particularly in the extreme ultra-violet and infrared spectral regions where the energy available is low, and the detectors are insensitive. The beam of radiation entering or leaving the instrument must pass through a small hole in a plane

[1] H. Ebert: Wied. Ann. **38**, 489 (1889). — W. G. Fastie: J. Opt. Soc. Amer. **42**, 641, 647 (1952).

[2] A. H. Pfund: J. Opt. Soc. Amer. **14**, 337 (1927).

mirror. This circumstance, and the use of paraboloidal mirrors for the concaves, limits the field of view to nearly a single spectrum line. The design provides a wider choice of positions for the dispersing element (grating or prism) than is the case for the Ebert system.

A limited extent of the focal surface of good image quality is a common characteristic of all plane grating spectroscopes in use for solar research. The range of wavelengths covered at adequate plate scale (not less than $\frac{1}{2}$ Å/mm for usual solar work) is not more than a few hundred angstroms at best, but for many types of observation of the Sun it would be desirable to record the entire photographic spectrum from 13 000 to 2900 Å at high dispersion and high resolving power. In principle, this problem can be solved by using the spectra of high order that are produced by a grating spectroscope. Since the efficiency of a

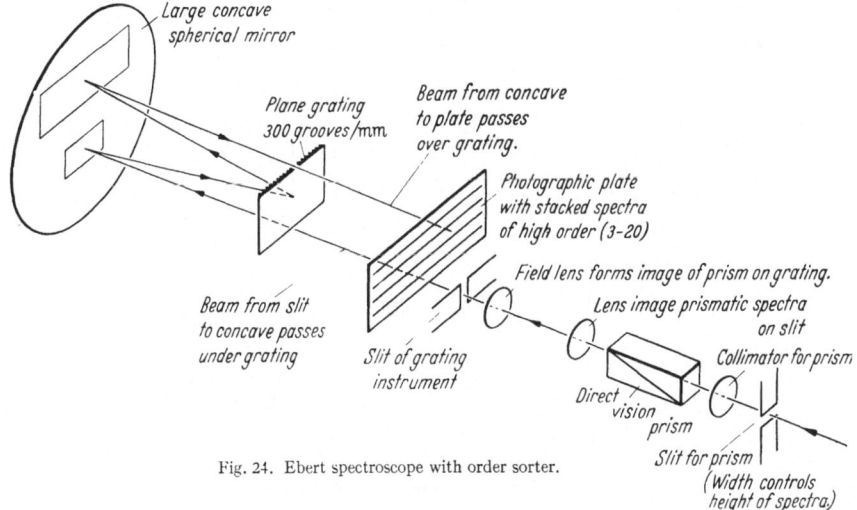

Fig. 24. Ebert spectroscope with order sorter.

grating spectroscope is proportional to the sum of the sines of the angles of incidence and diffraction, an additional gain is obtained through the use of high orders. These considerations have led to the design and construction of crossed-dispersion spectroscopes[1].

One of the crossed-dispersion instruments, which has been given the commercial name "Ebert spectrograph with order-sorter" in the United States (Fig. 24), appears to have many possible applications in solar research. A small direct-vision spectroscope is placed to project its spectral image on the slit of an Ebert plane-grating instrument, the slit direction being the direction of dispersion in the prismatic spectrum. On the focal surface of the grating instrument, adjusted for high spectral orders (10th to 12th at 4000 Å), will appear spectra of the several normally coincident orders separated one above the other. A number of spectra may be arranged one above the other in this way, each exhibiting the high dispersion and resolving power of the grating instrument. The spectra will be narrow (about one or two millimeters wide), since the slit is effectively shortened in the direction perpendicular to the dispersion of the grating. The narrow width of the spectrum is objectionable in some kinds of solar observation.

[1] R. MICHARD, R. SERVAJEAN and J. LABORDE: Astrophys. Journ. **127**, 504 (1958). — R. MICHARD, R. SERVAJEAN and J. LABORDE: Ann. d'Astrophys. **22**, 787 (1959).

Gratings with 300 grooves per millimeter, intended for use in the first order at 13 000 Å to twentieth order at 2900 Å, are employed in the "Ebert spectrograph with order-sorter". More coarsely ruled gratings with ten or fewer grooves per millimeter may be used in a somewhat similar way in the 100th to 500th orders[1]. The gratings with few grooves per millimeter have been called "*echelles*". The most compact arrangement for their use consists of a modified Wadsworth mounting of a concave grating of ordinary characteristics, 600 grooves/mm, focal length 3.2 m [2] (Fig. 25). After reflection from the collimator, the collimated radiation is reflected from the echelle to the concave grating which forms the image of the spectrum near the entrance slit. In this arrangement, the slit, echelle,

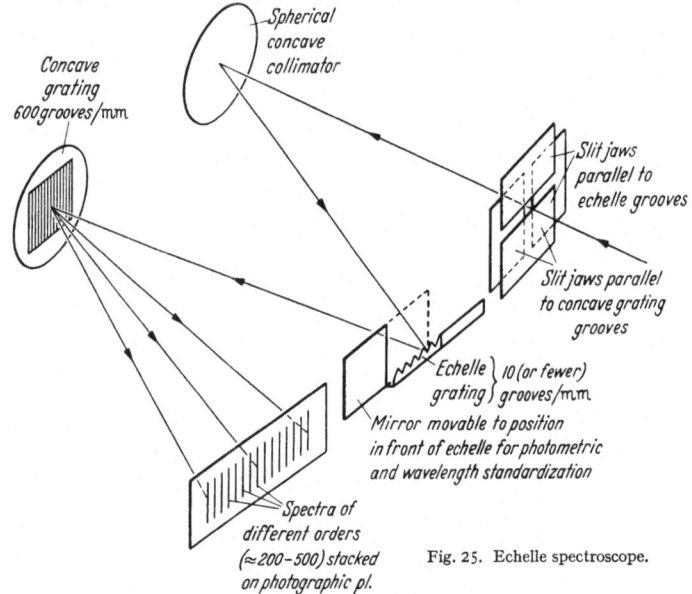

Fig. 25. Echelle spectroscope.

and the image of the spectrum are located near one end, the concave grating and concave collimator near together at the opposite end of the instrument. Much larger dispersion and resolving power are attained with the echelle than with the "order-sorter". At the present time, echelles do not achieve as high resolving power as the best plane gratings, ruled with 600 grooves/mm for use in the first-order direction of diffraction of 25 000 Å radiation. The best echelles produced up to the present time have 15 to 30% stray light in their spectra. Stray light as intense as this prohibits use for spectrophotometry of solar sources.

13. Spectroheliographs and spectroscopic monochromators. The modification of the spectroscope obtained with the placing of a slit in the focal surface so that the instrument transmits only radiation in a narrow band of wavelengths is called a monochromator. In this form, the spectroscope and a solar telescope are the image-forming instruments of the spectroheliograph. To complete the spectroheliograph, means are included for scanning the telescopic image in such a way as to form a monochromatic image of the Sun on a photographic plate, or, by increasing the rate of scanning, to make possible the visual impression of a mono-

[1] A. K. Pierce, R. R. McMath and O. Mohler: Astronom. J. **56**, 137 (1951).
[2] G. R. Harrison, J. E. Archer and J. Camus: J. Opt. Soc. Amer. **42**, 706 (1952).

chromatic image by persistence of vision. An instrument with a rapid scanning frequency for visual use is called a spectrohelioscope[1].

There are three possible scanning methods: (1) the monochromator is not moved and the image of the Sun and the photographic plate are moved in synchronism past the entrance and exit slits, respectively; (2) the image of the Sun and the photographic plate are fixed and the entire monochromator is moved; (3) since neither of methods (1) or (2) is practical for visual work in which the scanning frequency must form at least sixteen images per second, the equivalents of these motions are produced by optical and mechanical devices that permit the solar image, the eye (or other receiver), and the principal parts of the monochromator to remain stationary.

The spectroheliograph has proved to be one of the most powerful of all instruments for solar research, and, after a period of many years during which no new instruments of this type were built, several novel designs, intended to take advantage of the many advances in spectroscope construction, have been proposed, and some are in operation[2].

The spectroheliograph of the Crimean Astrophysical Observatory is a recently completed instrument that embodies some new features. The basic spectroscope is a Czerny type housed in a pit directly beneath the tower telescope (Fig. 13). Radiation enters slit S_1'' to concave mirror m_1'. A parallel beam from m_1' falls on grating G_2 (ruled area 150×150 mm^2, 600 grooves per mm) by way of plane mirror n_1, thence it is diffracted to m_2' and m_3' ,via n_2 and n_3 respectively. The concave mirrors m_2' and m_3' focus the beams on slits S_2'' and S_3''. The scanning motion for the solar image is provided by moving a reversing prism at half speed in front of slit S_1'' at the same time that a photographic film is moved full speed across slits S_2'' and S_3''. Simultaneous monochromatic images of the Sun, 50 mm in diameter, can be produced with this instrument, using radiation from any two spectral regions not closer together than 300 Å. Separate film transports are provided for each of the two regions of the spectrum so that long series of pictures can be made easily. The pass-band of the instrument is generally one-tenth of an angstrom.

14. Broad pass-band spectroheliographs and multiple-image spectrographs. A picture of the Sun made with a spectroheliograph, or any other monochromatic filter with a narrow pass-band, can record only those features whose line-of-sight velocities lie within a very narrow range. A series of spectrograms covering the same area gives complete complementary information about line-of-sight velocities, but rapidly changing objects cannot be observed often enough during their lifetimes in this way. It has been suggested frequently that a zero-dispersion monochromator, in principle like the type designed by van Cittert[3], would permit increasing the width of the pass-band of the spectroheliograph to include a wider velocity range without damaging the quality of the image. An operating instrument of this kind has not yet been constructed, but several other devices, designed to give information about line-of-sight velocities of monochromatic details, have been tried. The *spectroenregistreur* of Deslandres[4] and the Stone

---

[1] P. A. Wayman: Vistas in Astronomy, ed. by A. Beer, Vol. 1, p. 422. London and New York: Pergamon Press 1955.

[2] H. W. Dodson and O. Mohler: Astronom. J. **63**, 309 (1958). — R. B. Leighton: Astrophys. Journ. **130**, 366 (1959).

[3] P. H. v. Cittert: Rev. d'Opt. **5**, 393 (1926).

[4] H. Deslandres: Ann. Meud. Obs. **4**, 1 (1910).

Spectroheliograph[1] were early instruments designed to record spectra for many points within the area of a monochromatic photograph, but the rate of recording in these instruments was too slow to follow rapid motions.

The multi-slit spectrograph[2], used with great success on planetary nebulae, should be applicable to prominences, since they, like the nebulae, radiate chiefly in a few widely separated emission lines. The same technique can be applied to features of the solar disk if a narrow-pass filter, such as the polarizing monochromatic filter, is used to project a monochromatic image on the spectrograph slit. In a somewhat similar application, a monochromatic polarizing filter can be used to project two monochromatic images in the plane normally occupied by the slit jaws of a spectroscope. If one of the images is reversed left to right with respect to the other, the two images will be re-imaged in the focal plane of the spectroscope with line-of-sight velocity shifts appearing to give stereoscopic relief to moving features[3].

15. Auxiliary instruments: Guiding. Telescopes, spectroscopes, and monochromators are not only frequently modified, as the foregoing descriptions indicate, but auxiliary instruments of considerable complexity are often attached to them for the observation of specific phenomena. Brief descriptions of a few of these interesting devices follow.

Rather unusual demands for precision guiding are made on solar telescopes of all types. Measurement and analysis of long series of solar photographs are greatly facilitated if the telescope maintains identity of position from picture to picture. If, in other circumstances, a telescope of long focal length is used to project a small area of the Sun on a spectroscopic slit, the telescope must guide without attention for a period of minutes before any detectable error appears. Solar seeing is frequently at its best in the early morning hours with the Sun near the horizon, and moving at a changing rate in both declination and right ascension because of the effect of refraction. Under those conditions precise guiding cannot be obtained from the usual form of driving clock and more elaborate devices are generally used for solar telescopes.

The simplest of these consists of two pairs of photoelectric cells arranged with respect to a solar image so that one pair is affected by motion of the image in right ascension and the other pair responds to motion in declination. The cells may merely operate the equivalent of slow motion guiding buttons, adding corrections to a continuous approximate guiding motion supplied by a motor running at a constant rate[4]. Alternatively, the photoelectric cells may control the rate at which the driving motor runs, and all corrections to the telescope position may be made by changing the rate of the driving motor[5]. Driving systems of this sort will follow the center of the Sun accurately.

Corrections to the telescope position and to the driving rate may also be introduced manually by the observer and a variable rate drive will then provide a rate correct for a particular area of the Sun, which, because of the effects of refraction, rotation of the Sun, and motion of the object, may be rather different from the correct rate for the center of the Sun's disk[6].

[1] R. R. McMath: Publ. Obs. Univ. Michigan **8**, 141 (1943).

[2] O. C. Wilson: Science, Lancaster, Pa. **122**, 882 (1955).

[3] J. W. Evans: Unpublished, Sacramento Peak Observatory.

[4] A. E. Whitford and G. E. Kron: Rev. Sci. Instrum. **8**, 78 (1937). — H. W. Babcock: Astrophys. Journ. **107**, 73 (1948).

[5] W. O. Roberts: Electronics **19**, 100 (1946). — E. Fowler and S. Johnson: Electronics **24**, 118 (1948).

[6] R. R. McMath: Publ. Obs. Univ. Michigan **7**, 42 (1937). — L. P. Tabor: Publ. Cook Obs. Univ. Pennsylvania **4**, 3 (1939).

In some problems it is desirable to drive the solar image at a preselected rate along a line at a given position angle. To make this possible, solar telescope drives should have widely variable rates in both right ascension and declination and it should be possible to combine them in definite proportions. In studies of solar limb darkening, an additional and useful refinement is provision for a driving rate that will move the solar image rapidly near the center of the disk and then more slowly as the edge is approached.

Some types of solar observations require the use of auxiliary guiding devices to place a selected object in position with respect to the analyzing instruments. Often interesting objects are visible only in monochromatic light and it is unsatisfactory to try to locate them by offset from white light detail. An early solution of this problem[1] employed a spectrohelioscope as the guiding device. In a later solution, the spectrohelioscope served as both guiding device and spectroscope[2]. The first order spectra of the spectrohelioscope were used for visual observation and the guiding of monochromatic objects on the slit of the instrument, the higher orders were used to make photographs of the spectrum. More recently, polarizing interference monochromators have been employed to fulfill similar functions, either with auxiliary guiding telescopes (Fig. 13, F_1 and F_2) as in the solar tower telescope of the Crimean Astrophysical Observatory, or with light reflected from polished slit jaws (Fig. 21) as in the McGregor Tower Telescope of the McMath-Hulbert Observatory[3]. In both cases, a photograph taken through the filter is a detailed record of the position of the spectroscope slit in relation to the solar detail.

16. Auxiliary instruments: Methods of recording. The circumstance that the photographic method easily records simultaneously the relative positions and brightness of all points in a sizable area is a unique ability and it is the principal reason for the dominant use of photography in most types of astronomical recording. This property has been exploited for nearly a century by solar astronomers in the production of the daily records of the positions and areas of spots and faculae on the solar disk[4]. The successful use of time-lapse photographic recording at the McMath-Hulbert Observatory[5] has led to an ambitious extension of these methods on a world-wide basis that aims to provide a record of all transient solar phenomena at a minute-by-minute rate[6].

Some other considerable advantages of photographic methods are: (1) high sensitivity, that extends from x-ray wavelengths of a few angstroms into the infrared to a long wavelength limit near 13 000 Å; (2) ease in handling; (3) commercial availability; (4) simplicity in use with other instruments. However, the complexity of the process of translating photographic densities into measurements of radiation intensities (see WEAVER's article on Photographic Photometry, in this volume) makes other methods attractive for many problems.

Electrical detectors (photo-emissive and photo-conductive cells, bolometers, thermocouples, etc.) have gained steadily in the number of their applications

---

[1] G. E. HALE: Astrophys. Journ. **70**, 306 (1929).

[2] H. W. DODSON and S. E. A. v. DIJKE: Publ. Amer. Astronom. Soc. **10**, 122 (1941). — M. A. ELLISON: 7ième Rapp. Comm. pour l'étude des relations entre les phénomènes solaires et terrestres, p. 1. Paris: Hemmerle, Petit & Cie. 1951.

[3] The Solar System, Vol. 1: The Sun, ed. by G. P. KUIPER. Chicago: Univ. Chicago Press 1953.

[4] H. S. JONES: Sunspot and Geomagnetic Data derived from Greenwich Observations 1874—1954 compiled under the direction of H. S. JONES. London: H. M. Stationery Office 1955.

[5] F. C. McMATH, H. S. HULBERT and R. R. McMATH: Publ. Obs. Univ. Michigan **4**, 53 (1931).

[6] W. O. ROBERTS: Draft Reports, Internat. Astronom. Union, Dublin Meeting Aug. to Sept. 1955, report of sub-commission 11, p. 81. Cambridge: Cambridge University Press 1955.

to solar observation, partly because of their linear response curves over wide ranges of intensity (see Whitford's article on Photoelectric Techniques, in this volume) but also because, reliable and well-developed techniques of electrical measurements may be used. An electrical detector can be placed behind a slit, and the combination moved along the focal plane of a spectroscope to produce an intensity record of a region of the spectrum[1], or the detector may be fixed in position to receive radiation from the exit slit of a monochromator (Fig. 21) and an extensive part of the solar spectrum may be scanned by rotating the dispersing element[2].

Both of these scanning methods advance very slowly through the spectrum of a high dispersion spectroscope, generally not faster than at rates of one to four minutes per angstrom, and the instruments and observing conditions must be stable during the time required to record a region of the spectrum, or the records will be useless. The major causes of instability within the spectroscope are minimized by pumping the air out of the instrument, and several methods of making allowance for instability of the sky transparency and image unsteadiness have been developed.

One of the methods uses a second detector to monitor sky conditions. The output from the monitoring detector can be used to correct automatically for minor changes in sky transparency[3]. Unsteady images can be stabilized through the use of additional photoelectric servo-mechanisms[4].

Another method attempts to circumvent the difficulty by reducing the time required to make an observation. Recordings are made at very high speed with the aid of oscillographic recorders for following rapid changes in signal. A high speed scanning attachment is among the auxiliaries provided for use with the horizontal spectroscope illustrated in Fig. 13. A mirror rotatable about an axis perpendicular to the direction of dispersion, and parallel to the spectrum lines, is placed in the light path just before the focal surface at P, Fig. 13. The mirror reflects the spectrum to a focus on a fixed slit that is the entrance aperture for a photoelectric cell. Motion of the spectrum across the slit can be very rapid since only a small mirror must be rotated. The deflection of the light spot of an oscilloscope indicates the variations in spectral intensity. These are recorded on a photographic film moved in synchronism with the rotating mirror that scans the spectrum[5].

The monochromator-photoelectric cell arrangements just described provide excellent intensity data, and with small changes in their mode of operation may be used for the measurement of relative wavelengths. With the aid of a small prism monochromator similar to the one shown in Fig. 21, tracings of different ordered spectra may be recorded one after the other so that interpolation between the orders is possible. If the prism monochromator follows, rather than precedes the main instrument, the different orders can be recorded on independent channels continuously and simultaneously on a single paper tape and the relative wavelengths in the various orders can be established easily.

The method just outlined (essentially Rowland's method of overlapping orders) will, in the most favorable case, make possible determinations of relative

---

[1] T. Dunham: Phys. Rev. **44**, 329 (1933). — H. A. Brück: Monthly Notices Roy. Astronom. Soc. London **99**, 607 (1939).
[2] M. S. Lamansky: Phil. Mag., Ser. IV **43**, 282 (1872).
[3] F. Goos, and P. P. Koch: Z. Physik **44**, 855 (1927). — W. A. Hiltner and A. D. Code: J. Opt. Soc. Amer. **40**, 149 (1950).
[4] R. B. Leighton: Sci. Amer. **194**, 157 (1956).
[5] A. B. Severny: Izv. Astrophys. Obs. Crimea **16**, 12 (1956).

wavelengths in the solar spectrum to a precision of four or five parts in a million on a single tracing. This degree of precision is very near to the attainable maximum in the spectrum of the Sun in which the lines are not very sharp and are variable in wavelength from point to point on the solar surface. The long series of photographic measurements made with great care in the first half of this century used the Fabry-Pérot étalon or the Lummer-Gehrcke Plate (see Vol. XXIV, p. 232 to 233 and Vol. XXIX, this Encyclopedia) in conjunction with a stigmatic grating spectrograph (Fig. 26) [1] and they achieved a precision of two or three parts in ten million for the *average* values of the wavelengths of solar lines. Doubtless, this degree of precision can be obtained with the modern grating and photoelectric recording, if needed, but at the present time solar astronomy has shifted its emphasis from average values for the wavelengths of solar spectral lines to the study of wavelength variations within small areas of the Sun at particular times.

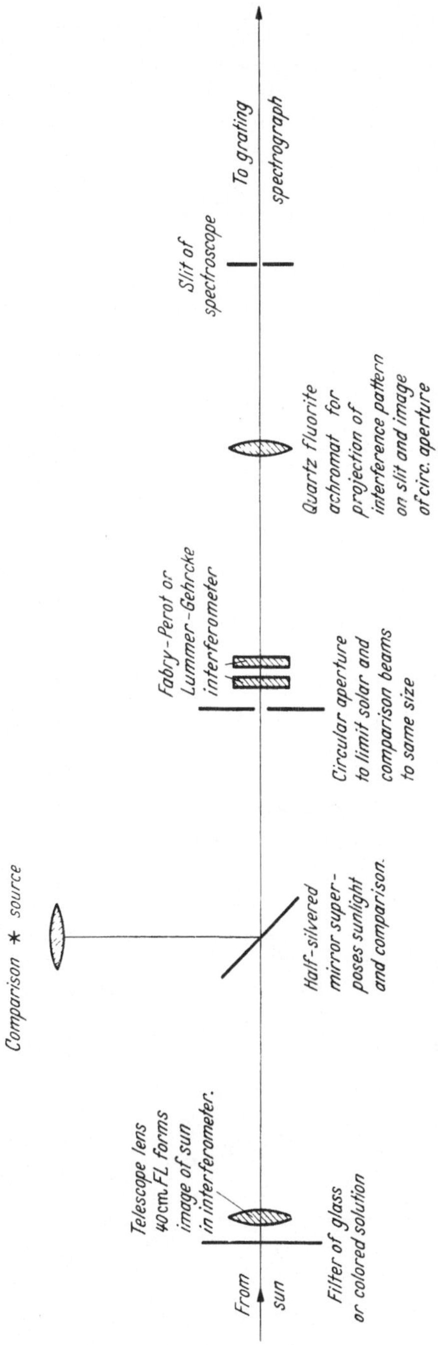

Nevertheless, the Fabry-Pérot étalon remains a very useful instrument in solar observation. It behaves like a light filter with very narrow transmission bands. It may be used to derive precise line contours [2] if one makes intensity measurements within the highly monochromatic channels obtained when the étalon is used with a spectroscope. The system of interference channels obtained with an arrangement like that of Fig. 26 can be swept across a line in the spectrum by varying the path length in the étalon, by mechanical means, or by changing the air pressure. In principle, a single channel can be followed across a spectrum line using the methods of direct photoelectric scanning. The data thus obtained reflect the high effective resolving power of the etalon, and are especially useful for the evaluation of the effect of stray light

Fig. 26. Interferometer in conjunction with grating spectrograph for wavelength measurement (see footnote 1, p. 31).

---

[1] K. BURNS and W. F. MEGGERS: Publ. Allegheny Obs. **7**, 105 (1929).

[2] C. D. SHANE: Bull. Lick. Obs. **16**, 76 (1932).

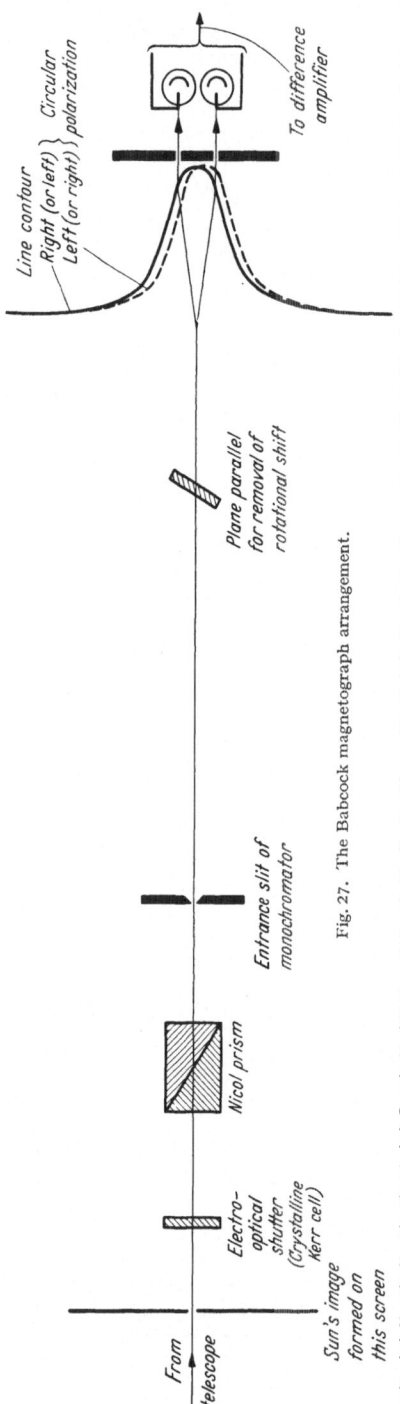

Fig. 27. The Babcock magnetograph arrangement.

upon the recorded spectrum. Such procedures have been applied in the derivation of precision line profiles an in the measurement of the sun's magnetic fields. Fields of sunspots (about 1000 gauss) are strong enough to split a sensitive spectral line into components separated by 3 or 4 millimeters on the scale of a solar spectrogram of high dispersion; hence, there is little problem in the measurement of these fields. But if it is required that fields of the order of one or two gauss be measured, such a difficult observational problem is presented that some of the most complicated collections of instruments to be found in astronomy are assembled in this connection[1].

One of the latest developments of a procedure for the daily measurement of small magnetic fields over the complete solar disk was carried out at the Hale Solar Laboratory of the Mount Wilson Observatory during the years 1950 to 1955 by H. D. and H. W. Babcock. The following are the essential elements of the method.

Light from a small area of the Sun is transmitted through an analyzer for circular polarization to the slit of a high resolving power spectroscope, which is arranged as a monochromator. The monochromator is adjusted to pass light from one side of a magnetically sensitive spectral line (Zeeman effect, Vol. XXVIII of this Encyclopedia). Radiation from the exit slit of the monochromator is received on a photoelectric cell. If the analyzer for circular polarization is shifted to transmit alternately the right- and left-handed circularly polarized components of the spectral line, and if circular polarization is present, the emergent radiation will be modulated at the frequency of shifting of the analyzer. Since the variation of intensity with wavelength is approximately linear for small distances along the contour of the edge of a spectral line, the amplitude of the modulated signal will vary linearly with field strength for small fields. A switching system following the scheme just outlined, which uses light from only one side of a spectral line, is very sensitive to many kinds of spurious modulation and the arrangement actually used is the one shown schematically in Fig. 27[2].

[1] H. v. Klüber: Vistas in Astronomy, ed. by A. Beer, Vol. 1, p. 751. London and New York: Pergamon Press 1955.
[2] H. W. Babcock: Astrophys. Journ. 118, 387 (1953).

Two slits are placed symmetrically on opposite sides of the spectral line (actually two spectral lines are used in the latest model of the instrument to increase the photoelectric signal) and the respective photoelectric cells for each slit are connected to a difference amplifier that is sharply tuned to the switching frequency of the analyzer for circular polarization. A shift in the position of the spectral line in synchronism with the switching frequency will produce a large signal, but a change in intensity common to the two positions of the line will cause no response. After amplification, the modulated signal is rectified synchronously with the driving circuit for the electro-optical analyzer so that the response of the system indicates which cell is the more brightly illuminated. (This feature indicates the polarity of the magnetic field producing the splitting of the line.) The telescope is driven to produce a stepwise scanning motion of the image perpendicular to the axis of rotation of the Sun. A correction for the rotational line shift is introduced by tilting a plane parallel plate (Fig. 27) in synchronism with the scanning motion. An oscilloscope displays the signal as a series of chords, and deviations from chords, of a circle that represents the Sun's disk. Usually about twenty-five parallel chords cover the Sun's disk. On the average over the Sun's disk, an instrument of this kind will indicate fields of the order of one gauss. The slit dimensions are approximately one and a half minutes by 2.7 seconds of arc, consequently, not very much fine structure on the Sun's surface can be investigated.

A circular polarizer can be introduced into the light beam before the electro-optical shutter and with this arrangement, the intensity of the transmitted light will be modulated 100%. In this condition, the instrument is exceedingly sensitive to very small displacements of the spectrum line. Displacements corresponding to line-of-sight velocities as small as ten or fifteen meters per second can be measured in this way.

Although most spurious signals are eliminated in the flicker method used in this instrument, and additional discrimination against unsought response is obtained by careful tuning of all amplifiers to respond to a single frequency (the frequency of switching of the two beams on the two photoelectric cells) with narrow pass-bands, convection currents within the monochromator can still cause trouble because they shift the lines in the spectrum in the same way as the flicker process. Removal of all air from inside the monochromator, as has been done in the instrument of the McMath-Hulbert Observatory, could improve stability of the spectral image, but mechanical vibration then still remains as a possible source of false signals. Delineation of extremely fine structure on the surface of the Sun will, as always, ultimately depend upon the excellence of the image of the Sun formed by the telescope.

Several experiments using the image orthicon as a receiver of radiation have been made with solar telescopes and spectroscopes in the hope that by increasing sensitivity, and thus reducing photographic exposure times, blurring caused by unsteadiness of the image might be reduced[1]. So far, all such experiments have failed to accomplish any improvement in the quality of the image. The best that can be attained at present introduces a net loss of approximately half the resolving power of a given instrument.

VI. Instruments for the observation of solar eclipses.

17. In the short time available for making observations at solar eclipses, it is almost impossible to collect data that would be considered of good quality

---

[1] P. B. FELLGET: Vistas in Astronomy, ed. by A. BEER, Vol. 1, p. 475. London and New York: Pergamon Press 1955.

in other types of solar research. One generally accepts observations that are at least one order of magnitude less precise than similar measurements made at non-eclipse times. Consequently, increase in sensitivity, with no sacrifice of the quality of the image, that results in decrease of the time required to record a given solar event would be most useful at solar eclipses.

Instruments for use at solar eclipses are, in many cases, merely observatory instruments transported to the track of the eclipse, but in nearly every instance many concessions must be made to the temporary character of eclipse observing sites. The principle distinguishing features of eclipse instruments arise from attempts to make the operation of the instruments semi-automatic so that the short duration of the eclipse may be used efficiently.

Geodetic instruments, intended to time precisely the instants of second and third contacts, need be nothing more than small telescopes equipped for motion picture recording of both the phenomena of the eclipse and accurate time records. The motion picture may record change in width of the crescent, the change in position-angle of the cusps of the crescent for stations located just outside the path of totality, or, with the help of a small spectroscope, the appearance and growth of the emission lines in the flash spectrum. Each of these aspects of an eclipse can be timed with great accuracy.

Astrometric observations at eclipse time are difficult only because the sky is never very dark and the exposure time required to photograph star images on the same plate with the eclipsed Sun generally results in very bad background fog. Since the star field around the Sun is distorted by relativity shifts, a scale field from a part of the sky far from the Sun is often reflected into the telescope by a half-silvered mirror so that exposure to the comparison field and the eclipse field can proceed simultaneously. Astrometric cameras used at eclipse time have not, as yet, been designed for this application and most of them could be improved with respect to control of stray light.

Although the problem of the control of stray light is made less difficult by the circumstances of an eclipse, some attention should be given to it, for the chromosphere and inner corona are bright objects that interfere with attempts to record the low light levels of the outer corona. The bright inner regions of the eclipsed Sun have been occulted by diaphragms placed at different points in the light paths of telescopes. In one ingenious arrangement[1], the occulting disks are placed on tracks some distance in front of the telescopic objectives. The disks are moved along the tracks at the correct rate for them to occult the Sun. In actual practice, however, it appears that the coronagraph system (Fig. 6), with the occulting disk near the focal plane of the telescope is the most effective in reaching low light levels.

Most often, orthodox equatorial mountings are used at eclipses to keep instruments directed toward the Sun, but instruments of unusual size may employ a siderostat, or coelostat with a fixed telescope. Fixed telescopes are also pointed to the position of the eclipsed Sun and the necessary guiding for direct photography is provided by moving a photographic plate in the focal plane of the telescope in the direction of motion of the image[2]. Such a telescope, producing an image of the Sun 150 to 300 millimeters in diameter, is ideal for the photography of the chromosphere, the prominences, and the inner corona.

Many of the spectroscopes employed at eclipses are also carried on equatorial mountings. The spectroscopic instruments are generally of intermediate disper-

---

[1] C. W. Allen: Monthly Notices Roy. Astronom. Soc. London **116**, 79 (1956).
[2] J. M. Schaeberle: Contrib. Lick Obs., Ser. I **4**, 1 (1895).

sion and resolving power (dispersion 0.20 to 1.0 mm/Å; 100000 to 200000 resolving power). Concave gratings, plane gratings, and prisms have been used in slitless spectroscopes for nearly all of the exploratory observations of the chromosphere. The slitless concave-grating spectroscope is in principle like the Wadsworth mounting[1] but since no collimator is necessary the grating receives the beam of sunlight directly and forms the spectral image. Similarly, the slitless plane-grating spectroscopes use no collimator, and sunlight is received directly on the grating. Light diffracted from the grating is focused by a lens into the spectral image. The slitless prism spectroscopes are identical in construction with the objective-prism instruments of stellar astronomy.

Rather elaborate plate (or film) moving mechanisms are used with all of these types of spectroscope in order to secure a number of photographs in rapid succession[2], or in a limiting form, to move the plate continuously perpendicular to the dispersion as the eclipse progresses[3]. In the instruments using the continuously moving plate, the height of the spectrum (perpendicular to the dispersion) is fixed by a slit parallel to the dispersion direction in a screen just in front of the photographic plate. The width of the slit frequently has been chosen to be one to three minutes of arc. The height of the spectrum in the instruments making individual stationary exposures in rapid succession is limited only by the length of the chromospheric arc.

A spectroscope with entrance slit and collimator is a necessity for all precision spectroscopy of the chromosphere, and since one wishes to form an image of details of chromospheric structure in the spectrum, the instrument must be stigmatic. The driving motion for the telescope that forms the image of the chromosphere on the entrance slit of the spectroscope should be provided with a variable rate so that utmost driving precision can be attained. A telescope equipped with a variable rate of drive in both coordinates can be rated to move exactly along the line of the Moon's motion at a fixed distance from the edge of the Moon. A complete record of limb darkening in a large number of spectral regions, free from many of the usual complications with stray light, as well as a record of the chromospheric spectrum, could be obtained in this way.

Eclipses are observed with radiation in the millimeter to meter wavelength range with exactly the same types of instruments that are used at solar radio-astronomy observatories; indeed, in many cases observations at eclipse time have been made with the observatory instruments moved to the eclipse site. Partial eclipses are of great use in solar radio-astronomy for locating small sources of radio emission. It is important for the interpretation of the radio record that accurately timed monochromatic (Hα or K) motion pictures of the eclipse be obtained on the site. Many of the objects most likely to be radio sources can be recorded in such pictures, and the details of their disappearance and reappearance made available for study.

VII. Some instrumental problems of solar observation.

18. The precision of nearly all types of observation of the Sun probably would be improved if the observations could be made from sites selected for astronomical excellence. Meteorological considerations indicate that, as far as can be

---

[1] S. A. MITCHELL: Astrophys. Journ. **38**, 407 (1913).

[2] D. H. MENZEL: Publ. Astronom. Soc. Pacific **44**, 356 (1932). — R. O. REDMAN: Monthly Notices Roy. Astronom. Soc. London **97**, 668 (1937). — R. O. REDMAN: Monthly Notices Roy. Astronom. Soc. London **102**, 134 (1942).

[3] W. W. CAMPBELL: Publ. Astronom. Soc. Pacific **10**, 129 (1898). — D. H. MENZEL: Publ. Lick Obs. **17**, 1 (1931).

predicted, nearly ideal conditions should occur on an isolated, high mountain peak rising from an ocean, far from land, within the latitude zone $\pm 35°$. The mountain should be high enough to rise above ordinary formations of cumulus clouds. The nearest approaches to such sites are the island of Tenerife[1] and the island of Hawaii[2], but little is known of the solar observing conditions at these sites.

Even if a perfect site could be found, all solar instruments so far constructed would continue to be handicapped seriously by air currents established in the neighborhood of the telescope, and even within the telescope, by the heat of the Sun. In principle, at least, much of the telescope and its supporting structure can be maintained by either cooling or heating at any temperature required to prevent establishment of air currents. Furthermore, the housing and auxiliary structures can be designed to minimize the turbulence in any residual motions of the air surrounding the instrument.

The solar telescope of the Kitt Peak National Observatory near Tucson, Arizona (U.S.A) is the first solar instrument to attempt complete control of the temperatures of the telescope. In the Kitt Peak construction an underground tunnel houses the telescope itself and its auxiliary instruments. A certain degree of thermal stability is automatically provided by such an underground location, but for additional control the tunnel will be surrounded by thermostatically controlled panels, and it will be tightly selaed to prevent air currents resulting from leaks. A polar heliostat with the mirror located 100 feet above the ground surface will direct sunlight along the direction of the polar axis into the underground telescope. Thermostatically controlled panels are to cover all of the buildings above ground (about 40000 sq. ft. in area). The panels will adjust the temperature of this structure as required to minimize thermal air currents.

The surfaces of the mirror and lenses in a solar telescope present more difficult problems for control than the telescope structure itself. A clean lens transmits and reflects nearly 99% of the total energy of the solar radiation incident on it, and if initially at air temperature requires only slight heating or cooling to follow the diurnal changes. However, no successful scheme for controlling the temperature of a lens has been developed. Despite lack of temperature control, lenses pointed directly at the Sun have defined the smallest structures yet reported on the surface of the Sun.

Mirrors perform less well than lenses because even the best mirror absorbs much radiation. A mirror in excellent condition, coated with aluminum, will absorb from 5 to 10% of the incident energy. Under prolonged exposure to the Sun, temperature differences of ten to fifteen degrees centigrade may be developed between the front and back surfaces of a mirror. A temperature difference of this size can cause serious degeneracy of the image in two ways. First: in glass of any kind the figure of the mirror surface is distorted if the disk is not uniformly heated. A disk of fused silica does not change figure detectably for small temperature differences of a few tens of degrees, and use of this thermally stable base for a mirror minimizes changes of figures. Controlled heating of disks of glass[3] might be applied to solar telescopes to correct thermal distortion of the surface. Second: air in nearly physical contact with the surface of the mirror is heated and flows slowly toward the top edge of the mirror[4]. This very thin layer that

---

[1] Rep. of the Council to the 37-th Ann. Gen. Meeting: Monthly Notices Roy. Astronom. Soc. London **17**, 107 (1857).

[2] Anon: Physics Today **9**, 42 (1956).

[3] A. Couder: C. R. Acad. Sci., Paris **231**, 1290 (1950).

[4] A. Couder: Vistas in Astronomy, ed by A. Beer, Vol. 1, p. 373. London and New York: Pergamon Press 1955. — W. H. Steavenson, W. H.: Vistas in Astronomy, ed by A. Beer, Vol. 1, p. 473. London and New York: 1955.

sticks tenaciously to the surface is one of the main causes of poor definition in reflecting telescopes. No successful method of dealing with the thin layer of air is known, but glass-on-metal mirrors[1], that offer the possibility of cooling or heating by circulation of a fluid through the metal, might be cooled or heated to the surrounding air temperature and this type of disturbance should then be suppressed.

All of the troubles arising from air currents can be eliminated by pumping all air out of the telescope, a procedure adopted with complete success in large spectroscopes and other instruments with long optical paths[2]. The manufacture of a completely sealed telescope and spectroscope from which air could be removed would be straightforward in the case of a refracting telescope, equatorially mounted, with the spectroscope carried on the telescope mounting. Any reflecting system would require at least one window, but in a meniscus type telescope, such as the Maksutov or Schmidt, the window might be used to improve the quality of the image at the same time that it acts as an air seal. Another type of telescope, in which air is replaced by helium, might also be insensitive to convection currents, but no tests of this proposal have been made. Any of the plane-mirror solar telescopes, such as the ordinary tower telescope using the coelostat and second mirror system, are difficult to protect from convection, either through the use of vacuum, or helium, or elaborate control of the temperature of the tower, telescope, and plane mirrors.

If the thermally stable telescope for solar observation planned for the Kitt Peak National Observatory is successful, the observation of the Sun will enter a new era. But equally significant advances may result from the development of techniques for observing at very high altitudes. Obviously, all troubles with absorption, convection, and turbulence in air can be eliminated completely by carrying telescopes and their accessories to the top of the atmosphere.

Manned balloons[3] have been used to reach high altitudes for many years, but the greatest success in solar astronomy has attended the use of unmanned free balloons[4]. The unmanned free balloons reach heights of about 25 kilometers, where the balloon may remain for several hours and carry out a rather elaborate program of astronomical observation.

Rockets are capable of carrying telescopes and accessories to extreme altitudes. Indeed, in principle these devices can be projected toward the sun and reach a point of closest approach that is limited only by the effect of the sun's radiation.

Since rockets enter regions in the atmosphere where absorption of radiation of wavelengths less than 3000 Å is negligible, solar observations concerning this part of the spectrum are especially valuable[5]. Radiation in the wavelength region 0—100 Å is recorded using filters to define narrow spectral regions. The detector preceded by the filter is exposed directly to the sun. An image can be formed by using a pinhole camera. For such observations guiding is relatively unimportant. However, for radiation in the range 100 to 3000 Å small grating or prism spectroscopes are used, and since the radiation is inherently weak and available exposure times are short, good guiding is essential. Photoelectric tracking controls[6] have been developed to the point where a small telescope can

[1] G. E. Hale: Astrophys. Journ. 82, 111 (1935).

[2] A. A. Michelson, F. G. Pease and F. Pearson: Astrophys. Journ. 82, 26 (1935).

[3] A. Dollfus: L'Astronomie 73, 345 (1959).

[4] M. Schwarzschild: Astrophys. Journ. 130, 245 (1959).

[5] A. Aboud, W. E. Bearing and W. A. Rense: Astrophys. Journ. 130, 381 (1959).

[6] R. A. Nidey and D. S. Stacey: Rev. Sci. Instrum. 27, 216 (1956). — H. D. Edwards, A. Goddard jr., M. Juza, T. Maher and F. Speck: Rev. Sci. Instrum. 27, 381 (1956).

Table. *Solar tower tele-*

Blank spaces in the table remain where information to fill them could not be obtained. — Only solar tower telescopes are listed. It is intended to supplement the information in the table by H. Gollnow [Naturwiss. **36**, 175, 213 (1949)], and to give data for solar tower telescopes not completed at the date of Gollnow's compilation. The telescopes are

a) Coelostats.

Number (Gollnow's Table)	Place	Year	Main mirror		Secondary		Material	Manufacturer
			⌀ cm	T cm	⌀ cm	T cm		
12	Lake Angelus 21.5 m tower University of Michigan	1940	47	5.8	41	10.5	Fused silica Pyrex	Perkin Elmer Company
15	Kanzelhöhe University of Graz	1947	30	5	30	5	Tempax-glass Schott	Zeiss, Jena
16	Rome Monte Mario	1948						
17	Capri Fraunhofer Institut	1953	30	5	30	5	Tempax-glass Schott	Zeiss
18	Co. Dublin Dunsink Obs.	1955	40.5		40.5		Fused silica	Cox, Hargreaves, and Thomson, Ltd.
19	P. O. Pochtovoye Crimean Astrophysical Observatory	1955	65	10	50	10	Tempax-glass Leningrad	Optical Institute Leningrad
20	Herestua Institute for Theoretical Astrophysics	1956	65	10	50	10	Tempax-glass	
21	Moscow Sternberg Astronomical Institute							Optical Institute Leningrad

c) Spectroscopes.

Number (Gollnow's Table)	Type	Collimator-Camera			Manufacturer
		⌀ cm	FL m	Type	
12	Auto-collimating spectroscope	15.2	7.6	Achromatic lens	Perkin-Elmer
	Pfund spectroscope	15.2	6.7	Concave mirrors and planes	Lloyd H. Sprinkle
	Echelle spectroscope	15.2 46.7	6.7 4.3	Collimator mirror Camera mirror	Lloyd H. Sprinkle
	Czerny vacuum spectroscope	31.0 31.0	15.2 15.2	Collimator mirror Camera mirror	Ferson Optical Company
15	Spectroheliograph and spectrohelioscope	10.0 10.0	2.0 2.0	Collimator mirror Camera mirror	Zeiss
16					
17	Spectroscope		6.0	Achromatic lens	Zeiss
18	Eagle concave grating spectroscope		7.0	Concave grating	
19	Spectroheliograph	17.0 17.0 17.0	5.0 5.0 5.0	Collimator mirror Camera mirror Camera mirror	Optical Institute Leningrad
	Spectroscope	23.0 30.0 30.0	10.0 10.0 20.0	Collimator mirror Camera mirror Camera mirror	Optical Institute Leningrad
20					
21					

scopes and accessories.

numbered as in GOLLNOW's table which includes nos 1—14 only.—Some of the major solar observatories do not use tower telescopes. Many of these are listed in Table 1, by R. COUTREZ in: The Solar System, Vol. 1: The Sun, ed. by G. P. KUIPER, p. 728. Chicago: Chicago University Press 1953.

b) Telescopes.

Type	∅ cm	FL m	T cm	Material	Manufacturer	∅ Solar image mm
1. Lens system, achromatic, parfocal C and K	a) 30 b) 23	14.9 −7.9 a+b: 29.0			Perkin-Elmer	139 270
2. Concave mirror and plane	31 31	15.2	5.7 5.7	Pyrex glass Pyrex glass	Rath and Brown Rath and Brown	142
3. Concave mirror and plane	41 31	31.0	7.0 5.7	Pyrex glass Fused silica	Ferson Ferson	288
4. Auxiliary plane	28		3.6	Fused silica	Perkin-Elmer	
Achromatic lens	20	5.0			Zeiss	47
Achromatic lens	20	5.0			Zeiss	47
Concave mirror	a) 35 b) 19	6.8 — a+b: 16.4		Fused silica	Cox, Hargreaves, and Thomson, Ltd.	153
Concave mirror Convex mirror Convex mirror Plane mirror Plane mirror	a) 40 b) 18 c) 14 d) 22 e) 30	a+d: 12.0 a+b+e: 21.0 a+c+e: 35.0		Tempax glass	Optical Institute Leningrad	112 195 326

d) Gratings.

Manufacturer	Ruled surface cm	Grooves per mm	Dispersion mm/A	Theoretical resolving power
University of Chicago, Gale	11.4 × 9.6	600	0.42	68400
Pfund spectroscope grating in vacuum spectroscope				
Bausch and Lomb, Richardson	15.0 × 7.5	7.8	2.5 400th order	470000
Mount Wilson Observatory, Babcock	20.3 × 13.3	600	6.0 fifth order	609000
Johns Hopkins University, Strong	13.3 × 13.3	600	0.87 second order	160000
Optical Institute Leningrad	15.0 × 15.0 15.0 × 15.0	600 600	0.33 first order 1.30 second order	90000 180000

be used to project an image of the sun on a spectrograph slit with sufficient stability to achieve an exposure time of the order of 30 seconds [1].

Similar guiding controls have been used with objective prism spectroscopes and objective grating spectroscopes to make monochromatic pictures of the sun with radiation from the Lyman-α line at 1215.7 Å [2].

Artificial earth satellites will extend the rocket observations by making it possible to obtain a record of the variation with time of various solar features. The problems encountered in the design of instruments for satellites and rockets are almost identical. Since the satellite, if successful, will continue to make observations for a long period of time special attention must be paid to the problem of providing instruments that will operate successfully for periods of years.

The two methods of attempting to improve the quality of solar observations: (1) the production of thermally stable telescopes of large size free from air currents for the use on the earth's surface; and (2) the use of balloons, either manned or unmanned, rockets, and artificial earth satellites as observing platforms high above the earth's surface, will result in extensive programs of instrumental development and may eventually provide the long-sought solar data of increased precision.

Bibliography.

ABETTI, G.: Handbuch der Astrophysik, ed. by G. EBERHARD, A. KOHLSCHÜTTER and H. LUDENDORFF, Vol. 4, Chap. 2, p. 61. Berlin: Springer 1929. — General descriptions of some typical solar intruments. In the same Encyclopedia see also Vol. 1, Part 1, Chap. 2, 3, 5. 1933.

ALLEN, C. W., and others: The Solar System, Vol. 1: The Sun, ed. by G. P. KUIPER, p. 592. Chicago: Chicago University Press 1953. — A review of the state of solar instrumental development. Tables 1, and 2, p. 728 and 734 list most of the active solar observatories and give some details about their telescopes.

AMBRONN, L.: Handbuch der Astronomischen Instrumentenkunde. 2 vols. Berlin: Springer. — Vol. 2, Chap. 12 of this work is especially directed to the description of the principal solar instruments.

BOWEN, I. S., and others: Vistas in Astronomy, ed. by A. BEER, Vol. 1, p. 400. London and New York: Pergamon Press 1955. — Several sections contain discussions by different authors of instruments and instrumental problems of solar astronomy. The section by H. VON KLÜBER: "Spectroscopic Measurements of Magnetic Fields on the Sun" is exceptionally complete.

DYSON, F., and R. v. d. R. WOOLEY: Eclipses of the Sun and Moon, Chap. 9, p. 57. Oxford: Clarendon Press 1937. — Brief descriptions of some of the important instruments used for solar observation at eclipse time.

GOLLNOW, H.: Naturwiss. 36, 173, 213 (1949). — A review of solar tower telescopes constructed before 1949. This article contains a valuable table of detailed information about solar towers.

HALE, G. E.: Astrophys. Journ. 70, 265 (1929); 71, 73 (1930); 73, 379 (1931); 74, 214 (1931); MITCHELL, G. A.: Astrophys. Journ. 88, 542 (1938). — A series of articles forming a connected account of the development and use of the spectrohelioscope. There are many discursion son tower telescopes, their construction and observing methods.

HALE, G. E., and S. B. NICHOLSON: Magnetic Observations of sunspots, 1917—1924, Part 1. Washington: Carnegie Institution of Washington 1938. — The original description of the 60-foot and 150-foot solar tower telescopes on Mount Wilson, California is contained in this work. The instruments for the measurement of the magnetic fields of sunspots are described.

---

[1] F. S. JOHNSON, H. H. MALITSON, J. D. PURCELL and R. TOUSEY: Astronom. J. 60, 165 (1955).

[2] R. MERCURE, S. C. MILLER, W. A. RENSE and A. STUART: J. Geophys. Res. 61, 571 (1956). — H. FRIEDMANN: J. Geophys. Res. 64, 1751 (1959).

HARTMANN, W.: Astronom. Abh. Hamburg Obs. Bergedorf **4**, 1 (1928). — The most detailed mathematical treatment of the plane mirror telescopes published so far.

KAYSER, H. G. J.: Handbuch der Spectroscopie, Vol. 1. Leipzig: S. Hirzel 1900. — The standard treatise on spectroscopic instruments and their uses. For a more recent account and bibliography of spectroscopic instruments see Vol. XXIX, this Encyclopedia.

KING, H. C.: The History of the Telescope. London: Charles Griffin and Company, Ltd. 1955. — Unusual details in the history of the development of solar astronomical instruments can be found only in this work.

LANGLEY, S. P., C. G. ABBOT, et al.: Ann. Astrophys. Obs. Smithson. Inst. **1900—1942**. — The Smithsonian observers describe in great detail their instruments and observing techniques. These accounts are the single most important publication dealing with the history of the solar constant.

REPSOLD, J. A.: Zur Geschichte der Astronomischen Meßwerkzeuge. Leipzig: W. Engelmann 1908. — A complete outline of the development of the heliostat and the coelostat, with descriptions of many other types of instruments used for solar observations.

STRÖMGREN, B.: Handbuch der Experimentalphysik, ed. by W. WIEN and F. HARMS, Vol. 26. Leipzig: Akademische Verlagsgesellschaft 1937. — An account of the instruments for the measurement of the solar constant. Bits of information about solar instruments are scattered throughout this volume.

VYAZANITSYIN, V. P., and O. A. MELNIKOV: Uspecki Astronom. Nauk **3**, 3 (1947). — A summary of solar instruments in use at the date of publication.

Radio Astronomy Techniques.

By

R. N. BRACEWELL.

With 56 Figures.

I. Introduction.

1. Outline of the chapter. The first two parts of this chapter, concerned respectively with receivers and aerials, represent a selection from a vast field of radio technology. It is true that the material has been selected for its relevance to radio astronomy, but it remains a body of information on practice in radio technology which is current in connection with quite other fields than radio astronomy. It will be recognized as such by the expert in radio and will appear to him to be profoundly specialized in places and shallow to the point of omission in others. But parts II and III have not been written primarily for the expert in radio, they have been written for the physicist or astronomer, engaged in radio astronomy, who would benefit from a collected and orderly exposition of those parts of radio technique which are especially relevant to his work. Whilst the compilation consisted merely in selection, it is felt that the result will serve a valuable end, as the corpus from which the selection has been made is too extensive for one not specifically grounded in radio to use efficiently without guidance.

Very much of the material of prime importance for radio astronomy is available only in technical papers and some of the topics dealt with in parts II and III are not properly covered in the literature at all. The theory of the radiometer is one item which is still in an incomplete state and the effect of errors on the performance of aerials is another.

Part IV lays the groundwork for studying those parts of radio astronomy technique which have not been borrowed from existing practice but have been developed recently by radio scientists engaged in astronomical observation. The developments which have been called forth by the challenging new problems are impressive and are already being adopted in communications, navigation, radar and other established fields of application of radio science. Some refined concepts are essential to an understanding of the way in which an aerial explores a radiation field, concepts which would correctly be classified under classical optics. Indeed it would seem that recent advances in optics, especially in diffraction theory, have been taking place to a considerable degree at the hands of men interested in radio wavelengths. Part IV introduces the quantities *brightness*, *brightness temperature*, and *flux density* in terms of which observations are expressed, and sets up the *aerial smoothing equation*, an integral equation which relates them to *available power*, the quantity which is actually measurable with a receiver and aerial. From this equation a number of important theorems are deduced. The spectral sensitivity theorem relates the resolution of an aerial to its aperture distribution, in full detail; that is to say, not in terms of the aerial's ability to resolve a pair of point sources but in terms of the response of the aerial to spatial Fourier components of all spatial frequencies. The spectral sensitivity function is shown to be equal to the normalized complex autocorrelation function of the aperture distribution. The aerial cut-off theorem then shows that no aerial whose width is S wavelengths in a certain direction has any response to a sinusoidal brightness distribution with more than S cycles per radian in that direction. The discrete interval theorem, which follows as a consequence, sets a limit to the number of independent measurements which can be made in a finite area of sky, and shows that in conducting a survey it is sufficient to observe at discrete intervals of angle. All the results of part IV apply also to interferometers, which are defined as aerials having two or more well separated parts.

Part V describes the simple two-element interferometer, the radio analog of MICHELSON's stellar interferometer, and interprets its function as a determination of the complex visibility of a temporal interference fringe system. Under conditions of symmetry the complex visibility of the fringes observed with infinitesimal isotropic aerials is equal to the complex correlation between the field phasors at the positions of the aerials, and this in turn is equal to the normalized two-dimensional Fourier transform of the distribution of brightness temperature over the sky. This latter quantity, the brightness temperature spectrum, proves to be very suitable for describing interferometers and their behavior, and is to a good approximation measurable. Interferometers composed of extended aerials and sensitive over an extended frequency band are discussed in terms of the idealized isotropic monochromatic case, and are shown to measure smoothed values of the brightness temperature spectrum. Several extensions of the simple interferometer have been developed. Multiple element interferometers with sharp fringes permit quicker and simpler operation, phase-switching and lobe-sweeping introduce technical advantages, and asymmetrical arrangements have special applications. Closely related devices, which should not perhaps be considered interferometers, include phase-switched crossed linear arrays which attain greatly enhanced pencil-beam resolution in exchange for collecting area, and two-aerial systems not preserving phase which permit extreme separations of the aerials. An optical analog of this latter device, which introduces a radically new departure in optics, has been demonstrated by BROWN and TWISS in a measurement of the diameter of Sirius. It is probable that among the many striking advances in instruments for radio astronomy there are others which have significance for practical technique in other fields.

Part VI of this chapter goes into details of the measurement of radiation fields from celestial sources by means of both pencil beam aerials and interferometers. All the components of the often complex apparatus which is immersed in the field, the concepts necessary for considering its interaction with the field, and the procedures for absolute calibration, are touched on in the preceding parts, and clarify the observing procedures. Some very interesting considerations bearing on the analysis of the data close the chapter.

The fascinating subject of radio astronomy is in a vigorous stage of development. It has called forth striking advances in instruments for use at radio wavelengths and as its instrumental needs continue, continued rapid development may be expected to emerge from the profound joint resources of electromagnetic theory and established radio technology on which radio physicists draw. It is hoped that the present chapter on current techniques in radio astronomy is sufficiently fundamental and restrained in its inclusion of ephemeral matter to provide a good basis for following the expected developments of the next few years.

II. Receivers.

a) Principal receiver parameters.

2. Bandwith, gain, noise temperature, and time constant. A radio receiver is a non-linear device which converts high frequency electrical oscillations applied to its input terminals into an output electrical potential with certain convenient properties. The primary desideratum is sufficient power to drive some indicating instrument such as a recording ammeter, an oscillograph, a loudspeaker, a magnetic tape recorder, or typewriter. The output potential must also have a spectral distribution of power suitable to the needs of the indicating instrument. Furthermore one relies on the receiver to select the narrow band of high frequencies to which attention is to be directed. Among the necessary functions of a receiver, therefore, are (i) amplification, (ii) frequency conversion, (iii) tuning, (iv) selectivity.

The simplest receiver is one which puts out a steady potential depending on the

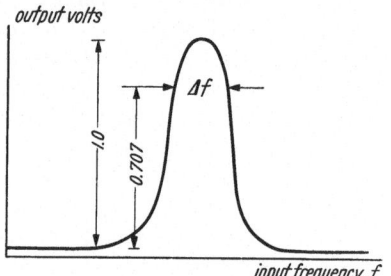

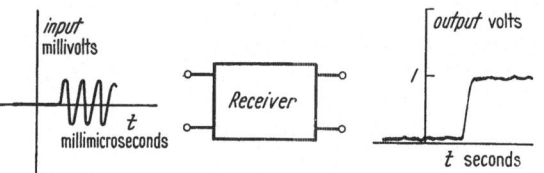

Fig. 1. Illustrating the amplifying and frequency converting properties of a receiver.

Fig. 2. Bandwidth of a receiver whose output voltage increases linearly with input voltage.

strength of the input oscillation. For example, the receiver in Fig. 1 puts out one volt with a monochromatic input signal of $10^{-3} \cos 10^{10} t$ volts.

In this case the voltage amplification is 10^3. The corresponding power ratio is referred to as the gain.

If input signals of other frequencies are considered the amplification becomes negligible outside a small range $\varDelta f$ centered on the nominal frequency f_0. Unless special effects, such as an unusually wide band, are wanted, the shape of the pass band will be approximately Gaussian since it is a product of the many similar characteristics associated with many successive stages of amplification.

The bandwidth $\varDelta f$ is generally defined as the difference between the frequencies at which the output voltage (or other output indication, such as meter deflection) falls to the value corresponding to a halved power input at the midfrequency. Fig. 2 illustrates a case where output voltage increases linearly with input voltage. Bandwidths in practice lie mostly between 10^4 and 10^8 Hz.

As Figs. 1 and 2 show, the output of a receiver contains some random noise which is generated in the receiver itself, and except at the lower frequencies, around 20 MHz, it is desirable to keep the receiver noise as low as possible. The receiver noise is measured by the receiver noise temperature T_R, a quantity which would be zero for an ideal receiver. The ability of a receiver to detect weak signals depends, among other things, on receiver noise temperature, which is therefore one of the important parameters specifying a receiver.

It commonly happens that the non-linear stages of a receiver cannot be described by a simple mathematical law, and a consequence of this is that the gain depends on the input level. It is therefore necessary to calibrate receivers and in the process of doing so it is often possible to use a wideband noise power source and to include the aerial, feeder, and indicating device so that the monochromatic gain of the receiver alone loses primary significance. It is replaced by the gain calibration factor, measured in units of output indication per unit increment of input power. A typical value might be 1 millimeter per degree Kelvin, where the output is pen deflection on the chart of a recording meter, and the input power source is thermal or equivalent.

Further operations on the output signal of a receiver may be carried out by circuits which are best considered separately. For example the noise fluctuations shown in Fig. 1 may be reduced by passing the output signal through a low-pass filter which discriminates against the fluctuations. A direct consequence of this is to impede the rapidity of response to sudden changes. The smoothing circuit which does this is often referred to as an integrator. It is essentially a low-pass filter and in one simple form the response to a voltage impulse is an exponential decay with a time constant which may range from a fraction of a second to several seconds.

The integrating time τ, since it also affects the ability to detect weak signals, is an important basic parameter. The principal receiver parameters are thus (i) nominal frequency f_0 and its range, if adjustable, (ii) bandwidth Δf (and shape of band if significantly non-Gaussian), (iii) gain calibration factor, or detailed gain calibration if, as is common, the output indication is not a linear function of input power, (iv) noise temperature T_R, (v) integrating time τ (or impulse response of integrator). Precise meaning is given to Δf, τ and T_R in Sect. 8 and 10.

b) Monochromatic operation of a noise-free receiver.

In order to simplify the description of receiver operation it is convenient to ignore receiver noise. The block diagram of Fig. 3 is applicable to most receivers.

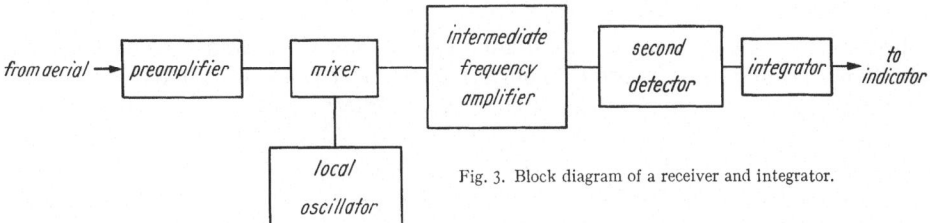

Fig. 3. Block diagram of a receiver and integrator.

It is also possible, by concentrating the amplification at the signal frequency, to omit the mixer and intermediate frequency amplifier, and it is advantageous to do this in the microwave range by cascading wide-band travelling wave tubes where large values of $\Delta f/f_0$, or special behavior of the receiver band pass characteristic, are required.

3. Preamplifier. A preamplifier means essentially an amplifier which amplifies at the original signal frequency, and may or may not be present in a receiver. If used, it may be situated away from the rest of the receiver. The main considerations in connection with a preamplifier are its effect on the noise temperature of the receiver, as discussed below, and the suppression of certain spurious responses. For frequencies greater than a slowly increasing limit, at present about 10000 MHz, it is an advantage to omit preamplification, since suitable amplifiers are only gradually being developed. There is a transition band of a few thousand MHz in which equally low noise temperatures are obtainable with or without preamplifiers (using triodes or travelling wave tubes), and at all lower frequencies triode or pentode preamplifiers are universal. If the attenuation between the aerial and the receiver is not negligible, preamplifiers must be reconsidered, and it has proved advantageous to use them at or near the focus of paraboloidal reflectors and at the extremities of extended interferometers (Sect. 60).

4. Mixer and local oscillator. The signal from the preamplifier and a strong signal from the local oscillator are combined in the mixer, a non-linear device such as a silicon crystal or suitably biased electron tube. Some energy now appears at the difference frequency, often 30 MHz, and is passed on to the intermediate frequency amplifier which accepts only a fixed band of frequencies centered on 30 MHz. Since the local oscillator frequency differs by 30 MHz from the nominal frequency f_0, any change in local oscillator frequency results in a change in f_0. This feature may be used for adjustment of f_0 but usually f_0 has to be kept fixed. The main requirement on a local oscillator is thus usually frequency stability. Other requirements are that it should give sufficient power at a steady level, be tunable as needed, and not absorb signal power. Provided the local oscillator power is strong, the envelope of the input signal is reproduced linearly.

5. Intermediate frequency amplifier. The intermediate frequency amplifier contributes all or most of the gain of the whole receiver and consequently must

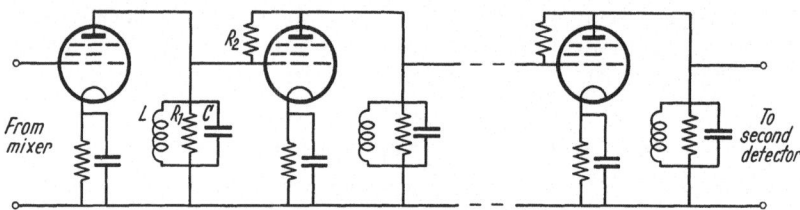

Fig. 4. An intermediate frequency amplifier.

meet the strictest demands on circuit design. Since the gain is typically one million, the minutest fraction of output power will cause instability if it leaks through to the input. For this reason intermediate frequency amplifiers have become highly developed along standardized lines. A great deal of skilled effort is needed to alter and readjust an existing design, hence the concentration on 30 MHz and other standard intermediate frequencies.

In Fig. 4 we see the basic pattern of an intermediate frequency amplifier[1]. A large number of amplifying stages are used, each with a selective tuned circuit LC. Since the effect of many tuned circuits in tandem is to produce bandwidths which are narrow, damping resistors R_1 may be provided. The negative feedback via R_2, if included, also tends to widen the band. The cathode resistor furnishes d.c. feedback which tends to stabilize the anode current and transconductance and should be chosen for the best time-stability. Inessential resistors, such as screen resistors, whose slow changes can influence the gain, should be omitted. Where the broadest bands are required, the frequencies to which the successive stages are tuned may be staggered.

Among the advantages of using a frequency change or superheterodyne receiver as described above are the ability to use a standard fixed-frequency intermediate frequency amplifier and the ability to adjust f_0 by a simple change in local oscillator frequency. Furthermore, the aerial structure is well adapted for picking up energy leaked from the receiver at the nominal frequency which makes a frequency change helpful if instability is to be avoided.

---

[1] G. E. VALLEY and H. WALLMAN: Vacuum tube amplifiers. New York-Toronto-London: McGraw-Hill 1948.

As a consequence of frequency changing the receiver is sensitive to two nominal frequencies each separated from the local oscillator frequency by 30 MHz. At the highest nominal frequencies there is little difference between what is received on these bands. However, the aerial pattern may be affected and for many purposes it is necessary to suppress one of the two. The one to be suppressed is called the image frequency. A selective preamplifier will usually suffice for image rejection. Otherwise an image rejection filter may be used. A superheterodyne receiver may also be sensitive to external signals at the intermediate frequency, and here again a preamplifier is effective.

6. Second detector. The second detector, so called because the mixer is regarded as the first detector, is a further non-linear frequency-changing device which performs the final step of converting the amplified signal to a steady output potential. It may consist of a diode rectifier (see Fig. 6) with suitable filtering to remove any traces of the intermediate frequency or its harmonics, or it may be some other standard detector utilizing a cathode follower or pentode. A linear detector is one whose output voltage or current is proportional to the amplitude of the voltage applied to it. The output of many detectors, however, is proportional to the square of the applied voltage amplitude, indeed at the lowest power levels virtually all detectors are square-law.

Before proceeding to the integrator which follows the second detector it is necessary to consider signals which are not monochromatic but are spread continuously over the spectrum.

c) Reception of a continuous spectrum.

7. Effect of non-zero bandwidth. Throughout the foregoing description of the parts of a receiver we have had in mind a monochromatic input signal and we have shown how it is converted to a steady output potential which measures its strength. However, the signals received from extra-terrestrial sources are not monochromatic; their energy is spread continuously over the spectrum. Let us consider such a signal as it passes through a receiver, and let us suppose that its spectral energy density is independent of frequency over the band of frequencies accepted by the receiver.

Fig. 5 (a) illustrates the time variation of electric field due to extra-terrestrial signals covering a wide spectrum as in Fig. 5 (b). The part of this field to which the receiver is sensitive is approximately monochromatic with frequency f_0 [Fig. 5 (c)] and has the spectral energy distribution shown in Fig. 5 (d). Because of beating between different components within the band, the waveform in (c) shows modulation at low frequencies which extend from zero to frequencies of the order of Δf. The important thing to notice is that the amplitude does not change much in times less than $(\Delta f)^{-1}$.

The output from the intermediate frequency amplifier, shown in (e), has the same envelope[1], but the mid-frequency has been shifted to 30 MHz as shown in (f). The second detector now suppresses or rectifies the negative half cycles and filters

---

[1] The probability of finding the envelope amplitude R between R and $R+dR$ is $p(R)\,dR$, where

$$p(R) = \frac{2R}{\langle R^2 \rangle}\, e^{-\frac{R^2}{\langle R^2 \rangle}}.$$

This is the Rayleigh distribution, and has the following properties: mean $= \frac{1}{2}\sqrt{\pi \langle R^2 \rangle}$, r.m.s. departure from mean $= \sqrt{(1 - \frac{1}{4}\pi)\langle R^2 \rangle} = 0.52 \times$ mean.

out higher frequency components to give an output (g) like the envelope of (e).
This is the receiver output potential, which in the case of a monochromatic input signal was steady, but which now varies about a mean value. However the
variations are no more rapid than the bandwidth Δf permits. For example,

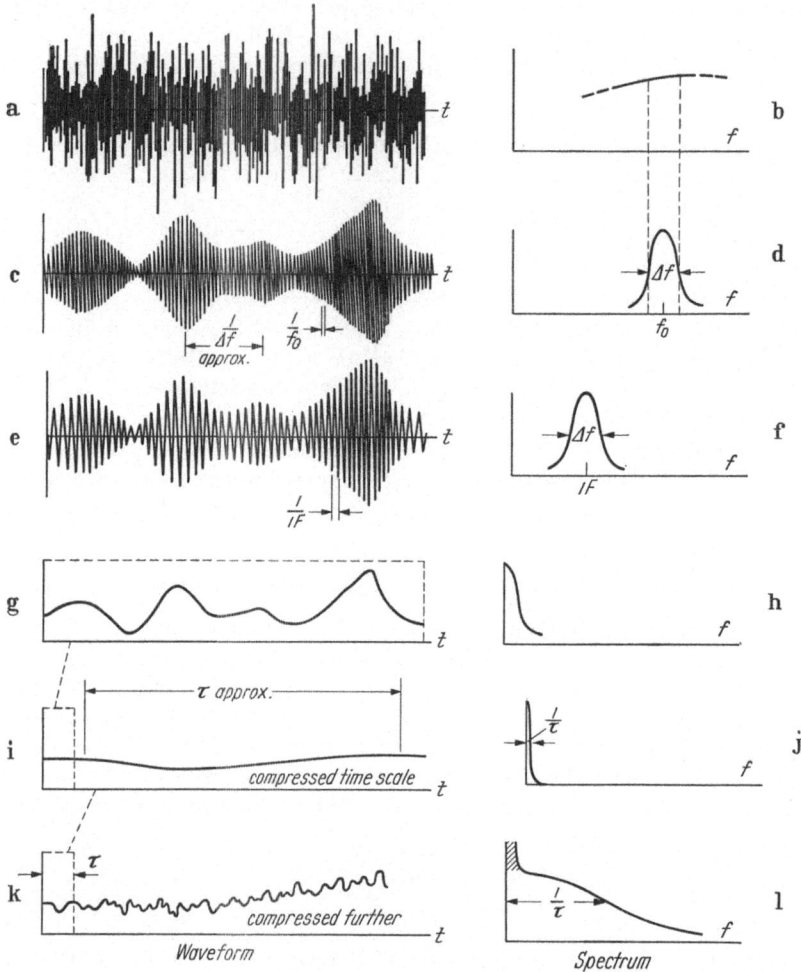

Fig. 5 a—l. Illustrating the changes in waveform and spectrum suffered by a signal with a continuous spectrum as it
passes through a noise-free receiver.

if $\Delta f = 1$ MHz then the output cannot change much in times small compared
with one microsecond.

The probability distribution of the receiver output voltage will depend on the
detector law. If the detector is linear, the output voltage V will be the same as
the envelope of (e) and thus have a Rayleigh distribution

$$p_1(V) = \frac{2V}{\langle V^2 \rangle} e^{-\frac{V^2}{\langle V^2 \rangle}};$$

more generally, the probability $p_2(V)\,dV$ of finding the output voltage between
V and $V+dV$ is the same as the probability $p_1(R)$ of finding the input envelope
between R and $R+dR$, i.e. $p_2(V)\,dV = p_1(R)\,dR$. For square-law detection

$V = R^2$ and $dV = 2R\,dR$; hence

$$p_2(V) = \frac{1}{\langle V \rangle} e^{-\frac{V}{\langle V \rangle}}.$$

This purely exponential one-parameter distribution has r.m.s. deviation from the mean equal to its mean value $\langle V \rangle$. Thus, as in the case of the linear detector, the r.m.s. deviation from the mean is proportional to the mean value.

To determine the input signal it is now only necessary to measure the mean value of the output.

d) Statistical limit to precision.

8. Standard deviation of a mean. We may assume the following theorem from statistical theory. If a large number N of independent variates are identically distributed (in almost any way) about a common mean M with r.m.s. deviation σ, then their mean is normally distributed about M with r.m.s. deviation $\sigma/N^{\frac{1}{2}}$, Regarding the receiver output as a succession of samples from a certain distribution we see that the uncertainty within which the mean can be determined depends on the length of time available for measurement and is proportional to the inverse square root of the number of effectively independent values assumed during the interval of measurement. Now independent values are separated by a time of the order of $(\Delta f)^{-1}$, so that about $\tau\,\Delta f$ independent values are assumed in an interval τ. Consequently, where the r.m.s. deviation of a single value from the mean is proportional to the mean, the uncertainty in the determination of the mean value is proportional to

$$\frac{\text{mean value}}{\sqrt{\tau\,\Delta f}},$$

which for $\tau = 1$ second and $\Delta f = 1$ MHz is equal to one thousandth of the mean value. This limit exists because of the random character of the signal to be measured, even in the absence of noise generated in the receiver.

One can assign precise definitions to Δf and τ to cover reception filters and smoothing filters with any frequency response. Let the power transfer characteristic of the reception filter be $R(f)$ and of the smoothing filter $S(f)$. The concept of "equivalent width" of a function proves to be basic, and is defined, for example for $S(f)$, by

$$W_S = \frac{\int_{-\infty}^{\infty} S(f)\,df}{S(0)}.$$

We also require the following pentagram notation:

$$R \star R \equiv \int_{-\infty}^{\infty} R(f')\,R(f'-f)\,df'.$$

We note that the output power spectrum from a square-law detector with input power spectrum $R(f)$ is

$$2R \star R + \left[\int_{-\infty}^{\infty} R(f)\,df \right]^2 \delta(f),$$

where the first term enumerates the number of ways a difference frequency f can be found within the spectrum $R(f)$, and the second term is the rectified component. This spectrum, when further limited by the smoothing filter, becomes the power spectrum of the receiver output

$$S(f) \left\{ 2R \star R + \left[\int_{-\infty}^{\infty} R(f)\,df \right]^2 \delta(f) \right\}.$$

The mean square fluctuation of the output is

$$2 \int_{-\infty}^{\infty} S(f) (R \star R) df = 2R \star R \big|_0 \int_{-\infty}^{\infty} S(f) df,$$

the mean value is

$$\sqrt{S(0)} \int_{-\infty}^{\infty} R(f) df,$$

and hence

$$\frac{\text{root mean square fluctuation}}{\text{mean value}} = \sqrt{\frac{2R \star R \big|_0 \int_{-\infty}^{\infty} S(f) df}{\int_{-\infty}^{\infty} R \star R \, df \, S(0)}}$$

$$= \sqrt{\frac{W_S}{\tfrac{1}{2} W_{R \star R}}}$$

$$= \frac{1}{\sqrt{\tau \, \varDelta f}}.$$

The precise definitions will now agree with previous custom if we take

$$\tau = \frac{1}{W_S},$$

$$\varDelta f = \tfrac{1}{2} W_{R \star R}.$$

The following tables, compiled with the assistance of Mr. R. Colvin, present a variety of cases for reference. For example, it is deducible from the tables that the r.m.s. fluctuation in the output of a receiver whose reception filter is Gaussian with half-power bandwidth B, and whose averaging is done entirely with a single RC circuit, is given by

$$\frac{\sqrt{\frac{\ln 2}{2\pi}}}{\sqrt{RCB}} \times \text{mean value} = \frac{0.56 \times \text{mean value}}{\sqrt{RCB}}.$$

Reception filter	$R(f)$	$\varDelta f$
Rectangular pass band	$\begin{cases} 1, & f_0 - \tfrac{1}{2}\varDelta < \|f\| < f_0 + \tfrac{1}{2}\varDelta \\ 0, & \text{elsewhere} \end{cases}$	$\varDelta$
Two rectangular non-overlapping pass band	$\begin{cases} 1, & f_1 - \tfrac{1}{2}\varDelta_1 < \|f\| < f_1 + \tfrac{1}{2}\varDelta_1 \\ 1, & f_2 - \tfrac{1}{2}\varDelta_2 < \|f\| < f_2 + \tfrac{1}{2}\varDelta_2 \\ 0, & \text{elsewhere} \end{cases}$	$\varDelta_1 + \varDelta_2$
Triangular pass band	$\begin{cases} 1 - 2\|f - f_0\|/\varDelta, & f_0 - \tfrac{1}{2}\varDelta < \|f\| < f_0 + \tfrac{1}{2}\varDelta \\ 0, & \text{elsewhere} \end{cases}$	$\tfrac{3}{4}\varDelta$
Single tuned circuit	$[1 + (\|f\| - f_0)^2/\varDelta^2]^{-1}$	$2\pi\varDelta$
Two isolated tandem tuned circuits	$[1 + (\|f\| - f_0)^2/\varDelta^2]^{-2}$	$\tfrac{4}{5}\pi\varDelta$
Gaussian pass band[1]	$\exp[-(\|f\| - f_0)^2/2\varDelta^2]$	$2\sqrt{\pi}\,\varDelta$

Smoothing filter	$S(f)$	τ
Takes running means over time T	$(\pi T f)^{-2} \sin^2 \pi T f$	T
Single RC circuit	$[1 + (2\pi RC f)^2]^{-1}$	$2RC$
Two isolated tandem RC circuits	$[1 + (2\pi RC f^2)]^{-2}$	$4RC$
Critically damped RLC circuit	$[(R/2L)^2 + (2\pi f)^2]^{-2}$	$8L/R$
Rectangular pass band	$\begin{cases} 1, & \|f\| < f_0 \\ 0, & \|f\| > f_0 \end{cases}$	$1/2 f_0$
Gaussian pass band	$\exp(-f^2/2f_0^2)$	$\sqrt{2\pi}\,f_0$

[1] The half-power bandwidth B is given by $B = \sqrt{8 \ln 2}\,\varDelta$.

9. Integrator. A method of averaging the fluctuating output is to pass it through a low-pass filter, e.g. a simple combination of one resistor R and one capacitor C with a time constant RC. In this way the measurement of the $\tau\,\varDelta f$ independent values is carried out automatically by the integrator, which gives out a potential approximately equal to the desired mean, but subject to small variations of order $(\tau\,\varDelta f)^{-\frac12}\times$mean which assume independent values at intervals of order τ. Electronic counters also furnish a convenient method of averaging by performing and printing running sums over consecutive intervals.

Fig. 6 gives the basic circuit of a diode second detector and integrator. The diode capacitor takes up a positive charge such that the steady leakage through the shunt resistor is balanced by the replenishment during moments when the noise peaks of the input voltage are sufficiently positive to pass current through the diode. The time average of the diode capacitor voltage then appears across the integrating capacitor.

This final electrical signal is the one which drives the recorder and is shown in Fig. 5 (i) on a time scale which is compressed so as to exhibit just one cycle of fluctuation. In recording, the time scale would be even more compressed, as in Fig. 5 (k), which shows the actual record

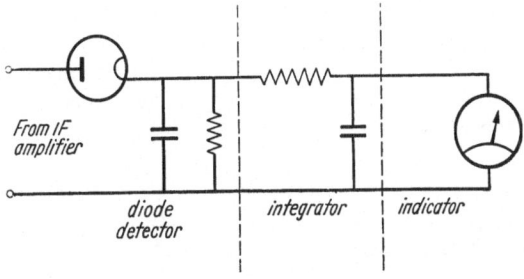

Fig. 6. Basic circuit of second detector and integrator in which $RC=\tau$

of the pen recorder indicating device. This device is often capable of following the most rapid variations passed by the integrator. Often, however, some further smoothing occurs in the recorder, which may for example be a moving coil galvanometer carrying a pen, and τ then assumes a larger effective value.

The constant of proportionality giving the amplitude of the final variations is calculable from the precise law of the second detector, the shape of the pass band of the intermediate frequency amplifier, and the shape of the impulse response of the integrator. Usually, however, the radio astronomer is unaware of the precise theoretical value but he does measure the strength of the variations and verifies that they are not unreasonably large. Many radio astronomical records reveal that the final minute recorded variations are non-Gaussian because of non-linear effects due to sticky ink, backlash, on-off servo-driven pens, etc. For these reasons, as well as the difficulty of determining the necessary parameters, the reasoning given above is most often resorted to for order-of-magnitude estimates.

10. Receiver noise. A noise-free receiver, such as has been assumed in the foregoing work can be approximated in practice by using sufficiently strong input signals. However, though the input power is reduced indefinitely, the output from the intermediate frequency amplifier will always give a positive indication due to noise generated internally in the receiver. Receiver noise is generated by statistical fluctuations in the rate of arrival of electrons at the anodes of the electron tubes induced by random times and velocities of departure of electrons from the cathode, thermal agitation in mixer crystals, and other causes. The first tube has by far the greatest effect since the noise contributions of later tubes are small compared with the already amplified signal.

4*

From a calibration such as that shown in Fig. 7 one can determine the gain g, the receiver noise output $g p_R$, and the amount of power p_R which, at the input, would account for the receiver noise output.

The quantity p_R is called the receiver noise power referred to the input terminals. It is a more suitable parameter than the noise output $g p_R$, which depends on gain, but is not free from dependence on Δf. To obtain a fundamental parameter specifying the noisiness of a receiver we refer p_R to the noise power available from a resistor at ambient temperature T_0 in the frequency band Δf, viz. $kT_0\Delta f$, where k is Boltzmann's constant. The receiver noise temperature T_R is defined by

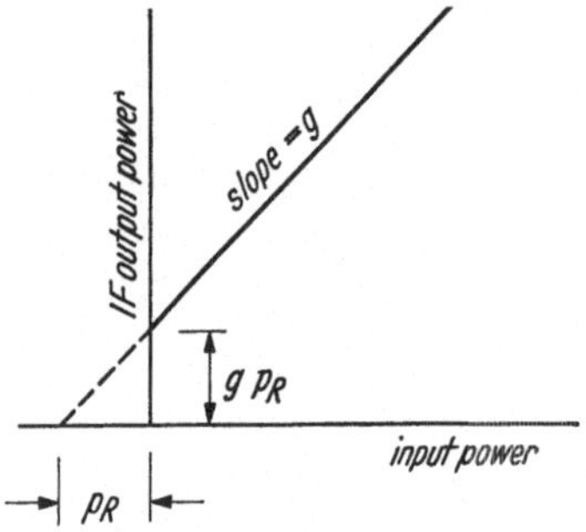

$$p_R = kT_R\Delta f$$

and the noise factor N is defined by

$$p_R = (N - 1)\, kT_0\, \Delta f.$$

The relation between the two parameters T_R and N is

$$T_R = (N - 1)\, T_0.$$

Fig. 7. Receiver calibration showing receiver noise.

In an ideal receiver such that $p_R = 0$, T_R would be zero and N would be equal to unity irrespective of T_0. Confusion over the choice of T_0 makes T_R clearer than N when ambiguity must be avoided.

The expression $kT\,\Delta f$ for the power delivered by a resistor at temperature T into a matched circuit was established by Nyquist[1] as follows. The thermal energy density per unit distance on a loss-free transmission line joining two resistors at temperature T is $2\,kT\Delta f/c$; hence the power flow each way is $kT\Delta f$. To establish the energy density, suppose some energy is trapped by the sudden creation of two perfectly reflecting short circuits a distance l apart. The field must be expressible as a sum of natural modes at frequencies that are multiples of the fundamental $c/2l$. In any range Δf there are $2l\,\Delta f/c$ such modes. Assigning energy kT to each mode in accordance with the principle of equipartition of energy, we find the stored energy $2\,kT\Delta f\,l/c$.

At the highest frequencies (or at low temperatures), where kT exceeds the quantum energy hf, and all the modes therefore cannot be excited, allowance for the Boltzmann distribution of energy over the modes gives an average energy per mode of $hf/[\exp(hf/kT) - 1]$. The general expression for available power from a resistor is thus

$$\frac{hf\Delta f}{\exp(hf/kT) - 1}\,,$$

which is equal to $kT\Delta f$ in the Rayleigh-Jeans long-wavelength regime where $hf/kT \ll 1$, but is otherwise less. This expression essentially describes the Planckian black-body emission from a body whose radiation is confined to one polarization and constrained to flow in one dimension by a transmission line. The resistor is "black" if it is matched.

Fig. 8 shows noise temperatures that are readily available today, but values down to 20 or 30° K over the whole spectrum are in sight, and even lower values

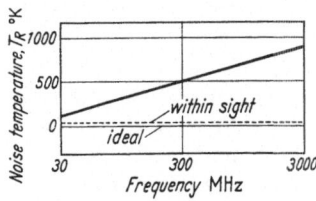

Fig. 8. [Receiver noise temperatures currently attainable.

[1] H. Nyquist: Phys. Rev. 32, 110 (1928).

with special attention. Bloembergen's maser [1] demonstrated an amplifying principle capable of yielding very low noise by operation at liquid helium temperatures, and the closely related parametric amplifier, which can be operated at room temperature, is another promising device. There is reason to think [2] that a theoretical lower limit hf/k may exist. But this is extremely small and it is apparent that we are entering an era of essentially noise-free amplification in which the thresholds of sensitivity will no longer be set by limitations of receiver technique but by external phenomena. Cosmic noise at meter wavelengths and microwave thermal emission from atmospheric molecules are examples of such phenomena.

e) Calibration.

11. Thermal noise and shot noise. Because of the complexity of a receiver, overall calibration using input signals of known strength is a necessity and it is advantageous to make the calibration with noise signals.

A good deal of care has gone into this phase of radio astronomy technique since it is only by careful absolute measurements that results can be compared between different equipments and different frequencies. The possibility of using selected areas of sky as secondary standards has been much discussed but hitherto absolute calibration has not been avoidable. Aerial calibration also affects final results and will be discussed separately.

Noise of known strength can be had from resistors and from temperature-limited diodes for each of which reliable theory exists. Thus the noise power available in a band Δf from a resistor at temperature T is $kT\,\Delta f$, and the mean square noise current in a temperature limited diode in the frequency band Δf is $2eI\,\Delta f$, where I is the mean diode current and e is the charge of the electron. The first of these is deduced from thermodynamics (Sect. 10). The second is based on the assumption that the current I is composed of charge carriers of amount e which cross the interelectrode space with random times of arrival and departure. Each electron carries a current $e\delta(t-t_k)$ with a flat power spectral density e^2, or $2e^2$ if negative and positive frequencies are combined. The power spectra of the I/e electrons per second, arriving independently, are thus additive, with a resultant $2e^2(I/e) = 2eI$.

12. Resistive noise sources. Resistive noise sources have the advantage of simplicity and accuracy, when used near room temperature. The available power is minute, of the order of 10^{-15} watts, but suitable for many purposes. In some applications to very faint sources, resistors have been cooled by liquefied gases,

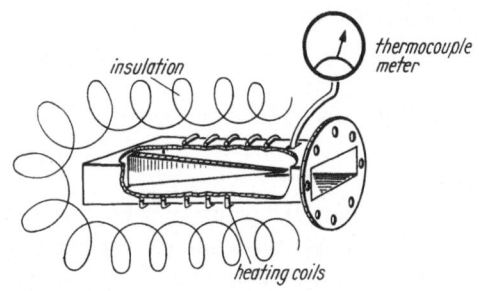

Fig. 9. A thermal noise source.

but more often hot resistors have been needed. The dependence of electrical resistance on temperature complicates the use of lumped resistors but it is convenient to use long lengths of slightly dissipative transmission lines. For example at wavelengths where waveguides are suitable a tapered length of resistive material is introduced into the guide (Fig. 9) in such a way as to absorb incident waves without appreciable reflection in the wavelength band of interest. The waveguide is then heated electrically, precautions being taken to ensure freedom

[1] N. BLOEMBERGEN: Phys. Rev. **104**, 324 (1956).
[2] K. SHIMODA, H. TAKAHASHI and C. H. TOWNES: J. Phys. Soc. Japan **12**, 686 (1957).

from temperature gradients, an exacting procedure if the highest precision is needed[1]. Finally the temperature is measured, usually with thermocouples.

By this means calibrating temperatures somewhat over 400° K are obtained. Higher temperatures have been attained by the use of a white-hot tungsten filament which is capable of rather higher but less accurately measurable temperatures.

13. Diode noise sources. Since calibration temperatures approaching one million degrees are wanted for some solar and galactic studies, something more intense than a thermal noise source is needed. For the longer wavelengths this is provided by the temperature limited diode.

By passing the mean square diode noise current $2eI\,\Delta f$ through a large resistor it is quite possible to produce noise powers approaching that available from a resistor at a million degrees. Fig. 10 illustrates a diode across which a potential difference high enough to draw the saturation current is maintained

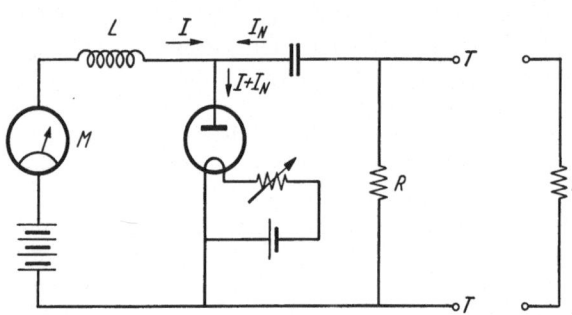

by a source shown as a battery. The diode current is $I + I_N$, I being the mean current and I_N the fluctuating part. A choke L prevents I_N from flowing in the battery circuit and a blocking condenser C ensures that the whole of I passes through the meter M. A resistor R defines the noise power which will be available from the terminals TT.

Fig. 10. A diode noise generator.

The noise power delivered to a load R_L is a maximum when the noise current i_N in the frequency band Δf is equally shared between R and R_L. Hence the available power is given by

$$\left(\frac{1}{2}\,i_N\right)^2 R = \frac{R\,e\,I\,\Delta f}{2}.$$

By writing

$$kT\Delta f = \frac{R\,e\,I\,\Delta f}{2}$$

we find for the equivalent source temperature T_s,

$$T_s = T_0 + \frac{e\,I\,R}{2k},$$

where the room-temperature contribution from the resistor R has been included.

In designing a diode noise source care is taken to make the blocking and choking elements adequate for the frequency band to be used, and adequate bypassing (not shown) for the meter is provided. A calculable correction may be allowed for. Too high a value of R will cause (i) leakage currents through conduction paths of unknown and perhaps variable resistance, (ii) fluctuations in potential comparable with the source potential, and (iii) problems of matching to loads. Stray capacity, including the interelectrode capacity, permits part of the diode current to bypass the resistor and therefore affects the accuracy of the calculated noise current. The source potential is not critical, but the current which heats the filament of the diode directly affects the noise output. It is necessary to stabilize this heating current but in addition it is usually made controllable as a simple method of adjusting the noise output.

---

[1] V. A. Hughes: Proc. Inst. Electr. Engrs. B **103**, 669 (1956).

The preceding considerations are all readily coped with up to frequencies of a few tens of MHz, beyond which the difficulties of measuring the resistance R, the current I, and above all eliminating stray reactance, cause the absolute accuracy to deteriorate. At higher frequencies, up to about 1000 MHz, diodes specially designed to fit without discontinuity into a coaxial line are available.

It is not necessary to include the resistance R but for practical reasons it is often desirable. For example if it is omitted small changes in R_L cause proportional changes in noise power whereas with R included, and R_L equal to R, poor adjustment of R_L has no effect to a first order, for $R = R_L$ maximizes the noise power delivered to R_L.

14. Gas discharge noise sources. In the microwave range gas discharge noise generators are available with noise temperatures above 10000° K. It appears that the velocity distribution of the free elec-

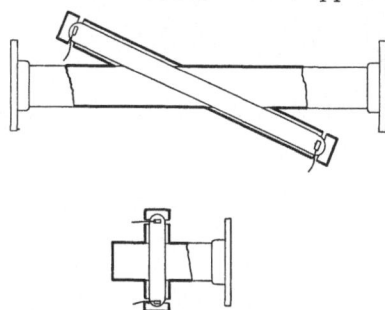

trons in an ordinary mercury-filled fluores-cent tube corresponds to a temperature of about 11000° K, and this temperature proves to be to a large extent independent of the discharge current and the gas pressure. To couple the thermal radiofrequency radiation resulting from collisions out of the discharge into a waveguide it is necessary to introduce the discharge tube into the waveguide in such a way that energy incident from a signal generator will be fully absorbed by the discharge. This presents no difficulty,

Fig. 11. Gas discharge noise generators in waveguide.

the arrangement introduced by MUMFORD[1] in his original work being to place the tube perpendicular to the axis of the guide. Matching arrangements are then necessary, but the scheme[2] illustrated in Fig. 11 (a) gives good match-ing over a wide frequency range. The perpendicular arrangement [Fig. 11 (b)] is suitable for precise measurement and also for fixed frequency operation, where it results in considerable space saving. At frequencies where waveguide is cum-bersome, a suitable arrangement is to place the discharge tube on the axis of a special coaxial transmission line whose inner conductor is helical.

Because of their ready availability and range of sizes commercial mercury-vapour lamps with fluorescent coatings have been widely used. These tubes contain mostly argon, though nearly all the radiation emitted is from the small amount of mercury. Tubes are now

Gas	Noise temperature degrees K	Noise power db above $kT_0 \varDelta f$
Mercury . .	11000	16
Argon . . .	11000	16
Neon . . .	18000	18

built especially for noise emission which contain pure argon and have clear glass tubes. Their noise temperature is almost the same but is more stable. Neon fillings are also obtainable with available noise 2 or 3 decibels higher than for argon. Careful work already done indicates that gas discharge tubes can perform as substandards to a precision of about one per cent. To achieve this accuracy in absolute level the very greatest care in calibration against a thermal source is at present required, but factory standardization is imminent. Meanwhile, the gas discharge source is very satisfactory to use, the absolute level being obtain-able to about 10% from the accompanying table.

[1] W. W. MUMFORD: Bell. Syst. Techn. J. **28**, 608 (1949).
[2] H. JOHNSON and K. R. DeREMER: Proc. Inst. Radio Engrs. **39**, 908 (1951).

f) Sensitivity to weak signals.

15. The least detectable signal. It was shown earlier that

$$\Delta T \propto \frac{T}{\sqrt{\tau \Delta f}},$$

where ΔT is the uncertainty in the measurement of a temperature T, τ is the integrator time constant, and Δf is the overall receiver bandwidth, including aerial selectivity. If T is constant for the duration of many time constants, higher precision could have been achieved by the use of a longer time constant.

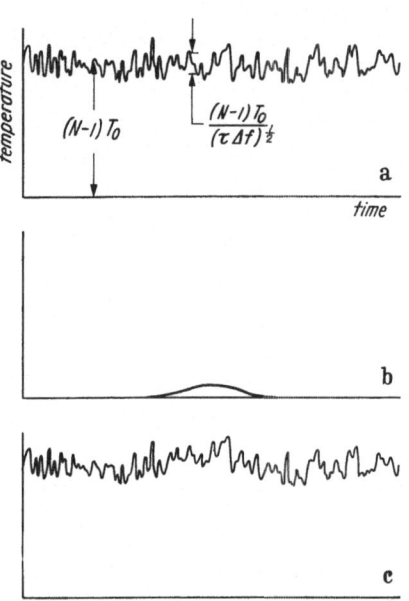

Fig. 12 a—c. Showing how a signal (b) is readily detectable in equally strong noise (a) because of its duration.

In practice one often gains this extra precision by scanning a pen record with the eye and smoothing it mentally, hence in discussing sensitivity to weak signals this aspect of τ must be borne in mind.

Suppose that the record shown in Fig. 12 (a) is obtained as an aerial sweeps over a part of the sky from which negligible radiation is received. Only receiver noise of amount T_R is recorded, and the fluctuations will be given by $T_R/\sqrt{\tau \Delta f}$. Now let the aerial sweep a part of the sky from which comes the faint signal shown in Fig. 12 (b); then the record obtained will be that shown in Fig. 12 (c). The question is how faint a signal will be detectable. In the figure the strength of the signal has been made roughly equal to the strength of the fluctuations and it will be seen that there is no doubt that this signal is detectable, for example by looking obliquely along the record.

Let the duration of the signal be $m\tau$. Then the value of T_R is determinate, over a length $m\tau$ of record, to $T_R/\sqrt{m\,\tau\,\Delta f}$, which is only a fraction of the strength of the recorded fluctuations. It is generally assumed that a signal of this strength is just detectable and hence that the sensitivity to weak signals is

$$\frac{T_R}{\sqrt{m\,\tau\,\Delta f}} = \frac{(N-1)\,T_0}{\sqrt{m\,\tau\,\Delta f}}.$$

If the efficiency of the aerial and feeder is η (Sect. 25), this expression becomes

$$\frac{\left(\dfrac{N}{\eta}-1\right) T_0}{\sqrt{m\,\tau\,\Delta f}}.$$

The exact criterion of detectability would involve the profile of the signal to be detected, the percentage reliability of detection, the experience of the observer and his prior knowledge of the profile to be expected. Such a study would be important since some of the most interesting and controversial results of radio astronomy are drawn from faint signals on the limit of detectability, but it has not yet been undertaken. Instead, the effort so far has gone into building more and more sensitive equipment.

Where the utmost sensitivity is wanted superposition of small numbers of independent records may prove feasible as in the attempts to detect 21 cm radiation from the cluster of galaxies in Coma and thermal radiation from Mars.

If one considers using very large time constants of hours or days extreme requirements on stability of equipment are encountered. For example in Fig. 12 the recorded level was assumed constant on the average, but receiver output is proportional to receiver gain, a parameter which is notoriously prone to drift.

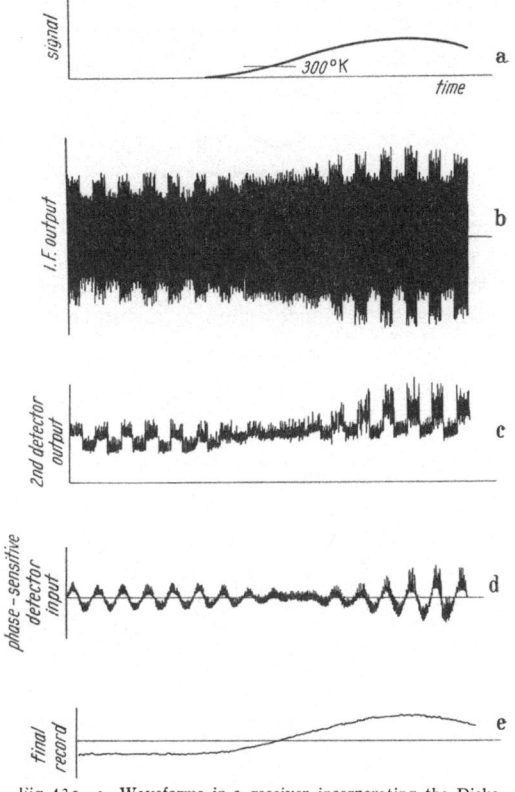

In the example illustrated a signal amounting to about 15% of the receiver noise $(N-1) T_0$ was readily detectable, but we assumed that the receiver gain remained stable to much better than 15%. In the illustration the fluctuations were deliberately exaggerated and in practice one per cent of the mean noise would be more representative, implying gain stability to one part in 10^3 over times of the order of seconds. A faint signal of hours' duration, obtained say by tracking a planet, would require higher stability over a longer time. However, a stability of one part in 10^3 is itself no small achievement. Methods of alleviating the stability requirements will now be considered.

Fig. 13 a—e. Waveforms in a receiver incorporating the Dicke system of switching between the signal and a reference noise source.

g) Comparsion systems.

16. Suppression of steady deflections. The intolerance of the direct system to changes in gain is essentially due to the largeness of the receiver noise relative to the signals to be detected. In the system due to DICKE[1], the steady component of receiver noise is cancelled out by switching the receiver between the signal coming from the aerial and a constant signal from a laboratory source. Only the difference is recorded, and whilst this still alters if the gain alters, the system ignores the accompanying change in steady component of receiver noise.

Fig. 13 (a) shows a signal which increases from 0° K to a value well over 300° K. The output from the intermediate frequency amplifier is shown in Fig. 13 (b), which should be compared with Fig. 5 (e). The time scale of Fig. 13 is so compressed that the 30 MHz oscillations merge together and the fluctuations of characteristic period $(\Delta f)^{-1}$ become narrow spikes. At regular intervals the receiver switches between the aerial and the reference source with a period which is long relative to $(\Delta f)^{-1}$ but short compared with τ (e.g. 30 Hz). The second detector takes the envelope, with some distortion depending on the law of the detector, as in Fig. 13 (c) [cf. Fig. 5 (g)]. An amplifier tuned to the switching frequency now removes the steady component, rounds off the square-wave modulation,

[1] R. H. DICKE: Rev. Sci. Instrum. **17**, 268 (1946).

and reduces the amplitude and bandwidth of the noise [see Fig. 13 (d)]. The next step is to extract the amplitude of the switch-frequency oscillation, before finally smoothing.

Now it will have been noted that the phase of the oscillation in Figs. 13 (c) and (d) undergoes a reversal as the signal level passes through the reference level, so that the final detector must be sensitive to phase. This is arranged by supplying the detector with a reference oscillation from the switch. A suitable circuit is shown in Fig. 14.

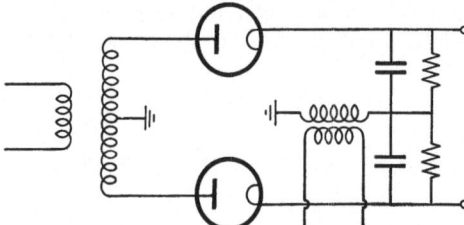

Fig. 14. A phase sensitive detector.

It will be perceived that the final deflection of the output indicator is a measure of the excess of the signal power over the thermal power received from the reference source, i.e. it is proportional to $T - T_{\text{ref}}$; in the direct system the deflection was proportional to $T + T_R$, where T is the signal noise temperature, T_{ref} is the reference noise temperature, and T_R is the receiver noise temperature. Hence when T is approximately equal to T_{ref}, the sensitivity of

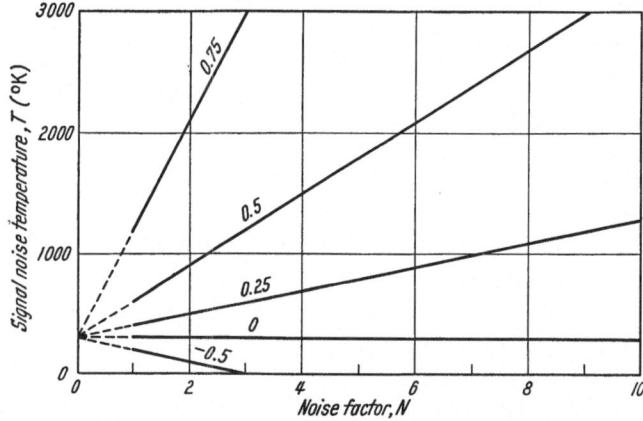

Fig. 15. The factors on the lines give the reduction in errors due to gain changes which results from adoption of the Dicke system.

the Dicke system to gain changes is negligible compared with that of the direct system. If a given fractional change in gain is interpreted as a signal level change then the ratio of the resulting error under the Dicke system to that using the direct system is

$$\frac{T - T_{\text{ref}}}{T + T_R} \, .$$

Taking T_{ref} equal to T_0 (290° K) and $T_R = (N - 1)\, T_0$, we show this reduction factor in Fig. 15.

It is notable that when N is around 10 or more and the signal to be measured is in the range from room temperature right down to absolute zero, the advantages of the Dicke system are very great. The detection of microwave radiation from the. Sun and Moon by Dicke and Beringer[1] was carried out under these conditions

However, unless $T = T_{\text{ref}}$, the dependence on gain changes is serious enough to require great care in stabilizing the receiver. Typical measures which are

[1] R.H. Dicke and R. Beringer: Astrophys. Journ. 103, 375 (1946).

taken include temperature control of the room, stabilizing the high tension and filament supplies, and stabilizing the mean anode current of the electron tubes. A further scheme is to apply gated automatic gain control to the intermediate frequency amplifier in a way which tends to maintain the long-term average second detector output constant as sampled during the half cycles when the reference source is connected.

If T is expected to remain relatively steady, for example when a source is being tracked, there is an advantage in seeking a reference source of corresponding temperature, as in the examples below.

When T is high the advantage of the Dicke system falls off and in fact it has not been generally used in meter-wave studies of the intense radiation from the Sun and Galaxy.

17. Reference sources. DICKE used a rotating disc which mechanically inserted a dissipative wedge through a slot in waveguide at a frequency of 30 Hz. This method has since been widely used and has proved satisfactory. The motor which turns the disc also drives a generator which provides a phase reference signal for the phase sensitive detector. The reference temperature is room temperature, which of course may be subject to variation.

It is possible to move the absorption band of a piece of ferrite[1] placed in a waveguide onto and off the operating frequency f by varying a magnetic field maintained in the ferrite. This can be done at a frequency of hundreds of Hz, without sound or vibration. Again the reference temperature is room temperature, the temperature of the ferrite. In the interesting experiments with this technique reported by MAYER[2] the effect of aerial mismatch on accurate measurements was studied and an application was found for the ferrite isolator in ensuring tolerance to mismatch.

Gas discharges in coaxial line and in waveguide have also been tried as high temperature reference sources.

h) The Ryle and Vonberg system.

18. Towards independence from receiver gain change. The sensitivity to gain changes and the dependence on ambient temperature would be removed if the reference source were always at the temperature of the signal. This is feasible at wavelengths where noise diodes are practicable and was applied by RYLE and VONBERG[3,4] to the measurement of meter wavelength radiation from the Sun.

The temperature of the diode filament is varied so as to keep the noise power output equal to the signal. The receiver is called on merely to detect inequalities, and need have only a high gain, not necessarily a stable gain. There is thus a great advantage in principle over the Dicke system. The quantity measured is the diode current I, which directly indicates the signal power. Two characteristic features are (i) the thermal lag of the filament which sets a lower limit to τ, and (ii) the limitation to the range of noise power output available from the diode.

The Ryle-Vonberg system has worked satisfactorily in the meter-wavelength range, at frequencies around 18 MHz, and is being extended to microwavelengths. The physical embodiments of the principle differ in these three wavelength ranges, especially as regards the switch, the noise source, and the means of calibration,

---

[1] Proc. Inst. Radio Engrs. **44** (1956). This volume is devoted to the subject of ferrites.
[2] C.H. MAYER: J. Geophys. Res. **59**, 155 (1954).
[3] M. RYLE and D.D. VONBERG: Proc. Roy. Soc. Lond., Ser. A **193**, 98 (1948).
[4] K.E. MACHIN, M. RYLE and D.D. VONBERG: Proc. Inst. Electr. Engrs. III **99**, 127 (1952).

and are not developed to an equally high pitch of perfection throughout. However the basic advantages offered are great.

In the example of Fig. 16 the reference source is a gas discharge G in waveguide whose output is attenuated by a ferrite F subjected to a variable magnetic field. Reference noise levels from 18000° K down to 298° K are thus available. For an application to solar work, a more satisfactory range would be from 1800 to 30° K; hence one tenth of the power is coupled through a directional coupler D into a waveguide W which ideally would be terminated with an absorber near absolute zero. An accessible approximation is to terminate with an antenna A pointing to the celestial pole. A signal received in antenna B passes to the ferrite switch S, together with the power from the reference source. This switch, which is operated electrically several hundred times a second, alternately connects the signal and the reference to the receiver. Any inequality produces an output from the phase sensitive detector which alters the attenuation at F in such a way as to bring the reference level into equality with the signal. The current taken by the ferrite attenuator (indicated at M) is then recorded.

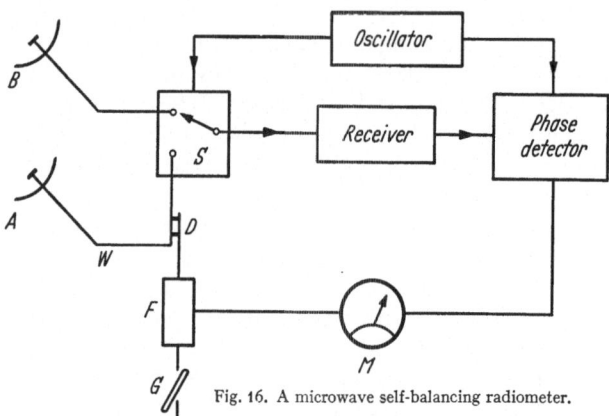

Fig. 16. A microwave self-balancing radiometer.

A modification for extending the range of operation below the minimum output of the noise generator was used by Brown and Hazard[1]. Locally generated noise was added to the signal noise and varied so as to maintain the sum constant.

III. Aerials.

Aerials are used in radio astronomy in much the same way as in other fields of radio, but with some differences. For example aerials may be used for transmitting or receiving but in radio astronomy, with the exception of astronomical radar, we are concerned only with receiving. For purposes of absolute measurement more attention must be paid to the determination of the effective collecting area of an aerial than is usually the case; polarization assumes a different importance; highly directional properties are called for; and methods of mounting and driving the aerial are quite distinctive. The various techniques practiced between 10^4 and 10^7 Hz do not concern us here.

The material which follows is a cross-section of aerial theory and practice selected for its appositeness to radio astronomy.

An aerial is a linear device which either (i) extracts power from a passing wave (often a plane wave) and delivers some of it to a pair of terminals or (ii) accepts power at its terminals and radiates some of it in all directions. The radiated field has a state of polarization which in general is elliptical and different in different directions; and a passing wave may always be split into two complementarily polarized components of which one is always fully rejected. The distribu-

---

[1] R. Hanbury Brown and C. Hazard: Monthly Notices Roy. Astronom. Soc. London **111**, 357 (1951).

tions of field and power flow around an aerial, and the currents flowing in it, are quite different in the receiving and transmitting cases. However, several reciprocal theorems closely bind the two uses, and we shall speak sometimes in terms of receiving, and sometimes in terms of transmitting, according to custom, even though the purpose is to receive.

a) Aerial parameters.

19. Radiation pattern. When power is fed into the terminals of an aerial currents are set up which produce magnetic and electric fields. Some of the energy of the fields flows into the conductors and surrounding ground and heats them, some ebbs and flows in the immediate vicinity of the aerial, and some is radiated away. Just as in the case of the diffraction fields produced by electric currents at optical frequencies, there is a division into Fresnel and Fraunhofer regions, only the latter being important in radio astronomy. In the Fraunhofer region, at great distances from the aerial, the flow of power is radially outwards and the electric and magnetic fields are transverse. The power per unit solid angle will however be a function of direction, $W(\alpha, \beta)$, where α is colatitude and β is longitude, referred to suitable axes in the aerial.

The function $W(\alpha, \beta)$, often pictured as a closed surface in space, or by means of representative cross-sections of this surface, is the radiation pattern. For a fuller description it would be necessary to add the state of polarization as a function of α and β and the departure of the isophase surface from spherical.

It is usual to normalize the radiation pattern in some way, especially by reducing the total radiated power or the maximum value of power per unit solid angle to unity.

20. Reception pattern. Consider a linearly polarized plane wave incident on an aerial from the direction (α, β), the plane of polarization making an angle $\psi(\alpha, \beta)$ with the meridian. Then the power delivered to a resistor connected across the aerial terminals will be a function of direction of arrival. This function of α and β is the reception pattern for the stated polarization and clearly other reception patterns exist for other conditions of polarization. For a full description one should include the phase of the current in the resistor under suitable assumptions of phase consistency of the plane waves from different directions.

21. Reciprocity theorems. The basic theorem stems back to work of HELMHOLTZ and RAYLEIGH and may be stated in terms of aerials and the intervening medium as follows. If a voltage generator G having zero internal impedance

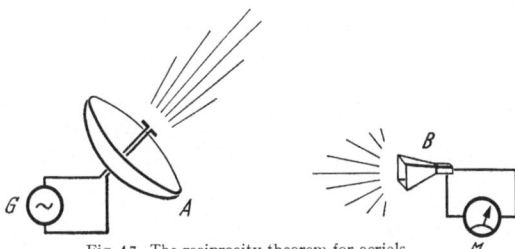

Fig. 17. The reciprocity theorem for aerials.

is applied to the terminals of an aerial A (Fig. 17) and produces a certain current in the zero-impedance meter M connected to the terminals of antenna B, then when the generator and meter are interchanged the current in the meter will be the same, under certain conditions. One of these conditions is linearity, both of the medium and of the elements of the antenna, another is freedom from the non-reciprocal effects associated with the presence of external magnetic fields, as in the ionosphere and in ferrites.

For theoretical purposes the reciprocity theorem is very important, as practical conditions for its applicability are readily realizable, though clearly also non-linearity and magnetic phenomena are commonly encountered in practice. Ground reflections, conductor loss, refraction, multiple paths, or transient conditions do not vitiate the reciprocity theorem.

A consequence of the basic theorem is that the radiation and reception patterns of an aerial are identical, provided the incident waves used for testing the reception pattern are suitably polarized. Another is that the impedance adjustment for maximum radiation is optimum for reception. These properties make it possible to test the reception of an aerial by using it to radiate, an arrangement which is often convenient.

In the usual pattern test arrangement the aerial under test and a second aerial are rotated relative to each other about an axis through the first. A generator is connected to one and the power received by the other is recorded. By the reciprocity theorem it will be immaterial which is used to receive, however the patterns so obtained refer not to the total power pattern $W(\alpha, \beta)$ but to a fraction of the power which is appropriately polarized. The remaining fraction, the cross-polarized component, is measured by rotating one aerial 90 degrees about the line joining the aerials and repeating.

22. Measurement of radiation patterns. The preceding discussion assumes a rigid aerial. When it is understood that the currents induced in the ground and objects surrounding an aerial are in effect part of the antenna, it will be seen that many rotatable and otherwise movable aerials are non-rigid. For this reason radiation patterns are usually virtually impossible to measure in their entirety.

When making measurements that are intended to apply to reception from celestial sources, it is necessary for the second aerial to be so far away that the path difference between any pair of rays joining the two aerials does not differ by more than a small fraction of a wavelength from the value it would have if the second aerial receded to infinity. Existing large aerials have already exceeded practical limits, but the partial check that is possible by flying a suitably oriented second aerial on a predetermined track at constant altitude through the distant field of the aerial under test has often been judged worthwhile. Direct observation of celestial sources such as the Sun and the strong discrete sources in Cassiopeia and Cygnus is helpful when the wavelength and beamwidth are such that only one source is important. While such observations may fall short of determining the directivity and details of the side radiation they may often have more pertinence than any other possible measurement.

23. Directivity. There is usually some direction which is of special interest in an aerial application such as a direction towards which most of the radiated power is concentrated. The directivity D of the aerial in this direction is defined by

$$D = \frac{\text{power radiated per unit solid angle in specified direction}}{\text{average power radiated per unit solid angle}}.$$

It is possible to calculate the directivity in terms of the radiation pattern. Thus if the specified direction is $\alpha = 0$, $\beta = 0$,

$$D = \frac{4\pi\, W(0, 0)}{\int\limits_0^\pi \int\limits_0^{2\pi} W(\alpha, \beta)\sin\alpha\, d\alpha\, d\beta}.$$

Since $W(\alpha, \beta)$ is difficult to measure, so likewise is D. However, in many cases it is possible to evaluate the integral to fair accuracy by combining one or two measured cross-sections of $W(\alpha, \beta)$, where it is large, with theoretical expectations.

24. Effective solid angle. The effective solid angle Ω is defined by

$$\Omega\, W(0, 0) = \int\limits_{0}^{\pi} \int\limits_{0}^{2\pi} W(\alpha, \beta) \sin \alpha\, d\alpha\, d\beta,$$

whence

$$\Omega = \frac{4\pi}{D}.$$

For example if an aerial has a directivity of 1000 it has an effective solid angle of $4\pi/1000$ steradians or 41 square degrees.

25. Gain. If the power radiated by an aerial were radiated equally in all directions in an amount W per unit solid angle the total radiated power would be $4\pi W$. Now when power $4\pi W$ is supplied to the terminals of an actual aerial the power $W(0, 0)$ radiated per unit solid angle in the direction $\alpha = 0$, $\beta = 0$ exceeds W by a factor g which is known as the gain. Thus

$$W(0, 0) = g\, W.$$

Of the power $4\pi W$ being supplied to the terminals of an aerial let a fraction $\eta\, 4\pi W$ be radiated, the remainder being dissipated as heat in the conductors of the aerial and the surrounding ground. The quantity η is called the efficiency, and it relates the gain to the directivity; thus

$$g = \eta\, D.$$

The efficiency of an aerial is not easy to measure but it is usually high and so can be deduced from relatively rough estimates of the losses of power in the feeders and in the ground.

The gain defined here is sometimes referred to as gain relative to a loss-free isotropic source and so is not directly measurable. The measurable quantity is gain relative to some standard aerial, such as a half-wave dipole, whose directivity in equatorial directions is known from the theoretical radiation pattern to be 1.64. One takes the ratio of the input powers necessary to produce the same effect at a distance, and allows for the factor 1.64. It is still necessary to know the efficiency of the standard. The loss in the dipole itself is normally negligible, ground loss can be made negligible by raising the dipole sufficiently, and the feeder loss can be accurately measured. The most difficult matter is to ensure that no significant amount of power is carried away back along the feeder by unbalance currents. In measuring high gains it is customary to use a standard aerial of moderate gain which may itself previously have been calibrated against a dipole. Alternatively a horn of calculable directivity may be used (SCHELKU-NOFF[1]) or the two-aerial method of absolute gain measurement in terms of the inverse square law of spatial attenuation (SILVER[2]).

26. Effective area. When a receiving aerial is placed in a radiation field the power which it abstracts depends on the load impedance which is connected to its terminals. By adjustment of the impedance a maximum transfer of power to the load can be arranged, and this power is referred to as the available power of the antenna. The effective area A measures the ability of an aerial to abstract energy from the field of a traveling plane wave coming from a specified direction and to deliver it to the aerial terminals. Since only one polarization can be

[1] S.A. SCHELKUNOFF: Electromagnetic waves, Chap. 9. New York: Van Nostrand 1943.
[2] S. SILVER: Microwave antenna theory and design, Chap. 15. New York-Toronto-London: McGraw-Hill 1949.

accepted, the wave is taken to have that polarization. Then

$$A = \frac{\text{available power at aerial terminals}}{\text{power crossing unit area of wavefront}} \ .$$

It will be perceived from the reciprocity theorem that A is proportional to g, in fact

$$g = \frac{4\pi A}{\lambda^2} \ .$$

This relation is normally proved by calculating both g and A for some one loss-free aerial (e.g. see Schelkunoff and Friis[1]) but it is in a way unsatisfactory to have to appeal to a special case with its own inevitable approximations to establish a general result. The following explanation is taken from Pawsey and Bracewell[2]. An aerial has connected to its terminals a resistor at temperature T which has been adjusted for maximum power transfer from a passing wave to the resistor. The reciprocity theorem enables us to say that the adjustment is also correct for maximum power transfer from a voltage generator in series with the resistor to the aerial. A black body also at temperature T subtending a small solid angle Ω in a direction in which the gain is g intercepts power $g\,kT\,\Delta f\,\Omega/4\pi$. Using Planck's formula for black body radiation the power delivered to the resistor is $\frac{1}{2}(2kT\,\Delta f/\lambda^2)\,\Omega A$, the factor $\frac{1}{2}$ resulting from the reception of one component of polarization. Under conditions of thermal equilibrium the principle of detailed balancing requires these two powers to be equal. We then find that $g = 4\pi A/\lambda^2$ without appeal to details of any one aerial.

b) Radiation patterns.

27. Fourier transform relation. One way of relating the physical structure of an aerial to its radiation pattern is to integrate the remote (and duly retarded) effects of its constituent current elements and space charges. Another way is to begin from the fields impressed on an infinite plane in or near the aerial and deduce the distant fields. For the purposes of the present section the latter approach is advantageous.

We know that twin plane waves equally and oppositely inclined to the normal to a plane as in Fig. 18(a) give rise to a cosinusoidal standing wave pattern as indicated by the parallel nodal lines of the figure. Conversely, such a standing wave pattern impressed on the infinite plane must launch two such plane waves into the semi-infinite space on the side opposite the sources of the field. When the nodal lines are closer together the pair of wave-normals are spread more widely apart as in Fig. 18(b), the sine of the angle ϑ between the wave normal and the $\eta\zeta$-plane being equal to the number of cycles of the standing wave pattern per free-space wavelength. With $\vartheta = 90°$ the two waves interfere with like nodes separated by the free-space wavelength λ; as ϑ approaches zero the nodes separate indefinitely. When the impressed field has more than one cycle per free-space wavelength no traveling waves are launched, the field merely evanesces as indicated in Fig. 18(c). The value of considering these simple distributions is that they constitute Fourier components of more general field distributions.

When there is periodic variation in two directions as in Fig. 18(d) each ray is further split into two, each making an angle φ with the $\xi\zeta$-plane. Each two-

---

[1] S.A. Schelkunoff and H.T. Friis: Antennas theory and practice, Chap. 6. New York: Wiley 1952.

[2] J.L. Pawsey and R.N. Bracewell: Radio Astronomy, Chap. 2. Oxford 1954.

dimensional Fourier component

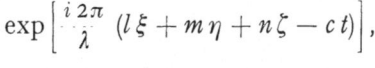

$$\cos \frac{2\pi\,\xi}{\lambda_1}\,\cos \frac{2\pi\,\eta}{\lambda_2}\,\cos \omega\,t$$

of a general impressed field distribution gives rise to four waves for which

$$\sin \vartheta = \pm \frac{\lambda}{\lambda_1},$$

$$\sin \varphi = \pm \frac{\lambda}{\lambda_2},$$

and the resulting disturbance is therefore of the form of the real part of

$$\exp\left[\frac{i\,2\pi}{\lambda}\,(l\,\xi + m\,\eta + n\,\zeta - c\,t)\right],$$

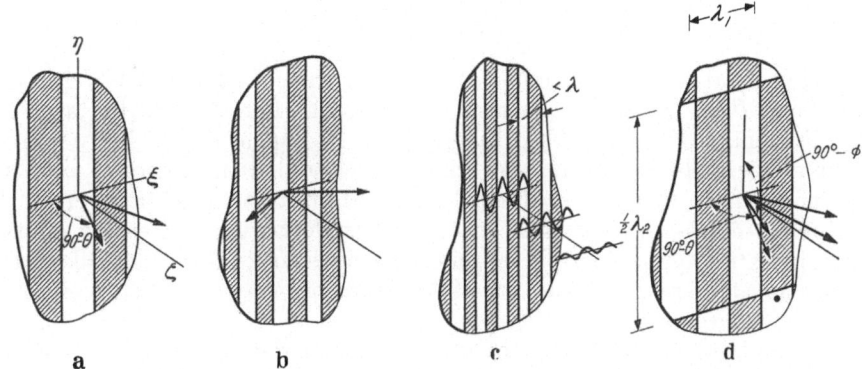

Fig. 18 a—d. Effects produced by a field distribution over an infinite plane aperture.

where l, m, n, the direction cosines of the wave normal, are given by

$$l = \pm \sin \vartheta,$$

$$m = \pm \sin \varphi$$

and

$$l^2 + m^2 + n^2 = 1.$$

The above wave function contains all the types of behavior illustrated in Fig. 18, including evanescent waves ($l^2 + m^2 > 1$).

The basic point of this section is the Fourier transform relation between an aperture distribution of field and the radiation which it launches. Let an aperture distribution of finite dimensions be analyzed into all its doubly periodic components [see Fig. 18 (d)]; then the field at a point in the direction (ϑ, φ) at a great distance from the aperture is fixed by the strength of the Fourier component of period $\lambda \operatorname{cosec} \vartheta$ in the ξ direction and $\lambda \operatorname{cosec} \varphi$ in the η direction.

Let

$$P(l, m) \exp\left[\frac{i\,2\pi}{\lambda}\,(l\,\xi + m\,\eta + n\,\zeta - c\,t)\right] dl\,dm$$

be the complex expression for the field strength at (ξ, η, ζ) due to waves in the interval l to $l + dl$ and m to $m + dm$. Then the field $F(\xi/\lambda, \eta/\lambda)$ over the aperture plane $\zeta = 0$ is given by

$$F\left(\frac{\xi}{\lambda}, \frac{\eta}{\lambda}\right) = \int\limits_{-\infty}^{\infty} \int\limits_{-\infty}^{\infty} P(l, m) \exp\left[\frac{i\,2\pi}{\lambda}\,(l\,\xi + m\,\eta)\right] dl\,dm.$$

This is the standard form of the two-dimensional Fourier transform (Sned-don[1]); consequently the inverse relationship

$$P(l, m) = \int\limits_{-\infty}^{\infty} \int\limits_{-\infty}^{\infty} F\left(\frac{\xi}{\lambda}, \frac{\eta}{\lambda}\right) \exp\left[-\frac{i\,2\pi}{\lambda}(l\,\xi + m\,\eta)\right] d\left(\frac{\xi}{\lambda}\right) d\left(\frac{\eta}{\lambda}\right)$$

follows immediately.

From the function P, which will be referred to as the angular spectrum, following Booker and Clemmow[2], the radiation pattern can be determined. The power radiated per unit solid angle in the direction (l, m) will be proportional to

$$P\,P^*\,\sqrt{1 - l^2 - m^2}\,,$$

since the element $dl\,dm$ subtends a solid angle

$$\frac{dl\,dm}{\sqrt{1 - l^2 - m^2}}\,.$$

When the aerial is highly directive, so that only values of l and m near zero are important, $P\,P^*$ represents the radiation pattern. Furthermore the aperture field distribution $F(\xi/\lambda, \eta/\lambda)$ then has negligible ζ-components, so that we arrive finally at the well-known statement that the field radiation pattern of an aerial is proportional to the two-dimensional Fourier transform of the transverse field distribution across its aperture. The preceding derivation will enable off-axis and polarization questions to be examined as necessary; meanwhile, taking the Fourier transform relation as a point of departure, one may immediately establish many aerial theorems.

The limits of integration $(-\infty$ to $+\infty)$ in all the double integrals which follow have been omitted, and we have taken $\sin\vartheta = \vartheta$ and $\sin\varphi = \varphi$.

28. Similarity theorem. By a change of variables,

$$P\left(\frac{\vartheta}{p}, \frac{\varphi}{q}\right) = p\,q \iint F\left(p\,\frac{\xi}{\lambda}, q\,\frac{\eta}{\lambda}\right) e^{-i\,2\pi\left(\vartheta\frac{\xi}{\lambda} + \varphi\frac{\eta}{\lambda}\right)} d\left(\frac{\xi}{\lambda}\right) d\left(\frac{\eta}{\lambda}\right);$$

hence compressing an aperture distribution, as measured in wavelengths, by factors p and q expands the associated field radiation pattern by the same factors. The similarity theorem underlies less precise rules of the form "beamwidth is inversely proportional to aperture size, and directly proportional to wavelength".

29. Shift theorem. By a different change of variable

$$P(\vartheta - \Theta, \varphi - \Phi) = \iint \left[e^{-i\,2\pi\left(\Theta\frac{\xi}{\lambda} + \Phi\frac{\eta}{\lambda}\right)} F\left(\frac{\xi}{\lambda}, \frac{\eta}{\lambda}\right) \right] e^{i\,2\pi\left(\vartheta\frac{\xi}{\lambda} + \varphi\frac{\eta}{\lambda}\right)} d\left(\frac{\xi}{\lambda}\right) d\left(\frac{\eta}{\lambda}\right).$$

Hence if an aperture distribution is modified by causing its phase to change linearly with distance across the aperture then the radiation pattern is shifted in the direction of lagging phase by an amount proportional to the phase gradient. A beam shift Θ requires a phase gradient of $2\pi\,\Theta$ radians per wavelength.

30. Array theorem. Let the convolution of the function $F(\xi/\lambda, \eta/\lambda)$ with some other function $L(\xi/\lambda, \eta/\lambda)$ be written

$$L * F \equiv \iint L\left(\frac{\xi}{\lambda} - \mu, \frac{\eta}{\lambda} - \nu\right) F(\mu, \nu)\,d\mu\,d\nu$$

---

[1] I. N. Sneddon: in vol. II of this Encyclopedia (Mathematical Methods), p. 198. Berlin 1955.
[2] H. G. Booker and P. C. Clemmow: Proc. Inst. Electr. Engrs. III **97**, 11 (1950).

and let the two dimensional Fourier transform of L be $A(\vartheta, \varphi)$, i.e.

$$A(\vartheta, \varphi) = \iint L\left(\frac{\xi}{\lambda}, \frac{\eta}{\lambda}\right) e^{\,i\,2\pi\left(\vartheta\,\frac{\xi}{\lambda} + \varphi\,\frac{\eta}{\lambda}\right)} d\left(\frac{\xi}{\lambda}\right) d\left(\frac{\eta}{\lambda}\right).$$

Then according to the two dimensional convolution theorem

$$A(\vartheta, \varphi)\, P(\vartheta, \varphi) = \iint (L * F)\, e^{\,-i\,2\pi\left(\vartheta\,\frac{\xi}{\lambda} + \varphi\,\frac{\eta}{\lambda}\right)} d\left(\frac{\xi}{\lambda}\right) d\left(\frac{\eta}{\lambda}\right).$$

One way of interpreting this theorem is as follows. Consider an array of elements each of which has aperture distribution F and pattern P, and let the elements be at the points $(\xi_i/\lambda, \eta_i/\lambda)$. The resultant aperture distribution can be expressed as the convolution $L * F$, where

$$L = \sum_i {}^2\delta\left(\frac{\xi - \xi_i}{\lambda}, \frac{\eta - \eta_i}{\lambda}\right),$$

and ${}^2\delta(\xi, \eta)$ is the two-dimensional impulse function with the properties $\iint {}^2\delta(\xi, \eta)\, d\xi\, d\eta = 1$ and ${}^2\delta(\xi, \eta) = 0$ where $\xi^2 + \eta^2 \neq 0$. Consequently the resultant field radiation pattern of an array of elements can be expressed as the product of the element pattern P and an array factor A which is the transform of the array arrangement L.

31. Directivity. In an earlier section the directivity was defined in terms of the field radiation pattern; we now deduce its dependence on the aperture field distribution. Consider a highly directional aerial consisting of a large but finite plane aperture across which is maintained a tangential electric field E. The field radiation pattern P is the Fourier transform of E; the radiation pattern $P P^*$ is the transform of $E * E^* (-)$, by the convolution theorem. The power radiated per unit solid angle in the direction $l = m = 0$ is $P P^*|_0$ and the total power radiated is, to a good approximation, $\iint P P^*\, dl\, dm$. Hence the directivity is given by

$$D = \frac{P P^*|_0}{\dfrac{1}{4\pi} \iint P P^*\, dl\, dm}.$$

Now the infinite integral of a function is the value of its Fourier transform at the origin; hence

$$D = \frac{4\pi \iint [E * E^* (-)]\, d\left(\frac{\xi}{\lambda}\right) d\left(\frac{\eta}{\lambda}\right)}{E * E^* (-)|_0}.$$

Since the infinite integral of the convolution of two functions is the product of their infinite integrals,

$$D = \frac{4\pi \iint E\, d\left(\frac{\xi}{\lambda}\right) d\left(\frac{\eta}{\lambda}\right) \iint E^*\, d\left(\frac{\xi}{\lambda}\right) d\left(\frac{\eta}{\lambda}\right)}{\iint E\, E^*\, d\left(\frac{\xi}{\lambda}\right) d\left(\frac{\eta}{\lambda}\right)}$$

or, in terms of the mean value E_m over the area $\mathfrak{A}$ of the aperture,

$$D = \frac{4\pi \mathfrak{A}/\lambda^2}{\dfrac{\lambda^2}{\mathfrak{A}} \iint \left(\frac{E}{E_m}\right)\left(\frac{E}{E_m}\right)^* d\left(\frac{\xi}{\lambda}\right) d\left(\frac{\eta}{\lambda}\right)}.$$

In taking the total power radiated to be $\int\limits_{-\infty}^{\infty} \int\limits_{-\infty}^{\infty} P P^*\, dl\, dm$ we have assumed the integrand to fall to negligible proportions while l and m are still small compared with unity. If this is not true then $\int\limits_{-\infty}^{\infty}\int\limits_{-\infty}^{\infty}$ should be replaced by $\iint\limits_{l^2+m^2<1}$ since the evanescent fields for which l^2+m^2 exceeds unity do not carry away radiation. These fields are associated with the Fourier components of E which have high spatial frequencies exceeding one cycle per free space wavelength and so, if necessary, the symbol E in the preceding formula may be interpreted as that part of the field distribution not containing high spatial frequencies. We do this in a later section which treats the effect of bolt heads and other irregularities.

32. Directivity factor. If the aperture is uniformly excited all over, E/E_m is unity and $D=4\pi\mathfrak{A}/\lambda^2$. Since we have proved that $g=4\pi A/\lambda^2$, it follows that the effective area A of a uniform loss-free aperture is equal to its physical area $\mathfrak{A}$. When the excitation over an aperture is not uniform, the directivity falls by a factor $\mathfrak{D}$ referred to as the directivity factor (sometimes gain factor); thus

$$\mathfrak{D}=\frac{1}{\dfrac{1}{\mathfrak{A}}\iint\left(\dfrac{E}{E_m}\right)\left(\dfrac{E}{E_m}\right)^*d\xi\,d\eta}\,.$$

c) Types of aerial.

A wide variety of aerials find application in radio astronomy. In this section we review many of the types briefly with comments on their behavior and considerations entering into their design.

33. Dipoles. As used in connection with aerials, the term dipole refers to a conductor about half a wavelength long, broken in the centre, as in Fig. 19(a).

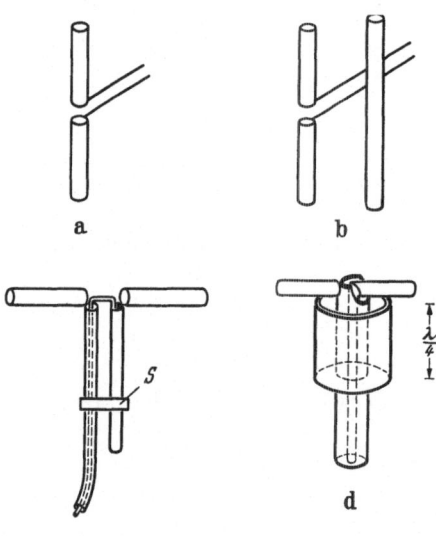

Infinitesimal dipoles are not ordinarily used. Power radiated from a dipole is concentrated towards the equatorial plane, equally in all longitudes. The radiation pattern takes the form

$$\left[\frac{\cos(\tfrac{1}{2}\pi\cos\alpha)}{\cos\alpha}\right]^2,$$

which differs only slightly from $\sin^2\alpha$, and the directivity in the equatorial plane is 1.64. As the operating frequency is changed the impedance of a dipole changes from its nominal value of about 70 ohms. However the need for tuning adjustments can be avoided over at least a two to one frequency range by fattening the conductors.

Dipoles are combined with other elements in many ways. The directivity may be doubled by placing a reflector behind the dipole in such a way that the reflected field reinforces the primary field in the forward direction. Fig. 19(b) illustrates one arrangement, but the mechanical embodiments of the idea are very numerous[1].

Fig. 19a—d. (a) A dipole, (b) a dipole with reflector, (c) and (d) baluns.

If it is necessary to connect a coaxial cable to a dipole a balance-to-unbalance transformer or balun is incorporated to prevent unbalance currents received on

---

[1] For an extensive treatment see G. H. Brown: Proc. Inst. Radio Engrs. **25**, 78 (1937).

the outside of the cable from contributing to the signal sent down the inside. The Pawsey stub [Fig. 19(c)] achieves this result by symmetry, at the same time providing an impedance matching adjustment by means of a shorting strap S, and the quarter-wave trap [Fig. 19(d)] inserts a high impedance in the path of unbalance currents.

34. Yagi aerials. A very convenient mechanical arrangement for increasing the gain of a dipole is afforded by the arrangement of Fig. 20, which was introduced by YAGI[1]. In addition to the reflecting element there are a number of "directors" so spaced and adjusted in length that the phase of the currents induced in them causes reinforcement of radiation in the direction of the arrow.

With care, the gain of a Yagi aerial containing n elements can be made about n times that of one dipole, the adjustment being carried out by trial. Since the lengths of the elements and their spacing depend on frequency, aerials of this type are essentially limited to a narrow band of frequencies.

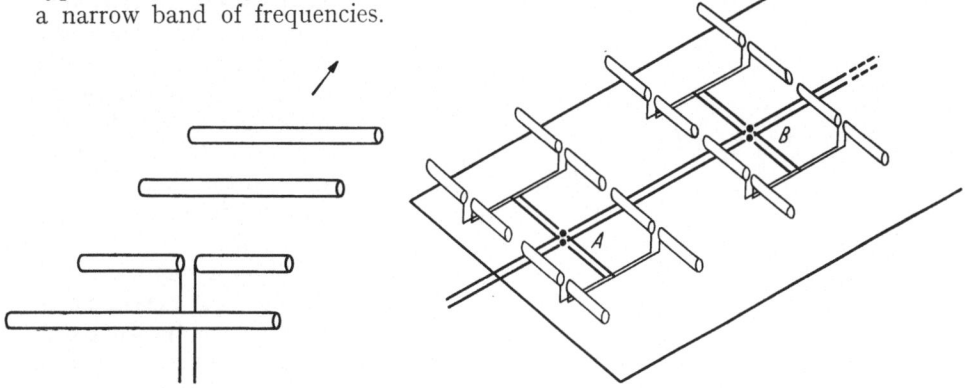

Fig. 20. A Yagi aerial.   Fig. 21. A broadside array of dipoles.

35. Broadside arrays. Arrays of dipoles such as in Fig. 21 are effective at meter wavelengths. The gain obtainable exceeds that of a single dipole by a factor roughly equal to twice the number of elements if, as is usual, a plane reflector backs up the array. The gain g per dipole-with-reflector is approximately 3, and the effective area per dipole, $g\lambda^2/4\pi$, is approximately $\frac{1}{4}\lambda^2$. This consideration illuminates the practice of spacing the dipoles at intervals of one half wavelength in both directions, but mechanical and electrical factors usually control this choice. If necessary the same gain can be obtained with substantially fewer elements. Feeding the dipoles all in equal amplitude and phase can be done by two essentially different methods both of which are illustrated in the figure. The group of dipoles fed from A are connected identically and therefore remain equally excited when the frequency is changed. But correct excitation of the group connected to B will require a certain relation between the length AB and the wavelength. Such frequency-dependent arrangements are often adopted for simplicity.

The effective area of a large broadside array can be made very nearly equal to the physical area, and the possibility of approaching the theoretical design in practice is an advantage where absolute measurements of radiation intensity are attempted.

Arrays of large size can be pointed in different azimuths but equatorial mounting, with few exceptions has not generally been deemed feasible. However

[1] H. YAGI: Proc. Inst. Radio Engrs. **16**, 715 (1928).

electrical beam-swinging is readily incorporated. Suppose for example that the section $A B$ in Fig. 21 had a different electrical length, i.e. the phase difference $(A B) \times (2\pi/\lambda)$ between A and B were different; then, as explained above in connection with the shift theorem, the beam would be shifted in proportion. Beamswinging is accomplished by actual physical adjustments or by changing wavelength.

36. Helical aerials. The aerial illustrated in Fig. 22 is akin to the Yagi, in that it is extended in the direction of radiation (indicated by the arrow), but has important differences which favor its application in radio astronomy. It is not as sensitive to changes in frequency, ranges of nearly two to one in frequency being not unreasonable. Careful adjustment is not required, a coaxial cable may be used without a balun, and high directivity is achieved with mechanical simplicity. The ground plane disc may be less than a wavelength in diameter.

Under correct operating conditions a wave travels around the helical wire with such a phase velocity that the phase difference between corresponding

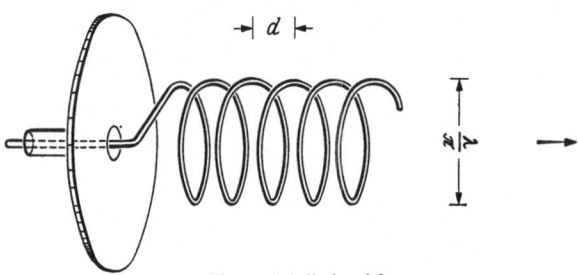

points on adjacent turns exceeds $2\pi d/\lambda$ by one cycle, d being the pitch of the helix. The distant fields of successive turns then reinforce in the axial direction and produce circular polarization.

Fig. 22. A helical aerial.

From a combination of theory and measurement KRAUS[1] gives the half power beamwidth in degrees as $52 (\lambda/C) (\lambda/L)^{\frac{1}{2}}$, where C is the circumference and L the length of the helix, provided the circumference is about one wavelength, the pitch about one quarter wavelength, and the number of turns greater than 3. From this we can say that the directivity will be approximately $15 (C/\lambda)^2 (L/\lambda)$ and the effective area $15 (L/\lambda) \times$ (projected area normal to axis). The input impedance is about 140 ohms.

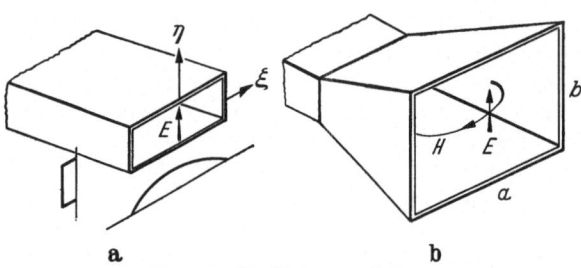

Fig. 23 a and b. Electromagnetic horns.

37. Horns. The open end of a rectangular waveguide radiates very much as one would expect from a plane aperture on which is maintained an electric field distributed as indicated in Fig. 23 (a). When the end of the waveguide is flared as in Fig. 23 (b) greater directivity results, and provided the aperture dimensions are less than about 3 wavelengths the electric field in the aperture can be written

$$E_\eta = \frac{\pi}{2} \cos \frac{\pi \xi}{a} \, \Pi \left(\frac{\xi}{a} \right) \Pi \left(\frac{\eta}{b} \right),$$

where $\Pi (\xi)$ is the rectangle function of unit height and width, i.e.

$$\Pi (\xi) = \begin{cases} 1 & (|\xi| < \tfrac{1}{2}), \\ 0 & (|\xi| > \tfrac{1}{2}). \end{cases}$$

[1] For details and further references see J. D. KRAUS: Antennas. New York 1950.

From the Fourier transform relation we have shown that the directivity is reduced, relative to a uniformly excited aperture for which

$$E_\eta = \Pi\left(\frac{\xi}{a}\right)\Pi\left(\frac{\eta}{b}\right).$$

by a factor $\mathfrak{D}$. Having normalized E_η to have unit mean value and to cover unit interval, we find succinctly,

$$\mathfrak{D} = \left[\int_{-\frac{1}{2}}^{\frac{1}{2}}\left(\frac{\pi}{2}\cos\pi\xi\right)^2 d\xi\right]^{-1} = \frac{8}{\pi^2}.$$

Hence the effective area is $8ab/\pi^2$, or about 0.8 of the physical area. The beamwidth in degrees between points where the radiation pattern is one tenth of the maximum is given by

$$\frac{88\lambda}{b} \quad (\eta\zeta\text{ plane}),$$

$$31 + \frac{79\lambda}{a} \quad (\xi\zeta\text{ plane}).$$

The value of these empirical formulae (SILVER[1]) will be seen below in connection with feeds for parabolic reflectors.

When the aperture of a horn is larger than about 3 wavelengths, there will be a phase inequality across the aperture depending on the length of the flared section. The directivity reduction integral corresponding to that given above can be evaluated and SCHELKUNOFF and FRIIS[2] present convenient charts for determining the directivity.

One of the few applications in radio astronomy of a horn alone was to the discovery of the galactic hydrogen line emission by EWEN and PURCELL. Usually a large horn requires too much volume for a given effective area. There are, however, some important uses. One is the provision of a calculable standard of gain intermediate between that of a dipole and that of an aerial under calibration. Another is for feeding paraboloidal reflectors. The high efficiency and relative freedom from wide-angle side radiation also suit them for use with future noise-free receivers.

38. Paraboloidal reflectors. By placing a source at the focus of a reflecting paraboloid of revolution one may produce extensive equiphase fields in planes normal to the axis. In comparison with a broadside array the means used are simple; there is only one radiator and one feeder. Furthermore, the frequency of operation can be altered by readjustment or replacement of the single feed system. Thus for many purposes a paraboloidal reflector offers advantages.

At the focus one commonly finds a dipole and reflector, or a horn; however any aerial may be used including small broadside arrays in the larger reflectors at longer wavelengths. The effective solid angle Ω_f of the feed antenna must harmonize with the solid angle subtended by the reflector at its focus. If Ω_f is too small, the outer part of the reflector is not fully utilized, and directivity is lost; while if Ω_f is too large there is "spillover", which has two effects. First there is a pure loss of energy in undesired directions and then, because of the large discontinuity in illumination at the edge of the reflector, the side radiation, though reduced, takes on a pronounced lobe structure. It is therefore customary to taper the illumination towards the rim, balancing the loss of directivity

S. SILVER: Microwave antenna theory and design. New York 1949.
S. A. SCHELKUNOFF and H. T. FRIIS: Antennas, theory and practice. New York 1952.

against the undesirability of distinct sidelobes. Equating the angle subtended at the focus by the rim of the paraboloid to the 10-decibel beamwidth of a feed horn, we have

$$4 \arccot \frac{4F}{D} = 88 \frac{\lambda}{b} = 31 + 79 \frac{\lambda}{a},$$

where F/D is the ratio of focal length to diameter of the paraboloid. These equations give suitable dimensions, a and b, for the feed horn. However, other proportions hould be considered for partic-

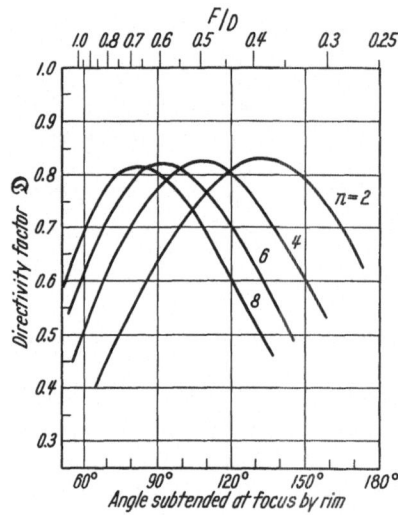

Fig. 24. Theoretical directivity factor as a function of focal ratio F/D for different values of the taper parameter n in the feed pattern $2(n+1)\,\Pi\,(\vartheta/\pi)\,\cos^n\vartheta$.

ular purposes, and some of the factors governing the design follow.

Whether the loss in directivity factor is quantitatively serious may be determined from Fig. 24 (adapted from SILVER), which gives the theoretical directivity factor for the normalized set of calculable feed patterns $2(n+1)\,\Pi\,(\vartheta/\pi)\,\cos^n\vartheta$. One of these functions will represent the actual feed pattern over most of the beam and Fig. 24 then gives the optimum focal ratio, and an approximate value of $\mathfrak{D}$. If desired, a better value of $\mathfrak{D}$ can then be calculated from an improved determination of the aperture field distribution which takes into account the actual feed pattern and allows for the inverse squares of the path lengths from the focus to the paraboloid.

The aerial feeding a paraboloid will normally have been adjusted to match its transmission line in the absence of the reflector, but when it is placed in the focus of a small paraboloid and excited by a generator it intercepts some reflected energy which then causes standing waves on the line. The fraction of power

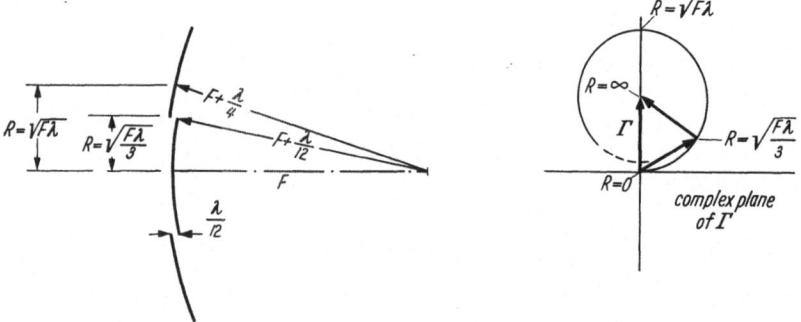

Fig. 25. Matching a paraboloid with a vertex plate.

intercepted will be approximately equal to A_H/A_P the ratio of the effective areas of the feed horn and the paraboloid and so the power reflection coefficient $|\Gamma|^2$ observed on the transmission line will approximately equal this ratio. Putting $A_P = F^2 \Omega_H$, where $4\pi/\Omega_H = g_H = 4\pi A_H/\lambda^2$, we have

$$|\Gamma| = \frac{g_H \lambda}{4\pi F} = \frac{A_H}{F\lambda},$$

alternative expressions for reflection coefficient which do not involve the aperture of the parabola. The reason for this is that the first few Fresnel zones around the vertex, which contribute the bulk of the reflection, occupy only a small fraction of the aperture. The radius of the first Fresnel zone, assuming a radius of curvature $2F$, is $\sqrt{F\lambda}$, and a cap of radius $\sqrt{F\lambda/3}$ contributes a reflection equal in magnitude to the full reflection but leading it in phase by a sixth of a cycle, as shown in Fig. 25. The region outside the cap contributes an equal amount which lags the full reflection by a sixth of a cycle. If now this cap is moved towards the focus by a twelfth of a wavelength, as in the illustration, the total reflection will be annulled. In practice a flat plate of the calculated radius is placed at the mean position of the cap, i.e. $\lambda/8$ from the vertex, any further adjustments being made by trial.

The foregoing considerations are complicated by mechanical details of the structure supporting the feed, which it may only be possible to take into account empirically. The following procedure nevertheless reveals some of the effects to be expected. Cancel the aperture distribution over the geometrical shadow cast by all the structures, including the feed horn, in front of the reflector. Then subtract from the field radiation pattern the pattern due to an aperture the same shape as the shadow.

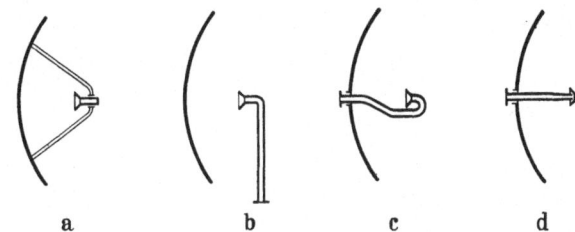

Fig. 26 a—d. Methods of supporting a feed horn at the focus.

The principal effects thus deduced are a reduction in effective area by the shadowing and an increase of side radiation in a distinctive arrangement associated with the shadow shape. The ultimate destination of the radiation impeded by the structure can also be considered, especially if it leads directly back into the horn.

Some basic arrangements are shown in Fig. 26. In (a) the horn is supported on four struts rising from firmly held points on the reflector, and the transmission line from the horn runs down one of the struts. In this very satisfactory arrangement a main concern is to minimize the shadow cast by parts of the structure near the horn. In (b) shadowing is small though asymmetrical, and the arrangement is ideal where access to the feeder near the rim is convenient. Arrangement (c) is suitable when access at the vertex is desired and it is simple mechanically. There is substantial asymmetrical shadowing and a direct reflection into the horn which must be minimized experimentally. In (d) an excellent mechanical arrangement is obtained at the cost of considerable empirical design and some restriction on bandwidth associated with the compact T-junction and pair of reflex horns at the focus.

Paraboloidal aerials receive some stray energy over and above that which is focused by reflection from the paraboloid. There is some side-reception by the collecting aerial at the focus from directions outside the cone defined by the rim of the reflector, and there may be some transmission through the reflecting surface itself (shine-through) if it is not a continuous metal sheet. There is also some scattering from the roughnesses of the reflector, and from the feed support and other incidental structures.

Let power W_{in} enter the terminals of an aerial such as is shown in Fig. 26. Some is absorbed in the transmission line; of that which emerges from the feed horn let W_{beam} be the amount intercepted by the paraboloid and focused into the main beam. We use the term main beam to refer to the radiation launched, ideally, with the characteristic Fraunhofer diffraction pattern of the circular aperture. Let W_{sky} be the total power launched towards the sky whether by the main beam, or by stray routes such as direct or ground-reflected spillover, scatter, or shine-through. Then the beam ratio $\mathfrak{B}$ is defined by

$$\mathfrak{B} = \frac{W_{\text{beam}}}{W_{\text{sky}}}$$

and, by the definition of efficiency η in Sect. 25,

$$\eta = \frac{W_{\text{sky}}}{W_{\text{in}}}.$$

We now calculate the aperture efficiency α, the ratio of the effective area A to the area $\mathfrak{A}$ of the circle bounded by the rim of the paraboloid. From Sect. 25 and 26, the gain g is given by

$$\frac{4\pi A}{\lambda^2} = \frac{W(0,0)}{\frac{1}{4\pi} W_{\text{in}}} = \frac{W(0,0)}{\frac{1}{4\pi} W_{\text{beam}}} \frac{W_{\text{beam}}}{W_{\text{sky}}} \frac{W_{\text{sky}}}{W_{\text{in}}} = \mathfrak{D} \frac{4\pi \mathfrak{A}}{\lambda^2} \mathfrak{B}\eta,$$

where the directivity factor $\mathfrak{D}$, referred to the rim of the paraboloid, is given as in Sect. 32 by

$$\mathfrak{D} = \frac{1}{\frac{1}{\mathfrak{A}} \iint \left(\frac{E}{E_m}\right) \left(\frac{E}{E_m}\right)^* d\zeta\, d\eta}.$$

(In evaluating this expression small-scale irregularities in E that scatter energy out of the main beam should be ignored.) Hence for the aperture efficiency α we have

$$\alpha = \frac{A}{\mathfrak{A}} = \mathfrak{D}\mathfrak{B}\eta.$$

The factor $\mathfrak{B}\eta$ is the beam efficiency β, the ratio of the power launched in the beam to that entering the aerial terminals:

$$\beta = \mathfrak{B}\eta = \frac{W_{\text{beam}}}{W_{\text{in}}}.$$

The application of these parameters is discussed in Sect. 69 and 70. Their measurement involves one step beyond the determination of g and η, such as, for example, a measurement of the shape of the main beam. Thus the presence of stray radiation will cause the effective solid angle Ω_{beam} of the main beam to be less than the expected value $4\pi/D = \lambda^2/\mathfrak{D}\,\mathfrak{A} = 4\pi\eta/g$. Hence

$$\mathfrak{B} = \frac{\mathfrak{D}\,\mathfrak{A}\,\Omega_{\text{beam}}}{\lambda^2}$$

and

$$\beta = \frac{g\,\Omega_{\text{beam}}}{4\pi}.$$

The effective solid angle of the main beam can be determined from careful non-absolute pattern measurement over a limited region. Then Ω_{beam} is the double integral of the limited reception pattern divided by the axial value. The integral is extended over a region sufficiently large, usually a few beamwidths in extent, that further extension makes an inappreciable difference for the purpose in hand.

Fig. 24 shows that $\mathfrak{D}$ will commonly be around 70 per cent. Beam ratios $\mathfrak{B}$ around 60 per cent are common and as the operating frequency is raised the surface irregularities of a given reflector introduce a factor ζ (Fig. 28) that leads to even smaller values. An efficiency η of 10 to 50 per cent is commonly accepted but much improved values are obtainable by eliminating the transmission line between the horn and the receiver. Further improvement in efficiency is obtainable by enlarging the horn to reduce spillover absorption by the ground; this reduces side-reception of thermal emission from the ground by the horn, but reduces $\mathfrak{D}$. The highest possible efficiency η is needed for sensitive observations in order to minimize the effective receiver noise temperature $(N/\eta - 1)\,T_0$ mentioned in Sect. 15, and to minimize objectionable changes of ground radiation as the aerial moves. This premium on efficiency calls for horn designs, for use with noise-free amplifiers, that accept a reduction in aperture efficiency α in favor of optimum performance with higher η.

d) Feeders.

39. Transmission lines and waveguides. The transmission line or feeder associated with an aerial may often conveniently be considered as a part of the aerial.

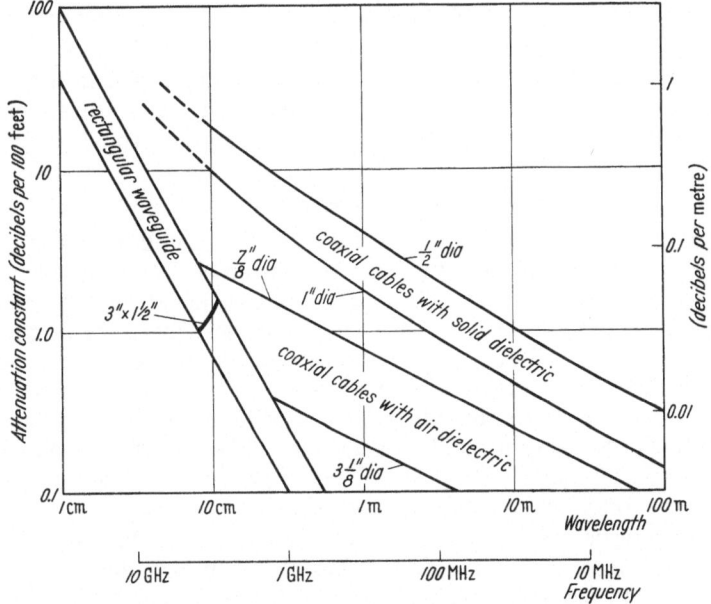

Fig. 27. Attenuation constant and range of use of various feeders.

It may serve an impedance matching function, be adapted for standing wave measurement, and enter into switching, coupling, and phasing devices.

In Fig. 27 the attenuation constant of various available feeders is shown. Flexible coaxial cables with solid dielectric are usable at all but the highest frequencies, where the dissipation in conductor and dielectric becomes prohibitive. Coaxial lines with a rigid outer conductor instead of braid, and with air dielectric, furnish lower attenuations especially in the larger diameters. However, when the mean circumference of the air space exceeds the wavelength the appearance of a second transmission mode renders the line unusable. There is therefore an upper limit to both diameter and frequency. Beyond the useful range of coaxial lines, and in a transition region, rectangular waveguide may be used. Unlike the coaxial line, which is satisfactory over a semi-infinite spectrum, each waveguide size is restricted to less than one octave. The figure shows a curve for the waveguide used at 10 cm; similar curves for the waveguide appropriate to each wavelength fall within the zone indicated.

Some other lines, not shown in Fig. 27, are important. Two wire transmission line has about the same conductor losses as coaxial line of similar dimensions and is far better mechanically in many ways. It cannot be used where complete shielding is essential, is subject to radiation loss at the highest frequencies, and is affected seriously by dew and other natural phenomena. It has been used successfully in very long runs for experimental purposes as high as 1400 MHz.

Ridged waveguide[1] permits operation over a range of two octaves, strip line[2] offers many convenient features, and single-wire surface wave transmission line[3] offers the possibility of extremely long straight runs for which waveguide would be too costly.

An important consideration in radio astronomy is stability of the phase length of a feeder, a quantity which may be affected by humidity, temperature, mechanical effects, and frequency change. Measures to achieve stability include sealing off from the air, pressurizing, burying or shading, heating, tensioning, and broadbanding connectors, bends, junctions, supports and other discontinuities in the line.

A large fraction of the essential technique in the development of a successful instrument for radio astronomy is spent on the design and adjustment of the transmission system, more than the length of this section would imply.

e) Tolerances.

Consider an aerial whose design calls for definite amplitudes and phases in a certain plane. When the structure has been assembled and ajusted there will be departures from the desired amplitudes and phases due to the following causes. In the case of a paraboloidal reflector there will be departures of the surface from the true paraboloid and the feed horn will not be perfectly located and oriented. In the case of a broadside array there will be small reflections on the feeders and discrepancies in phase length to the different elements. The location and orientation of the elements and the reflector will be imperfect, and in addition, the sizes of all the mechanical elements will be subject to variation from the design. There will be variable effects due to wind, gravity, and temperature.

It is therefore necessary to know what tolerances may be allowed in order to limit the deterioration in the final behavior to an acceptable degree. This branch of aerial theory is in a primitive state, but it is clear that the large and complex aerials of radio astronomy will benefit from a suitable theory of tolerances. It has been customary to measure the gain and side radiation of actual aerials and to make empirical adjustments, but the adjustments become more laborious with increasing aerial size and the measurements themselves are becoming infeasible.

In the following discussion it will be assumed that the amplitude and phase errors are known and the effect on the radiation pattern, and especially the directivity and beamwidth, will be studied.

---

[1] Very-high-frequency Techniques. New York-Toronto-London: MacGraw-Hill 1947.

[2] See various papers in Transactions of the IRE Professional Group on Microwave Theory and Techniques, vol. MTT 3, March 1955.

[3] G. GOUBAU: Electronics 27, 180 (1954).

40. The directivity possible of achievement. It was shown earlier that the directivity factor $\mathfrak{D}$ of an aperture of physical area $\mathfrak{A}$ is given by

$$\mathfrak{D} = \frac{\mathfrak{A}}{\iint \left(\frac{E}{E_m}\right)\left(\frac{E}{E_m}\right)^* d\xi\, d\eta}.$$

It is convenient to express the aperture illumination E relative to the mean value E_m over the aperture because the type of maladjustment which would cause the whole illumination to be say ten per cent low or ten degrees lagging in phase does not affect the directivity. Therefore, by first normalizing with respect to the mean we can concentrate separately on the quantity of interest.

Consider the case where the departures from the mean illumination E_m are εE_m. Then

$$\mathfrak{D} = \frac{\mathfrak{A}}{\iint (1+\varepsilon)(1+\varepsilon)^*\, d\xi\, d\eta}$$

$$= \frac{1}{1 + \frac{1}{\mathfrak{A}}\iint \varepsilon\,\varepsilon^*\, d\xi\, d\eta}$$

$$= \frac{1}{1 + \operatorname{var}\varepsilon},$$

where var ε is the variance, or mean square modulus, of the complex fractional departures ε. This basic formula, which is valid for both systematic and random departures in both phase and amplitude, has valuable applications.

If a non-uniformly illuminated aperture is perturbed so that the fractional departures from the perturbed mean are ε' and the perturbed directivity factor is $\mathfrak{D}'$, then the factor ζ by which the perturbation reduces the directivity is given by

$$\zeta = \frac{\mathfrak{D}'}{\mathfrak{D}} = \frac{1 + \operatorname{var}\varepsilon}{1 + \operatorname{var}\varepsilon'}.$$

Conversely defined, ζ is the factor which measures the extent to which $\mathfrak{D}'$ achieves the design value $\mathfrak{D}$; thus

$$\mathfrak{D}' = \zeta\mathfrak{D}.$$

We shall refer to ζ as the directivity achievement factor. In terms of a minimum acceptable value, such as 0.9, we can determine tolerances which are permissible.

Let us consider first the case of an aperture designed to be uniformly illuminated with a mean field E_m, but which is subject to an undesired phase error δ which varies from point to point over the aperture in any way. This is similar to the situation described in optics by a wavefront which departs from the ideal sphere, and it engenders all the types of aberration familiar in optical systems (MARÉCHAL[1]). The perturbed illumination $E_m \exp i\delta$ has a lowered mean value $E'_m \approx E_m(1 - \frac{1}{2}\operatorname{var}\delta')$, for small δ', where δ' is referred to the new mean phase, should it be altered, i.e.

$$\delta' = \delta - \frac{1}{\mathfrak{A}}\iint (1+\varepsilon)\,\delta\, d\xi\, d\eta;$$

but the mean square modulus of the perturbed illumination is unchanged. Therefore

$$\zeta = \left(\frac{E'_m}{E_m}\right)\left(\frac{E'_m}{E_m}\right)^* \approx 1 - \operatorname{var}\delta'.$$

---

[1] A. MARÉCHAL: in vol. XXIV of this Encyclopedia (Fundamentals of Optics), p. 44. Berlin 1956.

Alternatively we may say that the change in absolute mean value is small and that $E'_m(1+\varepsilon')=E_m \exp i\delta \approx E'_m(1+i\delta')$, whence $\varepsilon \approx i\delta'$. Then, with $\varepsilon \approx 0$

$$\zeta = \frac{1+\text{var } \varepsilon}{1+\text{var } \varepsilon'} \approx \frac{1}{1+\text{var } \delta'}.$$

Both lines of reasoning are useful for obtaining approximations in non-uniform cases; however, we can state without approximation that an aperture which is intended to be uniformly illuminated and which in fact has fractional departures ε' from the mean, of any magnitude, whether in amplitude or phase, has

$$\zeta = \frac{1}{1+\text{var } \varepsilon'}.$$

In a non-uniform case it may be possible to calculate the directivity factor in both the perturbed and unperturbed situations. If not, various approximate formulae may be evolved; for example if there are small phase errors δ, then by expanding the exponential factor in

$$E'_m = \frac{1}{\mathfrak{A}} \iint E \, e^{i\delta} \, d\xi \, d\eta$$

we find

$$\zeta \approx 1 - \text{var } \sqrt{1+\varepsilon} \, \delta'.$$

41. Squint. Let there be a progressive phase error across the aperture of amount α radians per meter in the x direction. If the extreme phase error is small, we have by the preceding equation, a reduction in directivity given by

$$\zeta \approx 1 - \alpha^2 \, \text{var } \sqrt{1+\varepsilon} \, x.$$

More precisely

$$\zeta = \left|\frac{E'_m}{E_m}\right|^2 = \left|\frac{1}{\mathfrak{A}} \iint (1+\varepsilon) \exp i\alpha\xi \, d\xi \, d\eta\right|^2$$

$$= \frac{PP^*|_\Theta}{PP^*|_0}$$

where $\Theta = \alpha\lambda/2\pi$; i.e. the whole reduction is due to a displacement of the radiation pattern PP^* through an angle $\alpha\lambda/2\pi$ without change of shape. The angle of squint Θ is the same as that which would have been predicted by the shift theorem.

42. Quadratic phase error. Let $\delta = \beta\xi^2$. Then different effects will result according as the aperture is uniformly illuminated or not. In the case of a uniformly illuminated rectangular aperture the mean phase of the perturbed illumination is one third of the phase error Δ at the edge of the aperture: Hence

$$\zeta = 1 - \text{var } \delta'$$

$$= 1 - \text{var}\left(\beta\xi^2 - \frac{\Delta}{3}\right)$$

$$= 1 - \frac{4\Delta^2}{45};$$

and for $\zeta = 0.9$ evidently an extreme phase error of 1.1 radians can be tolerated in the uniform case and even more if the illumination is tapered.

Now let $\delta = \beta(\xi^2 + \eta^2)$ over a uniformly illuminated circular aperture. Then the mean phase is one half the phase error $\varLambda$ at the edge and

$$\zeta = 1 \cdot \operatorname{var}\left|\beta(\xi^2 + \eta^2) - \frac{\varLambda}{2}\right|$$

$$= 1 - \frac{\varLambda^2}{12},$$

which is approximately the same as before.

Non-uniform illumination may be handled as indicated already or treated in a way mentioned below.

43. Random errors. Since there is nothing in the preceding work which excludes random errors the formulae given may be adopted immediately. In fact, the expressions in terms of mean square departures from a mean are especially appropriate. For example, a uniform broadside array whose elements are subject to small phase errors δ' from the mean has an achievement factor $1 - \operatorname{var} \delta'$; thus for $\zeta = 0.9$ a root-mean-square phase error of 0.3 radians (root-mean-square path difference of $\lambda/30$) can be tolerated. If the elements are in phase but unequally excited, then a root-mean-square fractional error of 0.3 can be tolerated; and if both sorts of error are present then the tolerances are 0.2 radians ($\lambda/28$) and $\pm 20\%$. These tolerances are liberal, especially when it is considered that extreme errors, both positive and negative, may substantially exceed the root-mean-square value. Whilst the illustrations given here are taken from convenient approximate formulae, it is recalled that, whether the errors are large or small, recourse may be had to the accurate result $\zeta = (1 + \operatorname{var} \varepsilon')^{-1}$. This result, moreover, does not assume a normal distribution of the errors, or any particular spatial correlation. There is, however, an important point of interpretation in the case of continuous reflectors.

As pointed out in the section on directivity, evanescent fields do not normally have importance in the angular spectrum of a highly directional aerial, but if they do, the symbol E in the formula for directivity may be interpreted as the part of the field distribution remaining when Fourier components of spatial wavelength less than the free-space radiation wavelength have been filtered out. Now in a reflector having irregularities such as bolt heads, ribs, or dents, evanescent fields are set up which should be ignored in evaluations of the distant field. Those spatial components of the irregularities whose scale is finer than the wavelength should therefore be rejected before calculating ζ; this conclusion rests on the assumption that an irregularity merely introduces a corrugation of the same shape in the wavefront, which seems appropriate to this first order discussion of smaller effects. Starting from a given corrugated wavefront one would first filter out the fine detail numerically, then evaluate $\operatorname{var} \delta'$. Alternatively, one might start from the (unnormalized) autocorrelation function[1] $\delta' \star \delta'$ of the phase corrugations referred to their mean. Then $\delta' \star \delta'|_0$, which is proportional to $\operatorname{var} \delta'$, is equal to $\iint F \, d\mu \, dv$, the infinite integral of F, the Fourier transform of the autocorrelation function of δ'. Then the variance of the filtered part of δ' will be proportional to the integral of that part of F for which $\sqrt{\mu^2 + v^2}$, the number of cycles per unit distance of wavefront, does not exceed λ^{-1}. An expression of

---

[1] We use the unobtrusive notation

$$f \star g = \iint f(u - \xi, v - \eta) \, g(u, v) \, du \, dv$$

for the unnormalized cross-correlation function of f and g as distinct from the convolution integral

$$f * g = \iint f(\xi - u, \eta - v) \, g(u, v) \, du \, dv.$$

this theoretical procedure, in terms of the rectangle function Π (Sect. 37), is

$$\zeta \approx 1 - \frac{\iint \Pi \left(\frac{1}{2}\lambda \sqrt{\mu^2 + \nu^2}\right) F \, d\mu \, d\nu}{\iint F \, d\mu \, d\nu} \text{ var } \delta'.$$

It may be applied to the problem of the variation of ζ of a uniform reflector as λ diminishes. Let the departures from the mean be h, then var $\delta' = (4\pi/\lambda)^2 \text{ var } h$ and for frequencies high enough to perceive the irregularities in full ζ will fall off parabolically as in the broken curve of Fig. 28. At low frequencies the exact form of $h \star h$ will determine the extent to which var δ' is diminished but in general a result such as the heavy curve indicates would be obtained. The point P marks the transition from mainly un-resolved irregularities to nearly complete resolution.

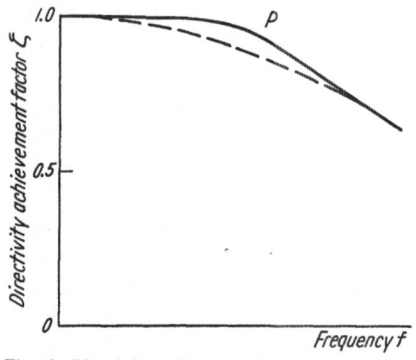

The effect of random errors on aerials has not been widely known. Ruze[1] has made an elegant experimental study and clearly explained the principal effects. He quotes the approximate expression $1 - \text{var } \delta'$, attributing it to R. C. Spencer, and examines an extension of the formula to a particular case of partial resolution of random irregularities. When an aerial has been designed for fractional de-partures ε from the mean uniform illumination and the departures are

Fig. 28. Directivity achievement factor for a reflector. The broken curve corresponds to complete resolution of the irregularities.

in fact $\varepsilon + \varepsilon_r$, where the ε_r are randomly distributed about zero mean value, then $\overline{\text{var}(\varepsilon + \varepsilon_r)} = \text{var } \varepsilon + \overline{\text{var } \varepsilon_r}$, where bars indicate the average over a batch, provided that ε_r is not correlated with ε on the average over a batch. Then we find the following general result:

$$\bar{\zeta} = \frac{1 + \text{var } \varepsilon}{1 + \text{var } \varepsilon + \overline{\text{var } \varepsilon_r}}$$

$$= \frac{1}{1 + \mathfrak{D} \overline{\text{var } \varepsilon_r}}.$$

When $\quad \varepsilon_r \approx \dfrac{i 4 \pi h}{\lambda},$

$$\bar{\zeta} = \frac{1}{1 + \left(\dfrac{4\pi}{\lambda}\right)^2 \mathfrak{D} \overline{\text{var } h}}.$$

In applying these formulae, and those below, ε_r or h must first be filtered.

It is not only the effect on ζ which is important in a study of tolerances; we are also concerned with the way in which the radiation pattern deteriorates, i.e. what happens to the shape of the main beam and what happens to the level of side radiation. A decrease in directivity implies by definition an increase in effective solid angle $\Omega = 4\pi/D$, and this may be interpreted as a widening of the beamwidth to half power in the case of errors of very low spatial frequency. However, by resolving the errors into spatial Fourier components, as described by the function F, we shall see precisely how much power is scattered into each off-axis direction. Should this scattered power be comparable with the intended radiation in a given direction; then the radiation pattern will be complicated. But if the scattered radiation predominates then we have an important property

[1] J. Ruze: Nuovo Cim. 9, Suppl. No. 3, 364, (1952).

of the undesired side radiation level immediately in terms of P, or $\delta' \star \delta'$, the autocorrelation function of the phase departures from the mean. For irregularities containing fine detail one would expect the side radiation level not far outside the main beam to be dominated by the irregularity contribution, but the shape of the main beam would not be much affected. For systematic errors such as those discussed in following sections, the side radiation would be mainly unchanged and the main beam would be fattened.

The following interesting result is obtained when numerous small roughnesses such as rivet heads scatter side radiation equally in all directions. Then the scattered side radiation level, expressed as a fraction of the aerial radiation level, is given by

$$2 \frac{\frac{1}{\zeta} - 1}{D} = \frac{2 \mathfrak{D} \operatorname{var} \varepsilon_r}{D} = \frac{2 \pi \mathfrak{D}}{A} \operatorname{var} h.$$

As examples of the application of these relations, suppose that an aerial has been designed for a maximum sidelobe level 20 decibels down on the main beam, and for a directivity $D = 100$. Then if the side radiation due to isotropically scattering roughnesses is to be kept negligible, say to one part in 1000 of the axial radiation level, then ζ must be kept up to 0.9. This is a lower limit for ζ since anisotropic scattering will mean stronger scattering than the average in some directions. If the directivity is much greater than 100, it will be seen that scattered radiation is unlikely to be important. Conversely stated, larger tolerances on roughness are permissible. The expression $\frac{2 \pi \mathfrak{D}}{A} \operatorname{var} h$ gives this in the case of a reflector and it will be noted that, provided the irregularities are fully resolved the side radiation level does not depend on wavelength.

44. Extrafocal errors. When the feed horn of a paraboloid is displaced from the focus there are two different effects according as the displacement is transverse or axial.

A given transverse displacement causes a progressive phase error across the aperture which can be written explicitly from the geometry, and which is approximately of the form

$$\delta = \alpha \xi - \beta \xi^3.$$

The first term causes a simple shift of the radiation pattern and the second term, which can be investigated as was done above for quadratic error, causes a loss of directivity. However this loss is small, because the cubic errors are strong near the edge, where the illumination is reduced.

A good deal of transverse extrafocal displacement can thus be tolerated since in most aerial applications, if not all, the axis of the radiation pattern is located by electrical observation, and not by reference to the mechanical axis.

The effect on ζ, if desired, can readily be evaluated for a given illumination and focal ratio. However by the time ζ falls appreciably below unity the displacement is so great that complications set in.

Axial displacement introduces a phase error approximately of the form

$$\delta = \alpha \xi^2 - \alpha_2 \xi^4.$$

Arguing from the quadratic term alone, which has already been studied, and neglecting illumination taper, we can say that an axial displacement of order $\lambda/3$ can be tolerated for a 10% loss in directivity.

45. Aparaboloidal aberration. It often happens that an intended paraboloid departs from the truth but remains a figure of revolution. If the departure is quadratic the shape remains paraboloidal but with a changed focal length. Therefore the effect is the same as for axial displacement of the feed from the focus and can be compensated for if the error is a permanent one such as might arise during construction, and not such as would be caused by thermal expansion or wind loading. If the departure is more general then the focus not only shifts but elongates into a line segment whose ends represent the foci of paraxial and extreme rays respectively. The best compensation would result from placing the feed between the ends of the segment as reminiscent of the position of best focus under conditions of spherical aberration in optics, a closely related phenomenon. Clearly one can set limits on allowable aparaboloidal distortions, including the astigmatic, along the approximate lines indicated here, or one can refer back to the actual geometry and apply the basic formula.

46. Side radiation. Away from the main beam the side radiation pattern exhibits numerous more or less distinct minima the regions between which are referred to as sidelobes. Unwanted natural or man-made point sources may give trouble by coming in on sidelobes, which therefore receive much attention. To calculate the side radiation one takes the Fourier transform of the aperture distribution and adds to it the direct radiation from the feed which is not intercepted by the reflector. In general no accuracy at all can be expected because small errors in assumptions regarding physical dimensions, etc., cause important effects at large path differences such as exist well away from the axis, where numerous large field components combine in a state of approximate cancellation Any method of reducing the off-axis response relative to the on-axis response essentially involves an increase in directivity. Sometimes however alleviation of sidelobes is sought by tapering the aerial excitation towards the edge which, as we have seen, reduces the directivity. Evidently this is only appropriate where o be structure of the side radiation, not its strength, is a cause of concern.

f) Mountings and drive systems.

47. Mountings. The principal mountings in use are meridian, altazimuth, and equatorial. A large meridian telescope at Pott's Hill is shown in Fig. 29, one meridian-mounted element of the giant Cambridge[1] interferometer is shown in Fig. 30 and Fig. 31 shows an extended array of meridian-mounted Yagis at Stanford.

Altazimuth mountings are illustrated in Figs. 32 to 34 which show the radio telescopes at Dwingeloo, Manchester[2], and Stanford.

For almost all astronomical purposes an equatorial mounting would be preferred, but for reasons of economy altazimuth mountings are accepted, even to the extent of embodying computing mechanisms, as has often been done, for converting coordinates. Meridian mountings are even more economical and have permitted larger aerials for a given effort in applications where the restricted motion has sufficed. Of course even larger reflectors are made possible by sacrificing all motion as in the case of the 212 foot fixed paraboloid at Manchester and the 80 foot fixed paraboloid at Sydney, both of which achieved discoveries depending on high resolution many years before the advent of steerable reflectors of comparable size.

---

[1] See M. Ryle and A. Hewish: Mem. Roy. Astronom. Soc. **67,** 97 (1955).
[2] See A.C.B. Lovell: Proc. Inst. Electr. Engrs. B **103,** 711 (1956).

Fig. 29. (Photo.) Large meridian telescope at Pott's Hill, Sydney. Its diameter is 35 feet, and at a wavelength of 21 cm it has a beamwidth to half power of 1.5 degrees.

Fig. 30. (Photo.) The Cambridge meridian interferometer which has four elements each 320 by 40 feet situated at the corners of a rectangle 1900 by 168 feet. The gain of each element is 950 at a wavelength of 3.7 meters.

Fig. 31.

Fig. 32.

Fig. 33.

Fig. 31. (Photo. A meridian array at Stanford. Length = 2020 feet, number of elements = 96, beamwidth to half power — 3 degrees at 23 MHz.

Fig. 32. (Photo.) Radio telescope on altazimuth mounting at Dwingeloo in the Netherlands. Diameter = 25 m, focal length = 12 m, beamwidth to half power at 21 cm = 0.56 degrees.

Fig. 33. (Photo.) The Manchester installation. Diameter = 250 feet, focal length = 62.5 feet.

Because radio telescopes are dish-shaped rather than pencil-shaped the various equatorial mountings for optical telescopes cannot be adopted unmodified. Most mountings abandon the feature of intersecting polar and declination axes, as a result of which the counterweights which bring the center of gravity of the reflector onto the declination axis have themselves in turn to be balanced by weights moving with the yoke to bring the center of gravity of the reflector plus yoke onto the polar axis. A further weight is then sometimes necessary to move the center of gravity along the polar axis to a point above the footing. A drawback associated with counterweights is the limit to motion imposed by their contact with the reflector, supporting structure, or ground; and modifications which increase the freedom of motion generally tend to increase the counterweights. A possibility occasionally adopted for small mountings is to make one of the counterweights variable with declination. Many ingenious variations can be worked out to suit special conditions, many no doubt still remaining to be invented. Fig. 35 shows some existing designs.

Fig. 34. (Photo.) One of a pair of 61-foot reflectors built by Stanford Research Institute, California.

48. Drive systems. Drive systems for tracking in sidereal time divide into those incorporating synchronous motors with suitable gearing, and others. Since the synchronous motor cannot change speed a gear change or second motor and differential gear must be provided for slewing. A remote position indicator, usually attached not to the reflector but to a gear wheel moving at higher speed, completes the system. A single non-synchronous motor can perform both tracking and slewing if the discrepancy between reflector time and clock time is used to control the motor speed. In one arrangement the clock turns the shaft of a synchro-receiver the error voltage from which excites one winding of a two-phase induction motor. In another the clock and reflector angles are subtracted mechanically and a switch turning the motor on and off is actuated by the difference shaft. An overriding control permits slewing at full speed.

Automatic returns from the western to the eastern horizon for long exposures on the one object, for example the Sun, require provision for keeping track of the object during the slewing. This is difficult when the synchronous element is the motor but is readily arranged when the synchronous element is the reference clock.

Special tracking rates such as required by the Sun, Moon, planets, etc., can be incorporated into non-synchronous drive systems by using appropriate reference clocks and into synchronous systems by adjusting the power frequency or by means of a further differential.

IV. Theory of aerial smoothing.

49. Definitions of brightness, brightness temperature, and flux density. The basic observation in radio astronomy is the determination of the strength of ratio waves coming from different directions over an area of sky. We may be concerned with a large part of the sky, as when making a galactic survey, or only a small part, as when studying the emission from discrete objects or sets of objects. We may also be concerned with the spectrum and the polarization and time dependence of strength, spectrum and polarization. In the case of those solar disturbances known as outbursts we have to deal with sources whose strength, spectrum, polarization, position and presumably size all change greatly in the course of minutes, but more usually the position is simpler and we begin with the measurement of strength as a function of position on the sky. This includes galactic surveys and the determination of position, size, and brightness distribution of discrete sources at a single frequency. All other types of measurement then reduce, at least in principle, to repetition on other frequencies, at other times, and with differently polarized aerials.

Three different quantities are customarily used in specifying the strength of celestial radio waves, viz. brightness and brightness temperature (in referring to extended sources) and flux density (in referring to discrete, but not necessarily point, sources). These quantities will now be defined.

Consider the energy ΔE which falls on a small area Δa at ground level in a time interval Δt and frequency band Δf, and which comes from directions within a cone of solid angle $\Delta \omega$ surrounding the point P on the celestial sphere through which passes the normal to the area. Let the epoch be t_1 and the mid-frequency f_1, and let us assume that there are no temporal, spectral, directional or spatial discontinuities[1] in the radiation at t_1, f_1, P or on the ground at Δa. Then we might expect that

$$\lim_{\Delta a, \Delta f, \Delta \omega, \Delta t \to 0} \frac{\Delta E}{\Delta a\, \Delta f\, \Delta \omega\, \Delta t}$$

would exist, thus defining a brightness b as a measure of the strength of radiation arriving at the receiving point from the direction of P and permitting us to write

$$dE = b\, da\, df\, d\omega\, dt.$$

In the meter-kilogram-second system of units, whose use in radio astronomy has been virtually universal, b is measured[2] in watts meters^{-2} Hz^{-1} steradians^{-1}. The precision with which one would know the strength of a signal occupying a band Δf and having a duration Δt is one part in $\sqrt{\Delta f\, \Delta t}$ for reasons already

---

[1] All these assumption can be expected to break down in practice in special cases.

[2] There is also an official movement under way to name this unit the jansky per steradian in honor of KARL JANSKY who first studied radio waves of extraterrestrial origin.

a

b

c

d

Fig. 35a—d. (Photo.) A group of equatorially mounted radio telescopes. (a) The 60-foot reflector at Harvard (photo by J. SHEAHAN, Boston Globe). (b) A model of the 140-foot reflector for Greenbank, West Virginia. (c) Christiansen array at Fleurs, New South Wales, forming one arm of a cross. (d) One of thirty-two 10-foot reflectors at Stanford, California.

explained in connection with the statistical limit to precision of noise measurement. Therefore it is necessary that $\Delta f\,\Delta t \gg 1$ and consequently it is not possible to proceed to the limit indicated above. Furthermore, the product $\Delta a\,\Delta\omega$ is subject to limitations, for suppose that the small collecting area Δa is to be realized by means of a horn or other aerial. Then $\Delta\omega$ will equal the "effective solid angle" $\lambda^2/\Delta a$ as defined earlier for an aerial of "effective area" Δa, and so the product $\Delta a\,\Delta\omega$ cannot be made less (or greater) than λ^2. This situation may not appear satisfactory as a basis for an observational science but it is accepted[1] and we can assert that the concept of brightness is useful in practice, the ratio $\Delta E/\Delta a\,\Delta\omega\,\Delta f\,\Delta t$ assuming a usably definite appearance while the factor $\Delta f\,\Delta t$ remains large enough to satisfy the inequality. The essential indefiniteness of b, as it is customarily conceived, reappears later.

Brightness is thought of as a scalar point function of direction (more generally a matrix of four quantities when polarization is included) specifying detail of the radiation field at a point. The use of brightness temperature T_b, as an alternative to b, derives part of its convenience from the proportionality at radio wavelengths between the brightness of the radiation from a black body and the temperature of the black body. Thus we may express PLANCK's radiation formula in the form

$$b = \frac{2h f^3}{c^2}\;\frac{1}{\exp\left(\dfrac{h f}{kT}\right) - 1},$$

where $b\,df$ is the power in the frequency range f to $f+df$ which is received perpendicularly per steradian on one side of unit area situated in an isotropic radiation field at temperature T[2]. When $hf \ll kT$, which is true for radio wavelengths

---

[1] In radiation theory, for example in the thermodynamics of stellar atmospheres, the concept of brightness is introduced along the lines of MILNE's well known treatment (Handbuch der Astrophysik, vol. 3, p. 65, 1930). MILNE did not mention the restriction on Δt and it continues to be overlooked in recent treatises (of which for example say A. UNSÖLD, Physik der Sternatmosphären, Berlin 1955); however it is known in connection with the description of signals in the time-frequency domain [see D. GABOR, J. Inst. Electr. Engrs. III **93**, 429 (1946), who also gives earlier references], and is an essential mathematical property according to which Δf and Δt, defined as the standard deviations of the spectral and temporal energy distributions about their mean abscissae, cannot, for a wide variety of signal functions, have a product less than $1/4\pi$. This mathematical fact assumes various embodiments in different branches of physics. Thus in quantum physics it appears in HEISENBERG's uncertainty relation [W. PAULI, this Encyclopedia vol. V/1, p. 20 (1957)], and in classical diffraction theory implies that $\Delta a'\,\Delta\omega' \geqq \lambda^2/(4\pi)$, where $\Delta a'$ and $\Delta\omega'$ are suitably redefined in terms of the standard deviations of the aperture intensity distribution and its power radiation pattern. This condition, together with the condition $\Delta f\,\Delta t \gg 1$, plays an important role in assessment of the efficiency of, and in setting a limit to, the rate of acquisition of information by an instrument. [In the strict proof that $\Delta f\,\Delta t \geqq 1/(4\pi)$, one defines Δf to be the standard deviation of the squared modulus of the Fourier transform of the signal waveform. For a quasi-monochromatic wave packet, $\Delta f \sim f$, and the inequality, through true, fails to be practical by a huge factor. If one defines Δf in the more natural way as the standard deviation of the positive-frequency part of the squared modulus of the transform of the signal, then $\Delta f\,\Delta t$ no longer possesses a strict lower limit of $1/(4\pi)$, as has been shown by E. WOLF, Proc. Phys. Soc. Lond. **71**, 257 (1958) and by I. KAY and R. A. SILVERMAN, Information and Control **1**, 64 (1957).]

[2] In m.k.s. units we have

PLANCK's constant, $h = 6.623 \times 10^{-34}$ joule-seconds,
BOLTZMANN's constant, $k = 1.3803 \times 10^{-23}$ joules degrees^{-1},
velocity of light, $c = 2.99792 \times 10^8$ meters seconds^{-1},

and

$$b = \frac{3.97 \times 10^{-25}\,\lambda^{-3}}{\exp\left(0.0143/\lambda T\right) - 1}.$$

except at the very lowest temperatures, we have the Rayleigh-Jeans approximation

$$b = \frac{2kT}{\lambda^2} \, .$$

This equation defines a temperature for every b, irrespective of whether the radiation field is of thermal origin or not, and this temperature is proportional to power. As we have seen in connection with the calibration procedure for receivers, absolute measurements are commonly made by comparison with the thermal radiation from a resistor at known temperature, the so-called available power $kT \, \Delta f$ of a resistor being itself an expression of the Rayleigh-Jeans approximation. Brightness temperature T_b is thus not only a convenient alternative to brightness b, for it is more often the actual datum.

The last of the three customary quantities, the flux density S of a source, may be defined as

$$S = \iint b \, d\omega ,$$

where the integral is taken over the whole solid angle subtended by the source and b is that part of the total brightness at the receiving point which is deemed due to the "source". Flux density is thus a scalar point function of position, as is b, but is not a function of direction. It is however a function of the possibly arbitrary outline deemed to define the source and therefore involves a subjective judgment by an observer. The way in which the subjective element enters is seen in a later section on the determination of flux density.

There is a fourth quantity, apparent disc temperature T_D, defined by

$$\frac{2kT_D}{\lambda^2} = \frac{S}{\Omega_D},$$

which is applicable to objects such as the Sun and planets which have definite optical discs or any other characteristic solid angle Ω_D. Since the optical disc may be unrelated to the size of the radio source (for example consider the corona and plages), the apparent disc temperature has deficiencies. But it is apposite, and has proved useful, in the absence of knowledge about the radio source.

50. The basic convolution relation. Let $T(C, R)$ be the distribution of true brightness temperature for the frequency and polarization accepted by an aerial, C and R being codeclination and right ascension respectively.

To specify the orientation of the aerial it is necessary to give not only (C_0, R_0), the direction of the beam axis,

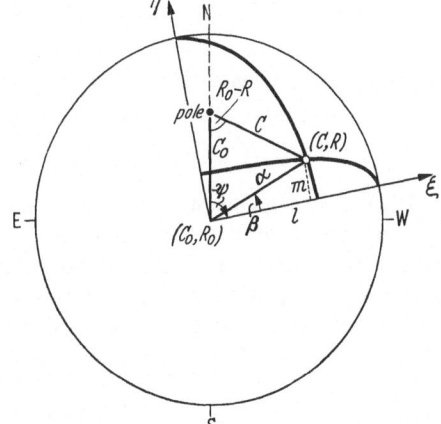

Fig. 36. Orthogonal projection of the sky on the aperture plane.

but also a position angle ψ which determines the rotation of the beam about this axis. The position angle ψ gives the direction of a transverse axis, fixed in the aerial, measured eastwards from north (Fig. 36).

$D(\alpha, \beta)$ is the directivity of the aerial in the direction (α, β), α and β being spherical polar coordinates relative to axes fixed in the aerial. The polar angle α is measured from the ζ-axis; the longitude β in the same sense as ψ from the great circle containing the ζ- and ξ-axes. From the definition of directivity, $D(\alpha, \beta)$ is normalized so that

$$\int_0^{2\pi} \int_0^{\pi} D(\alpha, \beta) \sin \alpha \, d\alpha \, d\beta = 4\pi.$$

We wish to calculate the available power, as defined in Sect. 26, when the aerial is pointed towards (C_0, R_0) with position angle ψ. This we can do with the aid of the principle of detailed balancing, quoting from Pawsey and Bracewell[1]. "Consider the situation of Fig. 37 in which it is required to find the available power at the terminals of aerial A due to thermal radiation from the body M which is at one temperature T. Let us tentatively connect to the terminals an impedance Z, for which the power transfer is maximum, and make the temperature of Z and of all the surroundings equal to T so that thermodynamic equilibrium may be realized. Then by the principle of detailed balancing the energy in the frequency range Δf transferred from M to Z equals that transferred from Z to M. Suppose a fraction α of the total radiation from the aerial is absorbed in M. The thermal power delivered by Z to the aerial is $kT \Delta f$, so that $\alpha kT \, \Delta f$ is absorbed in M. This quantity is also transferred from M to A. If Z is changed the actual power transfer may be altered but the available power from M will be as before. Also the temperature of the surroundings other than M may depart from T without affecting the contribution from M itself (if the changes do not alter its physical state or attenuate or deviate the radiation). The total available power P from all sources will be the sum of the contributions from all surrounding bodies. This is conveniently expressed in terms of the effective aerial temperature T_a, where

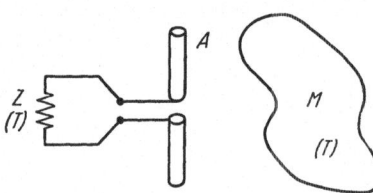

Fig. 37. An impedance Z in thermodynamic equilibrium with its surroundings at temperature T.

$$P = k \Delta f T_a = k \Delta f \sum \alpha_n T_n,$$

and the summation extends over all the surrounding bodies, α_n and T_n being the fraction of power radiated from the aerial absorbed in the body and the temperature, respectively, of the n-th body. This reduces to

$$T_a = \sum \alpha_n T_n,$$

so that T_a is a mean of the temperature of the surroundings weighted according to the fraction of power radiated by the aerial absorbed in each."

Let the direction (C, R) be the same as (α, β) when the aerial is placed in position angle ψ with its ζ-axis pointing in the direction (C_0, R_0). Then the fraction α of radiated power which would be absorbed by a black body subtending a solid angle $\sin C \, dC \, dR$ in the direction (C, R) is

$$D(\alpha, \beta) \frac{\sin C \, dC \, dR}{4\pi}.$$

Hence the effective aerial temperature T_a of a loss-free aerial surrounded by black bodies of temperature $T(C, R)$ in the direction (C, R) or surrounded by a brightness temperature distribution $T(C, R)$, is given by

$$T_a = \frac{1}{4\pi} \iint T(C, R) D(\alpha, \beta) \sin C \, dC \, dR.$$

[1] J. L. Pawsey and R. N. Bracewell: Radio Astronomy. Oxford 1955.

This integral may be taken over all directions in space or, with advantage, over the sky only, $D(\alpha, \beta)$ then being interpreted as the directivity of the aerial plus Earth. Since we have assumed that the total available power from all directions will be the sum of the separate contributions the conclusion will be invalid if there is coherence, such as would be produced by ground reflections, between rays arriving from different directions; and a good way to validate it is to deem the Earth to be part of the aerial.

Before the integral can be evaluated it is necessary to know C and R in terms of C_0, R_0, α, β, and ψ, for example by use of the relations

$$\cos C = \cos C_0 \cos \alpha + \sin C_0 \sin \alpha \cos (\beta + \psi)$$

$$\cot (R_0 - R) = \cos C_0 \cot (\beta + \psi) - \sin C_0 \cot \alpha \operatorname{cosec} (\beta + \psi).$$

We have now derived the basic formula of aerial smoothing in radio astronomy. The principal assumption, incoherence of the sources of radiation, is expected to prove widely valid for celestial sources, though it is also to be expected that the sky may contain examples of coherence caused by refraction or diffraction.

Before we can proceed further it is necessary to consider whether the function $D(\alpha, \beta)$ is unchanged when the ζ-axis points to different parts of the sky. In a great many practical cases it is accurate to assume that this is so, but there are important cases where it is not, for example movable aerials receiving significant ground reflections, and interferometers with aerials rotatable about axes not parallel to a line containing the elements. When the Earth is deemed to be part of the aerial as before, all these cases appear as non-rigid aerials whose parts may possess relative motion, and in the following development these cases are excluded; but the exclusion does not extend to the above aerials when they are caused to scan purely by the Earth's rotation.

It is desirable to specialize the basic equation to a simpler form and to distinguish two major branches of the theory according as ψ is retained or abandoned. By assuming that $\psi = \text{const}$ we restrict attention to radiation patterns subject only to "parallel" displacements; by allowing ψ to vary we generate the subject of strip integration which is taken up again in a later section.

Now placing the aerial on an equatorial mounting (an altazimuth mounting will allow ψ to vary) and causing the ξ- and η-axes of the aerial to fall respectively along declination and hour circles ($\psi = -\frac{1}{2}\pi$) we note that $-R \operatorname{cosec} C$ and C are approximate rectangular coordinates of points lying in a zone of declination containing the beam axis which is not too wide and not near the poles. Calling these coordinates x and y, and restricting attention to highly directional aerials within a certain zone of declination, we finally reduce the aerial smoothing equation to

$$T_a(x, y) = \iint A(x', y') \, T(x + x', y + y') \, dx' \, dy',$$

where x' and y', the coordinates of the element $dx' \, dy'$ relative to the ζ-axis of the aerial are essentially the same as l and m, $T(x, y)$ and $T_a(x, y)$ are the true and observed brightness temperatures respectively, and $A(x', y')$ is proportional to the directivity of the aerial but so normalized that

$$\iint A(x', y') \, dx' \, dy' = 1.$$

In these double integrals and those that follow, infinite limits are understood, but by the assumption of highly directional aerials the integrand differs from zero over only a small range of variables.

It will be noticed that in the notation introduced earlier (Sect. 43)

$$T_a = A \star T = \iint A(x', y') T(x + x', y + y') dx' dy'$$
$$= \iint A(x' - x, y' - y) T(x', y') dx' dy'.$$

We may force this result into the form of a convolution integral (denoted by $*$) by working in terms of $\mathbf{A}$, the response to a point source $^2\delta(x, y)$, instead of the conventional radiation pattern A. Substituting above,

$$\mathbf{A}(x, y) = A \star {}^2\delta = A(- x, - y).$$

Then

$$T_a = \mathbf{A} * T = T * \mathbf{A} = \iint \mathbf{A}(x', y') T(x - x', y - y') dx' dy'$$
$$= \iint \mathbf{A}(x - x', y - y') T(x', y') dx' dy'.$$

Because the operation of convolution is commutative ($f * g = g * f$), associative ($f * [g * h] = [f * g] * h$), and distributive ($[f + g] * h = f * h + g * h$) (see Doetsch[1]), it may be treated algebraically like ordinary multiplication and thus leads to simple mathematics, as below. On the other hand the operation of smoothing T with A (written $A \star T$) is perhaps more direct for some purposes than forming the convolution $\mathbf{A} * T$. In either case it is A which is plotted or tabulated when the calculation is performed. And, of course, A is the customary quantity in aerial physics, not $\mathbf{A}$. However, the smoothing process, being non-commutative and non-associative, proves to be not as convenient as convolution in the type of analysis presented below. But when the radiation pattern is symmetrical, the distinction between A and $\mathbf{A}$ disappears.

51. Formal solution by Fourier transforms. The Fourier transforms of functions related by convolution, as when

$$T_a = \mathbf{A} * T,$$

themselves have a simple product relationship, viz.

$$\overline{T}_a = \overline{\mathbf{A}} \, \overline{T},$$

where bars represent Fourier transforms, i.e.

$$\overline{T}(u, v) = \iint e^{-i 2\pi (u x + v y)} T(x, y) dx dy,$$

and conversely

$$T(x, y) = \iint e^{i \pi (u x + v y)} \overline{T}(u, v) du dv.$$

This is the two dimensional convolution theorem of which the array theorem introduced earlier is an expression.

In one dimension we write

$$\overline{T}(s) = \int e^{-i 2\pi s x} T(x) dx$$

and we shall often draw on this simpler form for illustration in what follows. The quantity s is the number of cycles per unit of x, or the spatial frequency, of a Fourier component of $T(x)$. In two dimensions a single Fourier component $e^{-i 2\pi (u x + v y)}$ may be regarded as a train of parallel crests and troughs in the (x, y) plane, proceeding in a direction inclined at an angle $\arctan (v/u)$ to the x-axis, with a spatial frequency $\sqrt{u^2 + v^2}$ cycles per unit distance in the xy-plane.

---

[1] G. Doetsch: Theorie und Anwendung der Laplace-Transformation. Berlin 1937.

The spatial frequency of a cross-section parallel to the x-axis is u, and parallel to the v-axis is y. We regard the true distribution $T(x, y)$ as composed of waves proceeding in all directions with all spatial frequencies, each wave of appropriate strength $T(u, v)$. Then the product formula tells us that $\overline{T}_a$ is derived from $\overline{T}$ by multiplication with a factor $\overline{A}$. Any values of $\sqrt{u^2+v^2}$ for which $\overline{A}$ is zero are particularly important, because then the modification is complete rejection. Knowing $\overline{T}_a$ and $\overline{A}$, we can partially infer $\overline{T}$; thus

$$\overline{T} = \frac{\overline{T}_a}{\overline{A}}$$

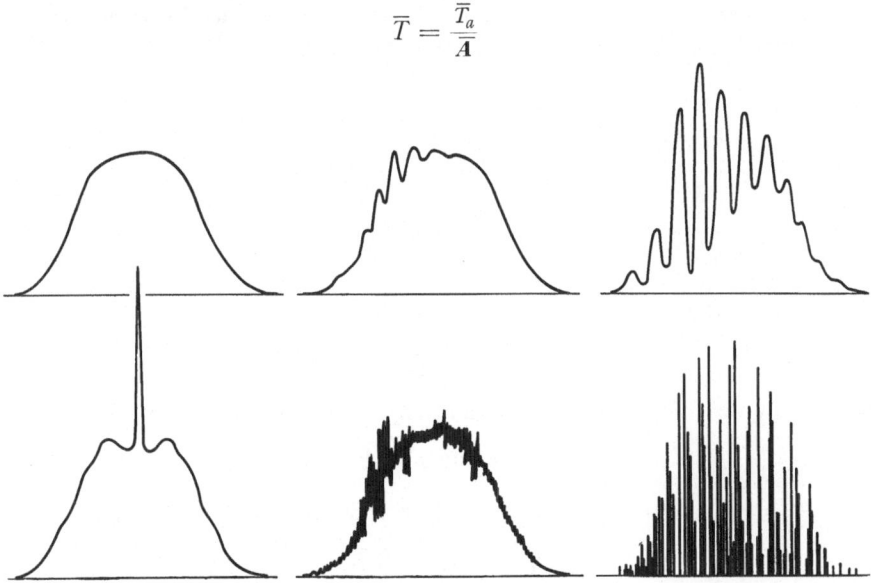

Fig. 38. A variety of distributions which when scanned with the same aerial all give the same result.

for values of $\sqrt{u^2+v^2}$ such that $\overline{A} \neq 0$. For other values of spatial frequency we can say nothing. Let $\overline{A}$ be zero where $u=u_k$, $v=v_k$. Then the product equation will be satisfied not only by $\overline{T}$ but also by

$$\overline{T} + \sum_k a_k \, {}^2\delta(u - u_k, v - v_k),$$

where the coefficients a_k are arbitrary and ${}^2\delta(u - u_k, v - v_k)$ is a unit two-dimensional impulse at $u=u_k$, $v=v_k$. It follows that T is not the only solution of the convolution equations; it is also satisfied by

$$T + \sum_k a_k \, e^{i 2\pi (u_k x + v_k y)}.$$

(If $\overline{A}$ is zero, not at discrete points, but over a continuous range, the summation is replaced by an integration.) The additive functions

$$\sum_k a_k \, e^{i 2\pi (u_k x + v_k y)},$$

which we term invisible distributions for the aerial, are obviously solutions of the integral equation

$$A \star T = 0.$$

They are of such a nature that it is impossible to detect them with the aerial in question, whatever their magnitude. Some one-dimensional examples of distributions containing invisible components are given in Fig. 38 taken from a paper by Bracewell and Roberts[1] which first discussed this aspect of the non-uniqueness of the solution of the aerial smoothing equation.

It is possible to use the formal solution in practice but only in simple cases and then only with caution and often with disappointing results for numerical reasons. However, as a theoretical basis for further discussion it is invaluable. It is clear that the zeros of $\bar{A}$ play a vital role and we now investigate from this point of view the Fourier transform of aerial patterns.

52. The spectral sensitivity theorem. Let an aerial specified by an electric field distribution $E_\eta = E(\xi/\lambda, \eta/\lambda)$, $E_\xi = 0$, give rise to an angular spectrum $P(l,m)$. Then we have shown that

$$ P(l, m) = \iint E\left(\frac{\xi}{\lambda}, \frac{\eta}{\lambda}\right) e^{-i2\pi\left(l\frac{\xi}{\lambda} + m\frac{\eta}{\lambda}\right)} d\left(\frac{\xi}{\lambda}\right) d\left(\frac{\eta}{\lambda}\right) $$

and

$$ E\left(\frac{\xi}{\lambda}, \frac{\eta}{\lambda}\right) = \iint P(l, m) e^{i2\pi\left(l\frac{\xi}{\lambda} + m\frac{\eta}{\lambda}\right)} dl\, dm. $$

We may write this latter equation

$$ E = \bar{P}. $$

For a highly directional aerial the radiation pattern A is given by

$$ A = \text{const } P^* P, $$

where P^* is the complex conjugate of P. By the convolution theorem

$$ \bar{A} = \text{const } \overline{P^*} * \overline{P} $$
$$ = \text{const } E^*(-) * E $$
$$ = \text{const } E^* \star E, $$

hence, $\bar{A} = \text{const } E \star E^*.$

In this proof we have used the fact that the transform of the conjugate of P, denoted by $E^*(-)$, is obtained from E^*, the conjugate of its transform, by reversing the signs of ξ and η.

The quantity $\bar{A}$ which we are investigating, and which we may call the spectral sensitivity function, is thus proportional to the complex autocorrelation function of E. The constant of proportionality follows from the normalization of $\bar{A}$; since $\iint A\, dl\, dm = 1$ it follows that $\bar{A}|_0 = 1$. Therefore, dividing $E \star E^*$ by its value for $u = v = 0$, we find finally that the spectral sensitivity function of an aerial is equal to the complex autocorrelation function of its aperture distribution, normalized in the conventional way, i.e. with a central value of unity:

$$ \bar{A}(u, v) = \frac{\iint E\left(\frac{\xi}{\lambda} - u, \frac{\eta}{\lambda} - v\right) E^*\left(\frac{\xi}{\lambda}, \frac{\eta}{\lambda}\right) d\left(\frac{\xi}{\lambda}\right) d\left(\frac{\eta}{\lambda}\right)}{\iint E\left(\frac{\xi}{\lambda}, \frac{\eta}{\lambda}\right) E^*\left(\frac{\xi}{\lambda}, \frac{\eta}{\lambda}\right) d\left(\frac{\xi}{\lambda}\right) d\left(\frac{\eta}{\lambda}\right)}. $$

We may now restate the basic relation of aerial smoothing in the form

$$ \bar{T}_a = \text{const } (E \star E^*)\, \bar{T}. $$

---

[1] R.N. Bracewell and J.A. Roberts: Austral. J. Phys. **7**, 615 (1954).

This is an expression of the spectral sensitivity theorem, i.e. it states the relation between the transforms of T_a and T in terms of the aerial aperture distribution E.

53. The aerial cut-off theorem. An aerial theorem of far-reaching consequences may now be proved. Since aerials are finite in extent the function E falls to zero for values of ξ (or ξ and η) greater than the finite values corresponding to the extremities of the aerial. When the function $E \star E^*$ is evaluated, the result, namely $\bar{A}$, must have this same property of falling to zero beyond a finite region; for if we examine the numerator in the expression for $\bar{A}$, we see that one or other of the factors is zero for all ξ and η when u and v exceed certain values depending on the exact extent of E over the $\xi\eta$-plane. A graphical procedure is described in Sect. 59.

For illustration consider an aerial consisting of a finite one-dimensional aperture of width w, across which is maintained a field constant in amplitude and phase. For this aerial the $A(x)$ for small x is approximately

$$A(x) = \frac{\lambda}{w} \left[\frac{\sin\left(\frac{\pi x w}{\lambda}\right)}{\pi x} \right]^2,$$

the numerical factor being chosen so that $\int A(x)\, dx = 1$. Fig. 39 shows $A(x)$ and $\bar{A}(s)$, which can conveniently be plotted against $|s|$ in this case since $\bar{A}(s)$ is real and even. The beam width of the main

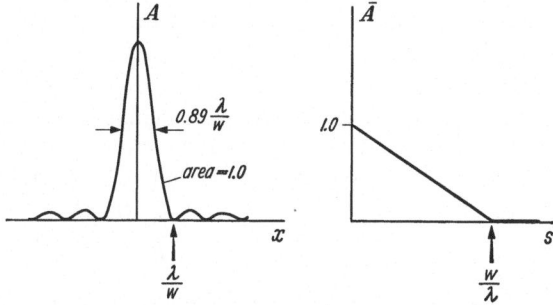

Fig. 39. The response to a point source, $A(x)$, and its Fourier transform $\bar{A}(s)$, for a uniformly excited aperture.

lobe between zeros is $2\lambda/w$ and the width to half power is $0.89\,\lambda/w$. The particular feature to note is that $\bar{A}(s)$ is zero for all values of s greater than the limiting spatial frequency $s_c = w/\lambda$.

The consequences of $\bar{A}$ falling to zero in this as in all other cases[1], is that $\bar{T}_a$ necessarily does the same, being derived from $\bar{T}$ by multiplication with $\bar{A}$. A powerful discrete-interval theorem then applies.

54. The discrete-interval theorem. A function $T_a(x, y)$ such that $\bar{T}_a(u, v)$ is zero for $|u| \geq u_c$ or $|v| \geq v_c$, is completely determined by its values at the points $(m/2u'_c + a,\ n/2v'_c + b)$, where m and n assume all integral values, a and b are arbitrary constants, and the spacing between points may be as wide as is compatible with $u'_c \geq u_c$ and $v'_c \geq v_c$.

The condition on $\bar{T}_a(u, v)$ may be expressed by saying that it is zero on and outside a rectangle which is centered on the origin of the uv-plane and set with its sides parallel to the axes; and we may note that this covers the case of $\bar{T}_a(u, v)$

---

[1] Our proof applies strictly only to plane apertures, but we know from experience that the directivity of a highly directional array cannot be improved much by rearranging its elements within the same overall dimensions. In any particular case the existence of the cut-off could be verified, and the value of s_c determined, by taking the Fourier transform of the radiation pattern. Low-gain apertures which are comparable with, or less than, a wavelength in extent, can be harmonized with the present treatment by considering only that part of the aperture distribution which does not generate evanescent fields. In fact strictly speaking the function $E \star E^*$ should be evaluated in all cases only after removal from E of those components which do not radiate. By not doing so one ignores an effective extension of the aperture distribution to a distance of about $\lambda/2$. To this extent, an aperture distribution never really cuts off sharply.

zero on and outside a circle or other region provided the rectangle is chosen sufficiently large.

It is sufficient to give a proof for the case where a and b are zero, i.e., where the origin of x and y is one of the sampling points. For, if the transform of $T_a(x, y)$ is zero on and outside a given rectangle, so also is that of $T_a(x+a, y+b)$ by virtue of the two-dimensional shift theorem, according to which the Fourier transform of $T_a(x+a, y+b)$ is $\overline{T}_a(u, v) \exp(i\, 2\pi(au+bv))$], which must be zero. Therefore, if the theorem is true for $T_a(x, y)$, it is also true for $T_a(x+a, y+b)$; but values of $T_a(x+a, y+b)$ at points of an array which includes the origin are values of $T_a(x, y)$ taken over an offset array.

To prove the theorem we use the bed-of-nails function ${}^2III(x, y)$ consisting of a two-dimensional array of unit impulses separated by unit distance. Thus

$$
{}^2III(x, y) = \sum_{m=-\infty}^{\infty} \sum_{n=-\infty}^{\infty} {}^2\delta(x-m, y-n).
$$

The bed-of-nails function is known to be its own two-dimensional Fourier transform (Bracewell[1]).

Proof of theorem. Let $F(u, v) \equiv (4u'_c v'_c)^{-1}\, {}^2III(u/2u'_c) * \overline{T}_a$, a function which may be pictured as an array of islands in the uv-plane, each the same as $\overline{T}_a$, spaced at intervals $2u'_c$ in the u direction and $2v'_c$ in the v direction. The islands will not overlap (but may touch) if $u'_c \geq u_c$ and $v'_c \geq v_c$. Under this condition, in the central region where $|u| < u_c$ and $|v| < v_c$, we have

$$
\overline{F}(u, v) = \overline{T}_a.
$$

Hence $\overline{T}_a$ may be recovered from $\overline{F}(u, v)$, and consequently T_a may be recovered from $F(x, y)$, the two-dimensional Fourier transform of $\overline{F}(u, v)$. But, by the two-dimensional convolution theorem,

$$
F(x, y) = 4u'_c v'_c\, {}^2III(2u'_c x, 2v'_c y),
$$

which contains values of T_a only at discrete intervals $(2u'_c)^{-1}$ and $(2v'_c)^{-1}$ of x and y. Hence T_a is completely determined by its values at discrete intervals of x and y which are equal to or less than $(2u_c)^{-1}$ and $(2v_c)^{-1}$. Since these intervals are peculiar to each aerial they will be referred to as the peculiar intervals.

This theorem is of great importance. For observational work it means that observations need not be more closely spaced than a limit determined by the aerial pattern. For computational work the property is equally important as it permits observed data to be represented exactly by a set of discrete values. An inverse form of the theorem, relating to celestial sources of finite angular extent, means that observations of coherence need not be made at spacings less than a limit set by the source size.

V. Interferometers.

a) Two-element interferometers.

As used in radio astronomy the term interferometer signifies an aerial with two or more well separated parts. When the aerial is used to transmit, the fields from the different parts combine in the distance to form spatial interference fringes and when it is used to receive from a moving point source the received power undergoes temporal variations which themselves are often called fringes. No confusion is caused by this transference of nomenclature, as the spatial fringes as such are not observed.

[1] R.N. Bracewell: Austral. J. Phys. **9**, 297 (1956).

Since interferometers are aerials, they are covered by the aerial theory which we have developed; they are distinguished from aerials in general by the possession of more or less periodic radiation patterns, a feature associated with two or more well separated parts.

55. Monochromatic theory. Consider first two identical aerials with directivity $D(\vartheta, \varphi)$ connected symmetrically together as shown in Fig. 40. Then the directivity of the combination can be deduced as follows. At a distant point in the direction (ϑ, φ), where ϑ and φ are the same angles as those in Fig. 18, two field components will combine with a phase differ-

ence $\delta = \dfrac{2\pi}{\lambda} a \sin \vartheta$ and amplitudes which are equal but $1/\sqrt{2}$ times less than would have been produced by a single aerial radiating as much total power as the interferometer. The flow of power into the direction (ϑ, φ) is therefore

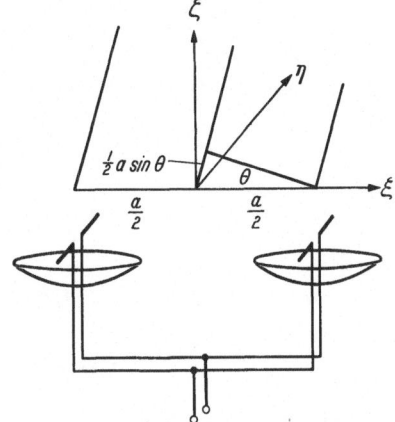

$$\left(\frac{1}{\sqrt{2}} 2 \cos \frac{1}{2} \delta\right)^2 = 1 + \cos \delta$$

relative to a single aerial and so the directivity of the interferometer is

$$D_i(\vartheta, \varphi) = D(\vartheta, \varphi) \left(1 + \cos \frac{2\pi a \sin \vartheta}{\lambda}\right).$$

Fig. 40. A two-element interferometer.

It will be noticed that the modifying factor is independent of φ and that the loci of constant ϑ are small circles with centers on the ξ-axis. For example a two element interferometer lying on an east-west line has a radiation pattern

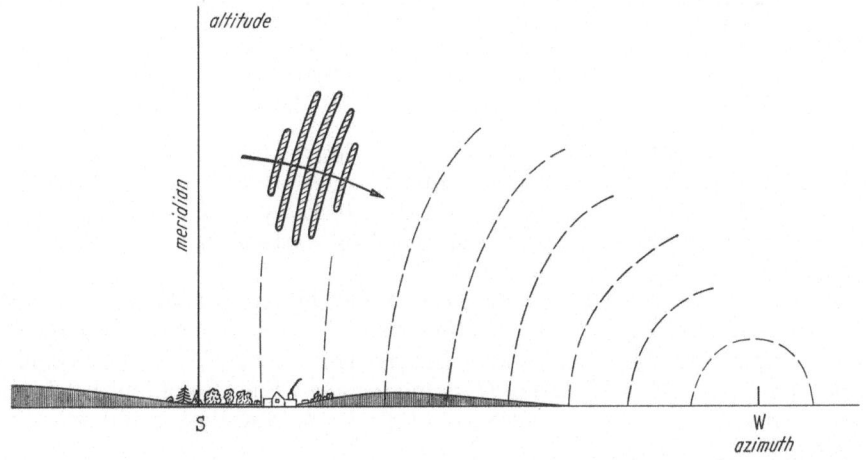

Fig. 41. View of the south-western horizon. The shading represents the reception pattern of a two-element east-west interferometer.

as illustrated in Fig. 41 which shows a view of the sky looking south. In this type of diagram, which the author commends for explanatory purposes, we shade those areas of sky where the directivity exceeds half the maximum. The envelope of the fringe system is the half-directivity diagram of a single aerial, and may be steered about in the sky by steering the aerials; but this does not move

7*

the fringes from their location on the small circles having the west point of the horizon as pole.

If a point source were to move through the reception pattern along the arrow shown in Fig. 41 the time variation of received power would be as shown in Fig. 42. If a single aerial were used the response would be the broken line.

Let the directivity $D(\vartheta, \varphi)$ and angular spectrum P be those which would result from an extended aperture distribution $E\left(\frac{\xi}{\lambda}, \frac{\eta}{\lambda}\right)$. Then a pair of such apertures may be represented by $II * E$, where II is a pair of impulse functions,

$$II = {}^2\delta\left(\frac{\xi}{\lambda} - \frac{a}{2\lambda}, \frac{\eta}{\lambda}\right) + {}^2\delta\left(\frac{\xi}{\lambda} + \frac{a}{2\lambda}, \frac{\eta}{\lambda}\right).$$

The angular spectrum P_i of the interferometer is given by

$$P_i = \overline{II * E}$$
$$= \overline{II}\, \overline{E}$$
$$= 2\left(\cos \frac{\pi a l}{\lambda}\right) P.$$

Then

$$P_i P_i^* = 4\left(\cos^2 \frac{\pi a l}{\lambda}\right) P P^*$$

and

$$D_i(\vartheta, \varphi) = D(\vartheta, \varphi)\left(1 + \cos \frac{2\pi a \sin \vartheta}{\lambda}\right).$$

This alternative derivation of the directivity of an interferometer illustrates an algebraic approach which is often useful for thinking out new systems. In

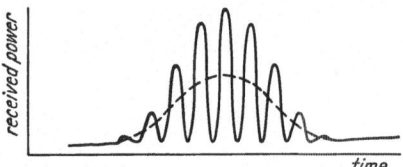

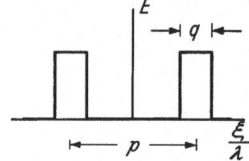

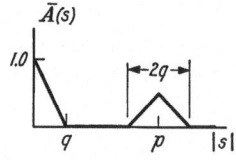

Fig. 42. The power received from a point source passing through the reception pattern of Fig. 41.

Fig. 43. Aperture distribution of an interferometer and its spectral sensitivity function.

effect we have said that a new system has been derived by convolution of a simple aperture distribution with a double impulse, which we know to be the Fourier transform of a cosine wave in the l-direction. Hence by the convolution theorem the old angular spectrum must be multiplied by a cosine variation and the old radiation pattern by cosine squared.

We now derive, under the conditions of applicability of the Fourier transform formula, the spectral sensitivity function for the one-dimensional interferometer whose aperture distribution is shown in Fig. 43. This distribution can be recognized as the convolution $II_q * II_p$ where II_q is rectangle function of unit height and width q, and II_p is the unit impulse-pair of spacing p. Now the spectral sensitivity function $\bar{A}$ is given by

$$\bar{A} = \text{const } E \star E*$$
$$= \text{const } E * E$$
$$= \text{const } (II_q * II_p) * (II_q * II_p)$$
$$= \text{const } (II_q * II_q) * (II_p * II_p)$$
$$= \Lambda_q * {}_1II$$
$$= \tfrac{1}{2}\Lambda_q(s + p) + \Lambda_q(s) + \tfrac{1}{2}\Lambda_q(s - p),$$

where $\Lambda_q = \Pi_q * \Pi_q$ is the triangle function of unit height and width $2q$, $_I\Pi_I =$ $\frac{1}{2}\delta(s+p) + \delta(s) + \frac{1}{2}\delta(s-p)$, and the constant has been adjusted to make $\bar{A}(0) = 1$. The important feature to notice is that the interferometer responds to a band of Fourier components of spatial frequency centered on p and of total width q, where p is the number of wavelengths between the centers of the elements and q is the breadth of each element in wavelengths. In addition, the interferometer responds to uniform brightness $(s=0)$ and to low-frequency spatial components up to q cycles per radian. The curve of Fig. 42 contains just such bands of Fourier components, one at low-frequencies and one centered on a high frequency; in fact, since the spectral sensitivity function (Fig. 43) and the response to a point source Fig. 42 are a Fourier transform pair they afford precisely equivalent descriptions of the behavior of an interferometer.

Having derived the properties of a two-element interferometer and considered a simple example, we calculate what is observed when an interferometer is used on a source distribution $T(x - Vt, y)$ which is moving with an angular velocity V in the x-direction past an interferometer whose radiation pattern is $A(x, y) \times (1 + \cos 2\pi S x)$. The aerial temperature T_a will be given by

$$T_a = \iint T(x - Vt, y) A(x, y) (1 + \cos 2\pi S x)\, dx\, dy$$
$$= T_{a_1}(-Vt) + \cos 2\pi S Vt \iint \cos 2\pi S x\, T(x, y) A(x + Vt, y)\, dx\, dy -$$
$$- \sin 2\pi S Vt \iint \sin 2\pi S x\, T(x, y) A(x + Vt, y)\, dx\, dy.$$

The term $T_{a_1}(-Vt)$ is what would be received on a single aerial. The remainder terms take on a simple form if the distribution T is compact relative to A and is mainly concentrated around (x_1, y_1). Then

$$T_a \approx A(x_1 + Vt, y_1) \left[\iint T(x, y)\, dx\, dy + \right.$$
$$+ \cos 2\pi S Vt \iint \cos 2\pi S x\, T(x, y)\, dx\, dy - \sin 2\pi S Vt \iint \sin 2\pi S x\, T(x, y)\, dx\, dy \right]$$
$$= A(x_1 + Vt, y_1) \iint T(x, y)\, dx\, dy\, [1 + N \cos(2\pi S Vt - \beta)],$$

where

$$N e^{i\beta} = \frac{\iint e^{-i2\pi S x}\, T(x, y)\, dx\, dy}{\iint T(x, y)\, dx\, dy}.$$

The remainder thus approximates to an oscillation whose phase and amplitude are given by that Fourier component of $T(x, y)$ having S cycles per unit of x in the x-direction, the whole modulated by the radiation pattern centered on the time of closest approach of (x_1, y_1) to the beam axis. The quantity N which measures the amplitude of oscillation relative to the mean is what in optics would be called the visibility of the fringes. By returning to the simple one dimensional illustration we can illustrate this Fourier-transforming action of an interferometer.

As the width q of the individual aerials approaches zero the aperture distribution can be represented by Π and the sensitivity function by $_I\Pi_I$. Hence the elementary aerial-pair spaced p wavelengths responds to source components of two spatial frequencies: (i) p cycles per radian, (ii) zero cycles per radian.

Since any aperture distribution can be regarded as composed of numbers of infinitesimally wide elements, it follows that an aerial will respond to all those Fourier components of spatial frequency p' such that the aperture contains a pair of elements p' wavelengths apart. Thus in the example of Fig. 43 the aperture distribution does not excite any two points whose spacing is greater than q and less than $p-q$, nor any whose spacing exceeds $p+q$. This is reflected by absence of response in the graph of spectral sensitivity function.

This concept of spaced pairs of elements is a very useful one and it also gives the strength of response. One simply counts the numbers of ways in which a pair of spacing p' can be found, allowing due weight for strength of excitation. Thus again referring to Fig. 43, the spacing $p-q+\varepsilon$ $(\varepsilon<q)$ can be found in a number of ways proportional to ε. All the linear segments of the graph of $\bar{A}$ can be explained in this way. Furthermore the height of the maximum at $s=0$ must be twice that of the maximum at $s=p$, for zero spacing can be found in twice as many ways as the spacing p. To prove these statements, which have been worded with a certain lack of rigor, one simply refers to the integral expression for $\bar{A}$ namely $E \star E^*(s)$, which prescribes just how the elements of spacing s are to be "counted" with "due weight". This view of the matter translates directly into two dimensions (Sect. 65) and throws immediate light on the effect of bandwidth (Sect. 57).

56. Coherence. We have measured the power arriving from the direction (x, y) in terms of a temperature $T(x, y)$, but we know that the instantaneous power is subject to random fluctuations. If attention is concentrated on a narrow band of frequencies, the distant field in the direction (x, y) may be represented by a complex phasor $f(x, y)$ whose modulus will have a Rayleigh distribution and whose phase drifts through all values equally. Let

$$\langle f f^* \rangle = T,$$

where

$$\langle \cdots \rangle \equiv \lim_{L \to \infty} \frac{1}{2L} \int_{-L}^{L} \cdots dt.$$

Since an interferometer composed of two small aerials at a certain spacing can measure one Fourier component of the distribution $T(x, y)$, it is in fact measuring one spatial component of $\langle f f^* \rangle$. The field distribution $f(x, y)$ produces in the $\xi\eta$-plane a Fraunhofer diffraction field $F\left(\frac{\xi}{\lambda}, \frac{\eta}{\lambda}\right)$ such that

$$F\left(\frac{\xi}{\lambda}, \frac{\eta}{\lambda}\right) \propto \iint f(x, y)\, e^{-i2\pi\left(x\frac{\xi}{\lambda} + y\frac{\eta}{\lambda}\right)}\, dx\, dy.$$

Hence the Fourier transform of $f f^*$ will be proportional to the spatial complex autocorrelation of the phasor F:

$$\iint f(x, y)\, f^*(x, y)\, e^{-i2\pi\left(x\frac{\xi}{\lambda} + y\frac{\eta}{\lambda}\right)} dx\, dy$$

$$\propto \iint F(\alpha, \beta)\, F^*\left(\alpha + \frac{\xi}{\lambda}, \beta + \frac{\eta}{\lambda}\right) d\alpha\, d\beta$$

$$= F \star F^*$$

$$= \langle F_1 F_2^* \rangle_{\text{spatial}}.$$

Taking the time average of both sides, and interchanging the order of spatial integration and time averaging, we have

Fourier transform of $\langle f f^* \rangle \propto$ spatial average of $\langle F_1 F_2^* \rangle$.

Since time averages such as $\langle F_1 F_2^* \rangle$ are independent of spatial position in the Fraunhofer region, it follows that

$$\bar{T} \propto \langle F_1 F_2^* \rangle.$$

Hence an interferometer measurement at a single spacing, which is known from Sect. 55 to be a measurement of a single value of $\overline{T}$, can also be regarded as a measurement of the time correlation $\langle F_1 F_2^* \rangle$ of the field phasors F_1 and F_2 produced at the two aerials by the celestial distribution T.

To show this directly, we let the distribution be *slowly* in motion as indicated by a distribution function $T(x - Vt, y)$. Then the field produced at one aerial is $F_1 e^{i\frac{1}{2}\varphi}$, where

$$\varphi = 2\pi S \sin Vt \approx 2\pi S Vt$$

and the received voltage at the terminals of the aerial pair will be

$$V = V_1 + V_2 = \alpha_1 e^{-i\delta_1} F_1 e^{-i\frac{1}{2}\varphi} + \alpha_2 e^{-i\delta_2} F_2 e^{i\frac{1}{2}\varphi},$$

where the factors $\alpha_1 e^{-i\delta_1}$ and $\alpha_2 e^{-i\delta_2}$ allow for the attenuation and phase delay in each arm of the transmission line. The instantaneous power will be proportional to VV^* and we have

$$VV^* = \alpha_1^2 F_1 F_1^* + \alpha_2^2 F_2 F_2^* + \alpha_1 \alpha_2 [F_1 F_2^* e^{i(-\varphi - \delta_1 + \delta_2)} + F_1^* F_2 e^{-i(-\varphi - \delta_1 + \delta_2)}].$$

As time elapses the mean value $\langle VV^* \rangle$ of the instantaneous power will measure the available power at the terminals, and under conditions where

$$\langle F_1 F_1^* \rangle = \langle F_2 F_2^* \rangle,$$

i.e. when behavior at one point is, on the average, the same as that at another point nearby, we have

$$\langle VV^* \rangle = \langle V_1 V_1^* \rangle + \langle V_2 V_2^* \rangle +$$

$$+ \frac{\sqrt{\langle V_1 V_1^* \rangle \langle V_2 V_2^* \rangle}}{\langle F_1 F_1^* \rangle} [\langle F_1 F_2^* \rangle e^{i(-\varphi - \delta_1 + \delta_2)} + \langle F_1^* F_2 \rangle e^{-i(-\varphi - \delta_1 + \delta_2)}]$$

$$= \langle V_1 V_1^* \rangle + \langle V_2 V_2^* \rangle + 2|\Gamma| \sqrt{\langle V_1 V_1^* \rangle \langle V_2 V_2^* \rangle} \cos(-2\pi S Vt - \delta_1 + \delta_2 + \text{pha}\,\Gamma),$$

where

$$\Gamma = \frac{\langle F_1 F_2^* \rangle}{\langle F_1 F_1^* \rangle}$$

and

$$|\Gamma| = \frac{|\langle F_1 F_2^* \rangle|}{\langle F_1 F_1^* \rangle}.$$

The final expression shows that as time elapses the available power rises and falls sinusoidally above and below a mean level $\langle V_1 V_1^* \rangle + \langle V_2 V_2^* \rangle$ with a depth of modulation characterized by the coefficient $|\Gamma|$, a dimensionless parameter which can be determined from

$$|\Gamma| = \frac{\langle VV^* \rangle_{\max} - \langle VV^* \rangle_{\min}}{\langle VV^* \rangle_{\max} + \langle VV^* \rangle_{\min}} \cdot \frac{1}{2} \left[\sqrt{\frac{\langle V_1 V_1^* \rangle}{\langle V_2 V_2^* \rangle}} + \sqrt{\frac{\langle V_2 V_2^* \rangle}{\langle V_1 V_1^* \rangle}} \right].$$

The first factor on the right hand side is the exact analog of the "visibility" or "contrast" of optical interference fringes as introduced by MICHELSON. The quantity Γ, which is the normalized time correlation of the field phasors at the two aerials, is the same as the "complex degree of coherence" of ZERNIKE[1]. Where the two aerials have equal gains and equally efficient transmission lines, so that $\langle V_1 V_1^* \rangle = \langle V_2 V_2^* \rangle$, the modulus of Γ is equal to the visibility of the fringes. In the case of unequal available powers from each of the aerials separately,

[1] F. ZERNIKE: Physica, Haag **5**, 785 (1938).

$|\Gamma|$ could be determined from an observation of fringe visibility when the proportions had been established by a further measurement. The phase of Γ depends on an observation of fringe epoch and is possible though difficult.

We have thus shown that a simple interferometer observation measures the normalized time correlation of the field phasors at the aerials, and we shall occasionally borrow the terms "complex degree of coherence" and "visibility", as already with the word "fringe". The visibility, which is a parameter principally of the electromagnetic field, depends also on inequality of the aerial gains and feeder losses.

The complex coherence appears to play a fundamental role in optics since it is an observable quantity (unlike the field vectors themselves). It has interesting properties[1] one of which is to remain unchanged on planes parallel to a radiating aperture, neglecting evanescent fields. We have already established the property that the complex coherence is the Fourier transform of the normalized brightness temperature distribution over the source.

57. Effect of bandwidth. When an interferometer consisting of two infinitesimal elements at a fixed spacing of p nominal wavelengths forms part of a system

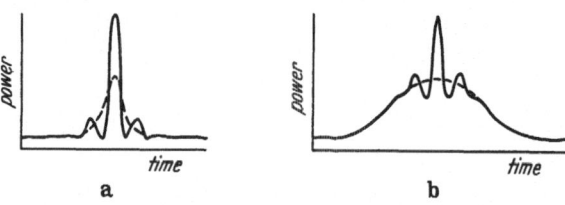

Fig. 44 a and b. Power received by the interferometer of Fig. 42 from a point source when the resolution is increased (a) by widening the aerials, (b) by widening the wavelength band.

sensitive to a band of wavelengths surrounding the nominal wavelength, then it can respond to a band of spatial frequencies surrounding $s=p$. If the spacing measured $p-q$ wavelengths at the longest wavelength and $p+q$ at the shortest, then the effect would resemble the result of making each element q wavelengths wide while keeping to monochromatic operation. In each case the effect is to extend the range of sensitivity to spatial frequency components, but with the difference that widening the elements widens the response around $s=0$ whereas broadening the wavelength band does not. Consequently the responses to a point source, though sharpened in each case, reveal differences. In Fig. 44 we see curves corresponding to that of Fig. 42 when conditions are modified so that in (a) the width of the elements is greatly increased and in (b) the band of wavelengths is opened out. The broken lines show what would be received on a single aerial. Case (a) is analogous to the monochromatic Fraunhofer diffraction pattern of a pair of slits and case (b) is connected with the white and colored fringes observed when two (narrower) slits are illuminated by white light from a point source.

It is rewarding to contemplate the Fourier transforms of curves (a) and (b), but there is a complication to be borne in mind when extended wavelength bands are considered, namely the wavelength spectrum of the aforesaid "point source". Changing to the domain of angles we can say in general that the response of a broad band system is given by

$$\int_0^\infty D_\lambda(\vartheta, \varphi)\, S_1(\lambda)\, S_2(\lambda)\, d\lambda,$$

[1] J.A. Ratcliffe: Rep. Progr. Phys. **19**, 188 (1956). — E. Wolf: Proc. Roy. Soc. Lond., Ser. A **225**, 96 (1954); **230**, 246 (1955). — A. Blanc-Lapierre and P. Dumontet: Rev. Opt. (theor. instrum.) **34**, 1 (1955).

where D_λ is the directivity at wavelength λ, and $S_1(\lambda)$ and $S_2(\lambda)$ are suitably normalized functions describing the source emission spectrum and the spectral sensitivity of the system.

Restricting attention to a system which is sensitive over a narrow band $\Delta\lambda$ and operates at a nominal wavelength $N\,\Delta\lambda$ we see that each wavelength in the band $\Delta\lambda$ produces a pattern such as that of Fig. 42, the central fringes coinciding but the outer fringes tending to cancel. The extent of the beating pattern will be of the order of N distinct fringes since the N-th fringe of wavelength $\lambda - \frac{1}{2}\,\Delta\lambda$ falls just between the N-th and $(N-1)$-th fringes of the nominal wavelength λ.

The precise way in which the obliteration of the fringes occurs depends on the function $S_1(\lambda)\,S_2(\lambda)$. The theory of this phenomenon is precisely that applicable to MICHELSON's attempts to determine the profile of spectral lines from the visibility of the fringes produced by two interfering beams.

58. The sea interferometer. The first application of radio interferometry to celestial objects was made by McCREADY, PAWSEY and PAYNE-SCOTT[1] who determined the angular size, and position on the Sun's disc, of those discrete sources of radio energy localized in the corona over sunspot groups, which are now referred to as noise storms. They did this with a radar aerial whose original purpose was for height finding of aircraft by means of an accurate range measurement combined with an accurate measurement of angle of elevation.

The requisite precision in angle of elevation was achieved by placing a single aerial on a high cliff overlooking the sea and allowing rays reflected from the sea surface to interfere with direct rays. In the application to height finding, ambiguities due to the fineness of the lobe structure were resolved with the aid of another aerial placed above the first. If one replaces the highly reflecting sea surface by a second aerial at the image of the first, one has a two-element interferometer and so it is not necessary to repeat the theory given by McCREADY, PAWSEY and PAYNE-SCOTT.

Some differences only need be noted. Because of a phase reversal on reflection the first fringe falls above the locus of zero path difference as in LLOYD's optical arrangement for producing interference at glancing incidence from a mirror. In actual fact the first fringe was a degree below the true horizon, mainly because of tropospheric refraction. The interference pattern has a sharp beginning and consequently permits higher angular resolution than the corresponding two element device; for example, otherwise unresolvable sources with different times of rising might be distinguished.

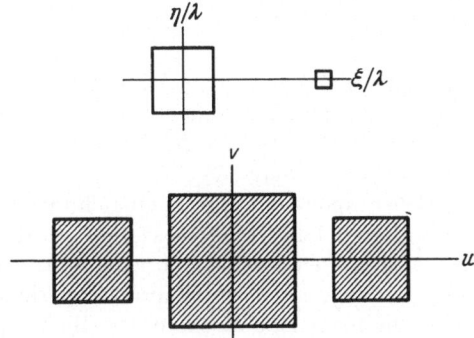

Fig. 45. Aperture distribution representing a large aerial and satellite (above) and the area of sensitivity in the uv-plane (below).

59. Unequal interferometer. A very large aerial, not readily movable may be accompanied by a smaller satellite aerial which combines with it to form an interferometer. This arrangement is becoming important and affords an opportunity to discuss a simple case in two dimensions. Fig. 45 shows a large square aerial supplemented by a small

---

[1] L. L. McCREADY, J. L. PAWSEY and RUBY PAYNE-SCOTT: Proc. Roy. Soc. Lond., Ser. A, **190**, 357, (1947).

square one represented by outlines in the $\xi\eta$-plane. The outlines in the uv-plane within which $\overline{A}$ is not zero are readily ascertainable graphically by copying the upper pattern on a sheet of transparent paper and sliding it about so that the copy is externally tangent to the original. A point moving with the copy then traces out the lower diagram.

Full evaluation of $E \star E^*$ requires a statement of the aperture distributions; for example, if they are uniform, the central island on the uv-plane is pyramidoidal with a central peak, and the adjacent islands are flat plateaux with steeply sloping borders.

The general conclusion is that the arrangement has the sensitivities of the separate aerials and in addition is sensitive to a set of spatial frequencies surrounding the value corresponding to the spacing between centers of the aerials. The extent of this latter set is considerable, being a little more than that corresponding to the dimensions of the large aerial.

Since interest may reside principally in the flanking islands, not in the central one, an advantage may accrue from reduction in sensitivity in the central island. The switched interferometer and the method of post-detection correlation, both discussed below, would have this effect. It might appear that attenuation of the signal from the large aerial would help, a procedure which can readily be studied by reducing the aperture function for the large aerial before taking the autocorrelation of the composite aperture distribution. An interesting optimum sensitivity property exists, but the attenuation reduces the power received from the components of higher spatial frequency.

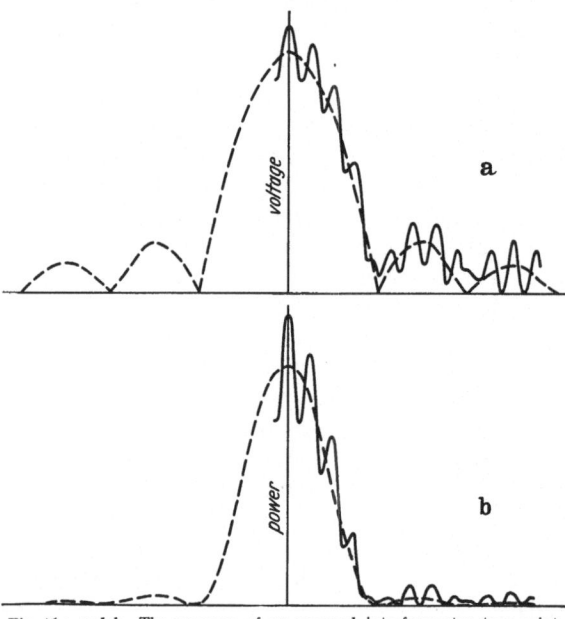

Fig. 46 a and b. The response of an unequal interferometer to a point source: (a) voltage, (b) power.

The response of the unequal interferometer to a point source is necessarily complicated but the cross section parallel to the ξ-direction is simple. Fig. 46 gives in broken outline the response of the large aerial in terms of the modulus of the voltage. The contribution from the small aerial, which is relatively constant over the main lobe of the other, then rapidly cycles through relative phases to produce the undulation of constant amplitude, shown as a full line (above). The power response (below) obtained by squaring, then shows how the high frequency band has been broadened and strengthened.

60. Payne-Scott and Little's interferometer. In Sect. 55 we described how the envelope of the radiation pattern of an interferometer could be steered in direction by rotating each aerial on an axis through itself; but the direction of the individual fringes, which is determined by the base line and the wavelength, does not change.

Now if the cables connecting the two aerials together are not of precisely equal electrical length, the central fringe will be displaced from the median plane by such an angle that the excess path from a distant source to the aerial with the shorter cable just compensates the cable defect. The electrical length of a cable can readily be controlled by changing its physical length or its phase velocity and a device for doing this may be called a phase-shifter. A phase shift changing progressively with time will cause the fringes to sweep across the sky, and to grow and decay in such a way that their envelope remains fixed.

An alternative explanation of this system can be given in terms of the interference between two waves of slightly differing frequency, since a steadily changing phase means a change in frequency.

Since the aerial pattern as a whole is not swept over the sky, the received power cannot be expressed as a convolution integral and the simple aerial smoothing theory does not apply. This was also the case with the pair of aerials on equatorial mountings and it was connected with the non-rigidity of the arrangement, an interpretation which can also be placed on the present case.

Let the radiation pattern of a single aerial be $A(x, y)$, let the true distribution of brightness temperature be $T(x, y)$, and let $T_a(x, y)$ be the distribution which would be observed with a single aerial. Then if the aerials are placed S wavelengths apart in the ξ-direction and a phase shifter in one arm introduces phase at the rate of Ω radians per second, the radiation pattern at time t is

$$A(x, y) \left[1 + \cos (2\pi S x - \Omega t)\right]$$

and the received power at time t when the aerial is pointed at (x, y) is given by

$$\iint T(x' + x, y' + y) A(x', y') \left[1 + \cos (2\pi S x' - \Omega t)\right] dx' dy'$$
$$= T_a(x, y) + \cos \Omega t \iint \cos 2\pi S x' T(x' + x, y' + y) A(x', y') dx' dy' +$$
$$+ \sin \Omega t \iint \sin 2\pi S x' T(x' + x, y' + y) A(x', y') dx' dy'.$$

The power received with the aerials pointed in a fixed direction, say $(0, 0)$, is

$$T_a(0, 0) + M \cos (\Omega t - \alpha),$$

where

$$M e^{i\alpha} = \iint e^{i 2\pi S x'} T(x', y') A(x', y') dx' dy' = \overline{TA}(S, 0).$$

Thus as time elapses, the power received with fixed aerials consists of a steady component equal to what would have been received on a single aerial plus a periodic component with period $2\pi/\Omega$. From the amplitude and phase of the periodic component one gets one Fourier component of $T(x, y) A(x, y)$, namely $\overline{TA}(S, 0)$. Since the steady component $T_a(0, 0)$ can be expressed as $\overline{TA}(0, 0)$, it follows that PAYNE-SCOTT and LITTLE's device is precisely equivalent to a simple two-element interferometer in the limiting case of small elements for which $A \rightarrow \text{const}$; it merely obtains the same information more quickly, and that is the purpose for which it was devised[1].

By increasing the bandwidth of the system one obtains greater sensitivity but at the same time the fringes of high order die out as described in Sect. 57. An advantage of narrow bandwidth in this instrument is the ability to work over a large area of sky without adjustment of the aerials. Increased bandwidth brings with it the need for a phase delay in one arm which will cause the zero-order fringe to follow the Sun; this is in addition to the physical following necessitated by the limited beamwidth of the aerials themselves.

[1] See Austral. J. Sci. Res. A **4**, 489 (1951); The Observatory **70**, 185 (1950).

Payne-Scott and Little introduced preamplifiers into their system at each aerial for the purpose of counteracting signal loss in the cables. This would have been impossible in a simple interferometer without exceptional equalization and stabilization of gains, phase shifts and noise figures, but in the phase-sweeping system, and in the phase-switching system discussed in the following section, moderate instability and inequality affect only the "steady" component, not the periodic component.

61. Ryle's interferometer. Of the two bands of Fourier components received by an interferometer, only the high frequency band contains information that could not have been obtained with a single aerial and so it would not be a loss if the low frequency band were filtered out. Such action would also bring advantages, for the steady background shown in Fig. 42 contains a contribution of galactic and extragalactic origin which rises and falls in strength as the day elapses and is embarrassing when large compared with the record of a desired faint source; this would be removed by filtering. So also would slow changes in receiver noise.

A method of doing this was introduced by Ryle[1]. By switching an extra half wavelength of cable in and out of one arm of the interferometer at a rapid rate such as 25 Hz, the fringe pattern can be switched back and forth, under its envelope, in a discontinuous version of the phase-sweeping scheme. Ryle describes it as follows.

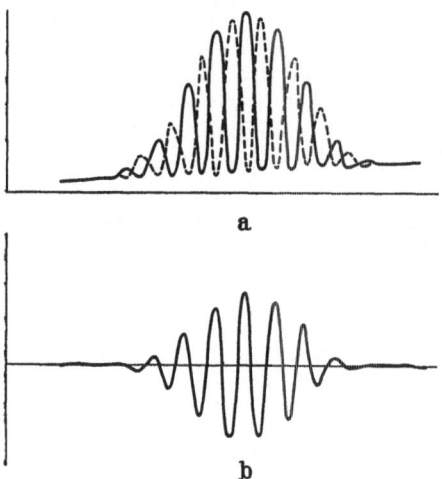

Fig. 47 a and b. (a) The two separate responses of a phase-switched interferometer. (b) The response of the system to a point source.

When the interferometer is in one of its two possible conditions, the response to a point source is

$$A(x) \left[1 + \cos 2\pi S x\right]$$

as shown by the heavy line in Fig. 47(a), and when in the other, the response, as shown by the broken line, is

$$A(x) \left[1 - \cos 2\pi S x\right].$$

When the interferometer alternates rapidly between its two conditions the received power has a steady component and a rapidly alternating component of amplitude

$$2A(x) \cos 2\pi S x$$

which may be rectified in a phase sensitive detector and recorded. The resulting record of Fig. 47(b) is in effect the difference between the two upper curves.

62. The Mills cross. Two aerials A_1 and A_2 (Fig. 48), one greatly extended in the ξ-direction, the other in the η-direction, form the two elements of an interferometer. The first receives from a narrow strip of sky a_1 and the second from a similar strip a_2 at right angles (not shown). When the two are connected together in phase their reception pattern is as shown at b, point sources in either a_1 or a_2 being received, and sources in both a_1 and a_2 being received more strongly.

---

[1] M. Ryle: Proc. Roy. Soc. Lond., Ser. A **211**, 351 (1952).

When the two aerials are connected together in phase opposition, which may be arranged by inserting a half-wavelength of cable into one arm, sources not in both a_1 and a_2 are received as before, but sources lying in the intersection of a_1 and a_2 are not received since their contributions to A_1 and A_2 are equal and combined in opposition [Fig. 48(c)]. Now if the system is caused to alternate rapidly between its two possible states, then the received power will comprise a steady part due to sources in only one beam and an alternating part due to sources lying in the intersection. The steady part may be ignored and the strength of the alternating component recorded by the use of a phase-sensitive detector.

Such a system was first constructed by MILLS and LITTLE[1] who thus achieved the angular resolution corresponding to a conventional aerial occupying the broken square outline, which would receive from an area of sky of size d. A photograph of the 1500 foot Mills cross for meridian use is shown in Fig. 49a. Fig. 49b

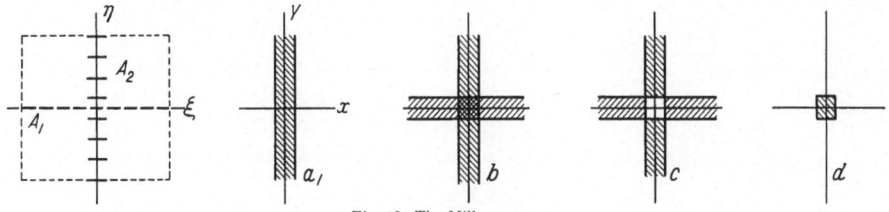

Fig. 48. The Mills cross.

shows an application [2] of the cross principle employing fully steerable elements, while Fig. 49c illustrates the quality of resolution obtainable on the sun.

The beamwidth of such a system is much less than would be achieved by a conventional aerial having the same collecting area, or same number of elements, and apparently violates the theoretical relation between the area and beamwidth of an aerial. However, while the system contains two aerials, it is not an aerial itself.

If the patterns represented schematically by a_1 and a_2 are written as $V_1(x, y)$ and $V_2(x, y)$, where V_1 and V_2 are respectively the phasors representing the voltages produced in the two separate aerials by a point source at (x, y), then the available power in the first condition of combination is proportional to

$$[V_1(x, y) + V_2(x, y)] \times [V_1(x, y) + V_2(x, y)]^*$$

and in the second condition,

$$[V_1(x, y) - V_2(x, y)] \times [V_1(x, y) - V_2(x, y)]^*.$$

The difference between these two expressions, which represents the amplitude of the alternating component, is proportional to

$$V_1(x, y)V_2^*(x, y) + V_2(x, y) V_1^*(x, y).$$

The power response of the system to a point source is thus the scalar product of the voltage responses of the separate aerials.

Such a pattern has sidelobes which are stronger relative to the main beam than is the case with the patterns of conventional aerials. In surveying the sky for discrete sources, there is a possibility that a strong source in the direction of a sidelobe may be mistaken for a faint source in the main beam. For this reason MILLS tapered his arrays to smooth out the lobe structure. The most

[1] B.Y. MILLS and A.G. LITTLE: Austral. J. Phys. **6**, 272 (1953).
[2] R. N. BRACEWELL: Inst. Radio Engrs. National Convention Record **5**, part I, 68 (1957).

suitable degree of taper in a given application will depend on a balance between
(i) spurious information on the one hand and (ii) lost information due to failure
of the widened beam to resolve and to obliteration by the increased general side
radiation on the other.

Arrangements for steering the beam on the meridian by introducing progressive phase shifts in a north- south dipole array have proved feasible in practice

b

Fig. 49 a, b. (Photo.) (a) The 1500 foot Mills cross at Fleurs, New South Wales, wavelength 3.5 m, beamwidth 50 minutes
of arc, (b) a cross of equatorial paraboloids at Stanford, California, wavelength 9 cm, beamwidth 3.1 minutes of arc.

and it is clear that the Mills cross represents a great instrumental advance in
those fields where resolution, not sensitivity, is the desideratum.

The cross principle lends itself to the use of aerials of very great physical
extent and, because of this, substantial effective area accrues even when aerials
A_1 und A_2 are only about a wavelength wide. For future astronomical observations requiring both high resolution and high sensitivity the aerials would
also have to be wide. Two distinct possibilities follow, each of which allows
construction in sizes much larger than the largest possible steerable giant paraboloid. For a meridian instrument, a wide east-west aerial could conveniently

be tilted and could assume the form of a parabolic cylinder with focal-line feed (as in Fig. 30). A wide north-south aerial could be a fixed parabolic cylinder steered by introducing progressive phase shifts along the focal line; alternatively the north-south aerial could be an array of more or less closely spaced tiltable segments. A second possibility, already being exploited both at Sydney and Stanford, is to split each aerial into an array of equatorially mounted segments (Sect. 64), thus relieving the restriction to transit observations (Figs. 35c and 49b).

Whilst the theory given above is quite valid there is a subtlety that may be overlooked. If the aerials A_1 and A_2 are represented by aperture distributions

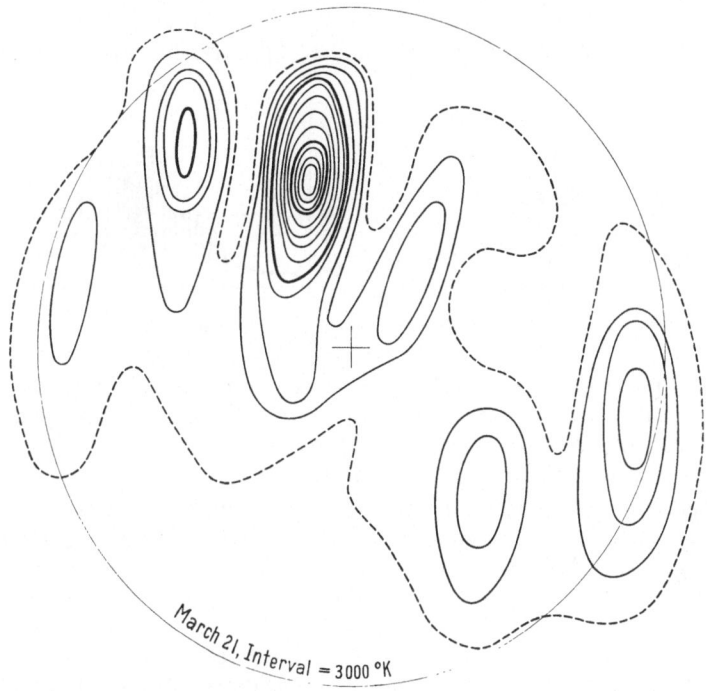

March 21, Interval $= 3000\ °K$

Fig. 49c. Picture of the microwave sun showing the quality of resolution obtainable at 9 cm.

they may not overlap since the energy falling at $\xi = 0, \eta = 0$ can only be absorbed once. It is therefore necessary that one or both aerials have a defect at the origin, let us say A_1. Then $V_1(x, y)$ comprises the narrow strip a_1 and also a shallow negative part distributed in accordance with the voltage reception pattern of the missing part. The response of the system to a point source then has negative parts which bring its integral to zero; which is necessary for a system so designed that it cannot respond to a uniform distribution. In Sect. 64 we refer to this matter again.

63. The Brown and Twiss system. As we have seen, the measurement obtained with a pair of spaced aerials yields a value of $\overline{T}$, normalized in the case of a simple interferometer, otherwise not. For a very small source $\overline{T}$ falls only slowly from its value at the origin, and so very great aerial spacings are required to reveal the appreciable change in $\overline{T}$ that will yield a size measurement. The large spacing in turn requires the preamplification which is only feasible with a phase-switched interferometer, and then careful absolute measurements at at least two spacings are necessary since a single phase-switched interferometer record does not by itself yield a value of fringe visibility. For these reasons difficulty was encountered in the first attempts to measure the diameter of the source in Cygnus, which though strong proved to have an extremely small angular diameter, of the order of one minute of arc. From the inverse theorem of Sect. 54 it follows that independent observations of $\overline{T}$ cannot be made at intervals less than about 3000 wavelengths. This places a considerable strain on technique and was faced in three ways. Smith[1] obtained measurements to 423 wavelengths ($\lambda = 1.4$ m) giving very great care to equalizing and stabilizing cable loss and preamplifier gains. Mills[2] substituted a radio link for the cable and obtained measurements out to 3000 wavelengths ($\lambda = 3$ m), taking particular care to preserve phase in an ingenious fashion. It appears, however, that it is not

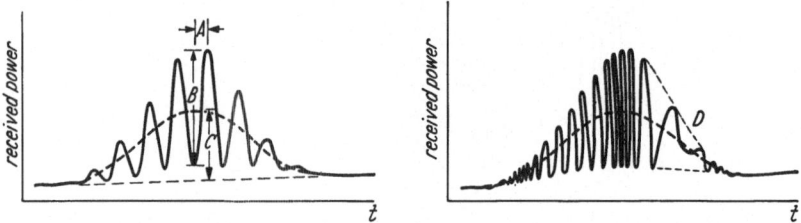

Fig. 50. Simple interferometer records of a discrete source in which phase is preserved (left) and allowed to drift (right).

necessary to preserve phase if only one shape parameter, viz. a second moment of T, is all that is sought. And it is clear that with the moderate spacings mentioned above only one parameter would be justified. Brown, Jennison and Das Gupta[3] based their measurement on this fact.

Consider a simple interferometer with which a single record at a large spacing has been obtained on a very small source, and from this record suppose that a measurement of complex degree of coherence Γ has been taken. Then $|\Gamma|$ yields the size of the source and pha Γ relates the center of gravity of the source to the median plane of the interferometer. It will now be shown that a record taken without care to preserve phase yields $|\Gamma|$ the desired quantity but loses pha Γ a quantity which in any case is normally considered difficult to measure. In Fig. 50 is shown on the left a record from the simple interferometer, which should be compared with the record of a point source shown in Fig. 42. From the depth of the minima it is apparent that the spacing is sufficiently great to reveal a fringe visibility $B/2C$ appreciably less than unity, which can be used as explained in Sect. 68 to determine the source width. The distance A, taken in conjunction with the fringe period and corrections for collimation gives pha Γ. On the right hand side of Fig. 50 we see the effect of an irregular phase drift. Clearly the envelopes from which fringe visibility are determined are not lost, except momentarily by an accident such as that indicated at D.

On this basic point regarding phase Hanbury Brown based a new system which is described by Brown and Twiss[4] and by Jennison and Das Gupta[5]. The independence of phase was to permit the extension of aerial spacings to 50 km

[1] F. G. Smith: Proc. Phys. Soc. Lond. B **65**, 971 (1952).
[2] B. Y. Mills: Nature, Lond. **170**, 1063 (1952).
[3] R. H. Brown, R. C. Jennison and B. K. Das Gupta: Nature, Lond. **170**, 1061 (1952).
[4] R. H. Brown and R. Q. Twiss: Phil. Mag., Ser. VII **45**, 663 (1954).
[5] R. C. Jennison and M. K. Das Gupta: Phil. Mag., Ser. VIII **1**, 55 (1956).

which would be necessary if, as was thought, the radio sources should prove to be of stellar dimensions. The system actually built incorporates other, inessential, departures from previous practice, viz. (i) the two received signals were detected before being combined, (ii) the multiplication effected in the phase-switched system by differencing the squares of the sum and difference of two voltages, was carried out by a special envelope multiplying circuit referred to as a correlator, (iii) the time-varying quantities whose cross-correlation was evaluated were the squares of the envelopes of the received signals.

For a full understanding of the Brown and Twiss system it is therefore necessary to examine the relationship between $\langle F_1 F_2^* \rangle$, the quantity involved in an observation of fringe visibility with a simple interferometer (Sect. 56), and $\langle F_1(t) F_1^*(t) \cdot F_2(t+\tau) F_2^*(t+\tau) \rangle$ the cross correlation between the squares of the envelopes of $F_1(t) e^{i\omega t}$ and $F_2(t) e^{i\omega t}$. In practice τ will be arranged to be zero by observing in the median plane, or, if the source under study is off to the side, by inserting a compensating time delay. It can be shown (BRACEWELL[1]) that

$$\frac{\langle (F_1 F_1^* - \langle F_1 F_1^* \rangle)(F_2 F_2^* - \langle F_2 F_2^* \rangle) \rangle}{\langle (F_1 F_1^* - \langle F_1 F_1^* \rangle)^2 \rangle} = \left[\frac{\langle F_1 F_2^* \rangle}{\langle F_1 F_1^* \rangle} \right]^2 = |\Gamma|^2$$

i.e. the fluctuating parts of the squared envelopes $F_1 F_1^*$ and $F_2 F_2^*$ are correlated; but not as strongly as are the phasors F_1 and F_2^*, since the first correlation coefficient is the square of the second.

Receiver noise must not be neglected in this discussion. When visibility is being measured from a record taken with a simple interferometer the presence of receiver noise may hinder the measurement but does not change its character. However, the correlation between the squared envelopes of two oscillations containing steady receiver noise components large compared with the contributions due to the fields F_1 and F_2 is not the same as in the absence of noise and proves to be equal to the correlation between the field phasors. The important point is that the correlation is not destroyed by demodulating the radio frequency signals. The essence of the proof for the noise-free case, has been concisely set out by WOLF[2] who shows that the correlation between the fluctuating parts of the instantaneous field intensities is equal to twice the square of the correlation between the instantaneous real fields at two points.

b) Multi-element interferometers.

64. CHRISTIANSEN's interferometer. Consider a long array of identical aerials, such as paraboloids, spaced at equal intervals of B wavelengths along the ξ-axis. The aperture distribution may be expressed as the convolution of the distribution over one aerial with the function

$$\sum_{n=1, 3, 5 \ldots}^{M-1} {}^2\delta \left(\frac{\xi}{\lambda} \pm \frac{nB}{2}, \frac{\eta}{\lambda} \right),$$

which is a row of M impulses in the $\xi\eta$-plane. Consequently the field radiation pattern is the product of the pattern for a single aerial with the Fourier transform of the above set of impulses. By taking the impulses in symmetrical pairs, the required transform can be expressed as a sum of $\frac{1}{2}M$ cosine functions

$$2 \sum_{n=1, 3, 5 \ldots}^{M-1} \cos \pi n B l,$$

[1] R. N. BRACEWELL: Proc. Inst. Radio Engrs. **46**, 97 (1958).
[2] E. WOLF: Phil. Mag., Ser. VIII **2**, 351 (1957).

which is the Fourier series for a periodic function consisting of overlapping functions of the form

$$\frac{\sin [M \pi B l]}{\pi B l}$$

repeated with alternation of sign at intervals $1/B$. This latter expression will be recognized as the field pattern of a uniform array $M B$ wavelengths long and of

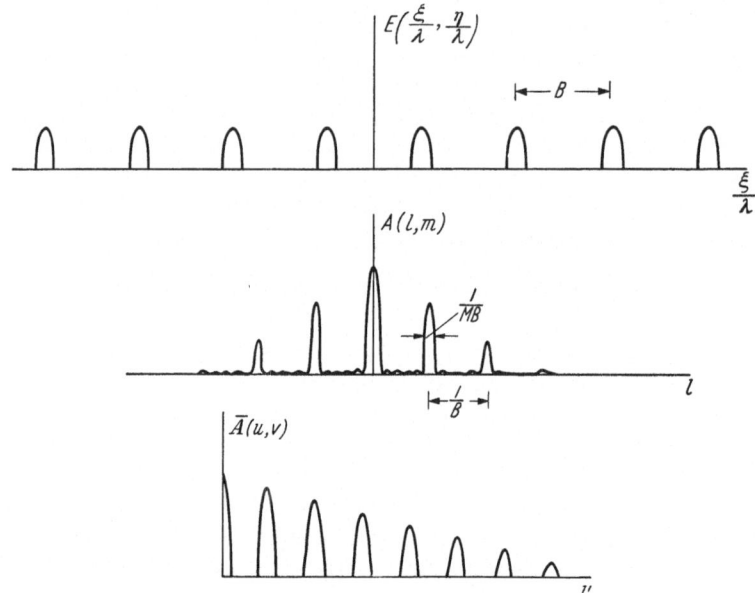

Fig. 51. The aperture distribution, radiation pattern, and spectral sensitivity function of a Christiansen array.

zero width. The power radiation pattern, which we obtain by squaring the field pattern in this case, thus consists, when $M B$ is large, of a set of parallel sharp fringes whose profile is of the form

$$\left[\frac{\sin \pi M B l}{\pi l}\right]^2,$$

the whole multiplied by the radiation pattern of a single aerial.

These conclusions are illustrated in Fig. 51 which also shows the spectral sensitivity function.

This aerial array is analogous to a diffraction grating with a total of M lines and an aperture of $M B$ wavelengths, in contact with a slit set perpendicular to the ruling. The repeated fringes correspond to the beams of various order, and the fringe profile corresponds to the diffraction pattern of the slit in the long direction. The envelope of the fringe pattern corresponds to the diffraction pattern of a single transparent element of the ruling, and for the closest analogy the width of the slit would be equal to the width of a single transparent element. The first array of this kind was built by Christiansen and Warburton[1] and had 32 elements each spaced approximately 30 wavelengths apart.

When a discrete source, such as the Sun, whose angular extent is less than the fringe spacing passes through the fringe system, a record is obtained which is virtually identical with what would be obtained with a uniform aperture a little longer than the array. The omission of aperture excitation between the

[1] W.N. Christiansen and J.A. Warburton: Austral. J. Phys. 6, 262 (1953).

elements thus results in no loss of resolution, though there is a loss in available power. This is another example of the inverse discrete interval theorem. The spectral sensitivity function shows that the spatial spectrum will be sampled at uniform intervals and nothing will be lost thereby if the object does not extend beyond a certain finite width. In the case of the Sun, which is approximately 0.01 radians in diameter, the gap between the aerials could not exceed 100 wavelengths, and should be less to allow for the fringe width.

The importance of a new idea of the present kind is that a great advance over previous technique becomes possible. Furthermore the idea is transferable to other systems which combine aerials, such as the Mills cross, and a future application to arrays of giant aerials of the size just now coming into existence can be foreseen.

65. Multi-element phase-switched systems. The use of phase-switching in conjunction with two general aerial arrays had been contemplated by RYLE[1] and various complicated systems are conceivable. One interesting example developed by COVINGTON and BROTEN[2] resulted in the achievement of twice the angular resolution which was believed possible at the time within the overall dimensions.

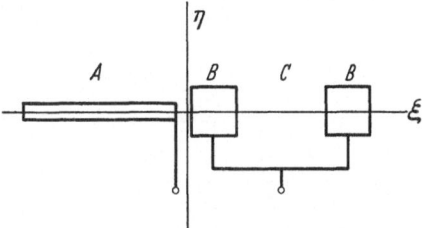

Fig. 52. A compound phase-switched system.

A long narrow existing aerial A (Fig. 52) was combined with a two-element interferometer BB by phase switching. The (power) response to a point source can be shown to be the product of the (voltage) responses of (i) the long array, (ii) an interferometer consisting of two isotropic aerials at BB, (iii) an interferometer of spacing AC, (iv) a single aerial B. The net result was a beam which in one dimension was about half that for an ordinary aerial of the same overall length in the ξ-direction.

Consider the general case of an aperture E_1 of any kind centered at $\xi = -\frac{1}{2}a$, $\eta = 0$ and another E_2 at $\xi = +\frac{1}{2}a$, $\eta = 0$, and let them be connected together through a phase-switch. Let

$$E_2 \star E_1^* = E_3$$

and

$$E_1 \star E_2^* = E_4.$$

Then the spectral visibility functions of the two states of connection are

$$\left[E_1\left(\frac{\xi+\frac{1}{2}a}{\lambda}, 0\right) \pm E_2\left(\frac{\xi-\frac{1}{2}a}{\lambda}, 0\right)\right] \star [\text{conjugate}]$$

$$= E_1\left(\frac{\xi+\frac{1}{2}a}{\lambda}, 0\right) \star E_1^*\left(\frac{\xi+\frac{1}{2}a}{\lambda}, 0\right) + E_2\left(\frac{\xi-\frac{1}{2}a}{\lambda}, 0\right) \star E_2^*\left(\frac{\xi-\frac{1}{2}a}{\lambda}, 0\right) \pm$$

$$\pm E_2\left(\frac{\xi-\frac{1}{2}a}{\lambda}, 0\right) \star E_1^*\left(\frac{\xi+\frac{1}{2}a}{\lambda}, 0\right) \pm E_1\left(\frac{\xi+\frac{1}{2}a}{\lambda}, 0\right) \star E_2^*\left(\frac{\xi-\frac{1}{2}a}{\lambda}, 0\right).$$

Consideration of the alternating part ($\pm$ signs), viz.

$$E_3\left(\frac{\xi+a}{\lambda}, 0\right) + E_4\left(\frac{\xi-a}{\lambda}, 0\right),$$

[1] M. RYLE: Proc. Roy. Soc. Lond., Ser. A **211**, 351 (1952).
[2] A. E. COVINGTON and N. W. BROTEN: IRE Transactions on Antennas and Propagation **AP—5**, 247 (1957).

will give the Fourier transform of the response of the system to a point source, subject to normalizing. When E_1 and E_2 are real and even functions of ξ, both E_3 and E_4 are equal to $P_1 P_2$, the product of the field patterns; hence the response of the system $A(\vartheta, \varphi)$ is given by

$$A(\vartheta, \varphi) = P_1 P_2 \cos 2\pi a\, \vartheta.$$

This result applies to the example quoted above, to the Mills cross, and to the phase-switched interferometer with identical aerials. The more general result, found by substituting for E_3 and E_4 and transforming, is

$$(A - \vartheta, -\varphi) = P_2 P_1^* \, e^{i\, 2\pi a\, \vartheta} + P_1 P_2^* \, e^{-i\, 2\pi a\, \vartheta}$$
$$= |P_1 P_2^*| \cos [2\pi a\, (\vartheta - \vartheta_1)].$$

Interesting special cases include (i) the long uniform linear array phase-switched against a single aerial at one or at each end, which gives a flat spectral

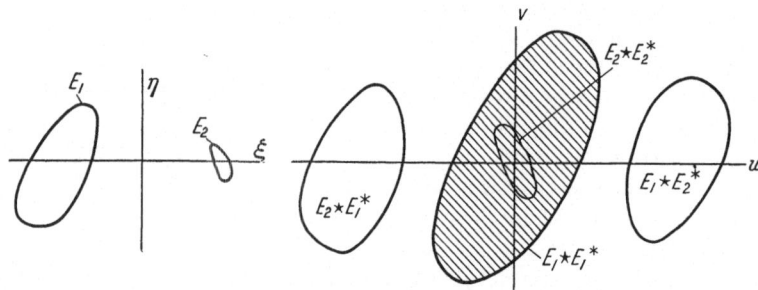

Fig. 53. Spectral sensitivity islands in the uv-plane for a phase-switched system comprising apertures E_1 and E_2. The shaded island, shown for completeness, is suppressed.

sensitivity function and doubles the resolving power of the array even if the single aerial has little directivity; (ii) two long perpendicular uniform arrays combined in a tee instead of a cross, thus doubling the resolution in one direction; (iii) a phase-switched unequal system which, referring to Fig. 45, would have the effect of suppressing the central island; (iv) a four element system comprising a pair of crossed variable-spacing two-element interferometers; (v) a long array with a single aerial at variable positions on the perpendicular bisector. In thinking about devices of this kind the two-dimensional spectral sensitivity functions are easier to obtain than the spatial responses, and are often more revealing. Fig. 53 illustrates a spectral sensitivity island diagram which was very readily drawn by sliding a copy of the $\xi\eta$ outline over itself as explained above. The shaded island is the one whose sign does not change when the sign of E_2 is reversed and which is therefore suppressed.

66. The 1100101 array. The spectral sensitivity functions so far illustrated exhibit bands (Fig. 43) or taper away towards their cut-off (Figs. 39, 51). The question may be asked whether an aperture distribution exists such that the spectral sensitivity function is constant out to its cut-off, i.e., can we have $\overline{A}(s) = \Pi\left(\dfrac{s}{2 s_c}\right)$, without resorting to phase-switching. This would imply that $A(x) = \dfrac{\sin 2\pi s_c}{\pi x}$ which is impossible since $A(x)$ cannot go negative. However $\dfrac{\sin 2\pi s_c x}{\pi x} + \dfrac{4 s_c}{3\pi}$ is non-negative and has a transform $\Pi\left(\dfrac{s}{2 s_c}\right) + \dfrac{4 s_c}{3\pi} \delta(s)$ which is an interesting possibility. In terms of equally spaced discrete elements of amplitude

0 or 1 we find the following arrays where the sequence of numbers $a_0\, a_1 \ldots a_n$

$E\left(\frac{\xi}{\lambda}\right)$	$\bar{A}(s)$
11	21
1101	3111
1100101	4111111

stands for $\sum_{i=0}^{n} a_i\, \delta\left(\frac{\xi}{\lambda} - i\right)$ in the left hand column and $\sum_{0}^{n} a_i\, \delta\,(s-i)$ in the right hand column.

The last entry in the table represents the 1100101 array, an array of four equal elements so spaced that there is one and only one pair of elements for each spacing from one unit up to six units, the full extent of the array. The spectral sensitivity function is therefore flat. However zero spacing occurs four times and so the array is unduly sensitive to the $s=0$ component, which is necessitated by the condition that $\bar{A}(s) \lessgtr 0$. An array of parabolas spaced in this way has been built and demonstrated by Arsac[1]. There are no further arrays with the property shown by the three tabulated above, and in two dimensions no corresponding property has been found.

It is clear that the economy and simplicity of the 1100101 array will fit it for consideration when giant aerials become sufficiently numerous to group into arrays.

VI. Observing procedures and analysis of observations.

a) Discrete Sources.

67. Position of a point source. If one observes with a pencil beam, the position of a point source is simply the direction of the beam axis when maximum power is being received. An accompanying calibration is then necessary to determine the beam axis in terms of readings on the setting circles on the aerial mounting, the Sun being one convenient source for this purpose. A direct observing procedure is to fix the beam on the meridian and determine the right ascension from the time of transit, as revealed by the maximum in the record. Then take a series of drift curves at discrete intervals of declination equal to the peculiar interval of the aerial (Sect. 54). From the maxima of the drift curves determine the declination by interpolation, using if necessary the exact method of Sect. 70. The judgment of the observer will be necessary in removing any trend in the background and any confusing effects of nearby sources.

If one observes with an interferometer, and the great bulk of the early positional work was done in this way, various combinations of right ascension and declination will be measured and the final coordinates will be obtained by elimination, unless the interferometer is on an east-west baseline. In this case the right ascension is given by the time of transit, and the declination δ is given by

$$\cos \delta = \frac{86400}{2\pi S \tau},$$

where τ is the average period of the fringes in sidereal seconds, and the interferometer spacing is S wavelengths. The result follows from the factor $\cos\,(2\pi S V t - \beta)$ in the expression for the signal received from a discrete source (Sect. 55), which

[1] J. Arsac: C. R. Acad. Sci., Paris **240**, 942 (1955).

shows that the period of a fringe is $\frac{1}{SV}$, where V, the velocity of the source through the beam, is proportional to $\cos \delta$.

The order of accuracy of position determination has been minutely studied by Mills and Thomas[1] and by Smith[2], the latter paper including results on the sea interferometer and many other arrangements. With great care absolute accuracy of one minute of arc was achieved in the location of the source in Cygnus, but only the most devoted observers will wish to resume this type of work since there are many reliably identified extragalactic sources whose optical positions may be used with confidence as points of reference for absolute positions within a few minutes of arc.

With the advent of aerials of great effective area serious confusion between adjacent sources has been encountered at the longer radio wavelengths. Let us assume that any continuous background is uniform and assess the limit to the number of resolvable sources set by random clustering of the sources themselves and the finite resolving power of the instrument. Since independent measures can be made only at the peculiar interval a survey of the whole sky will yield only a finite number of data equal to $4D/\zeta$, where D is the directivity and ζ the achievement factor of the aerial. Now if a saddle point in an observed distribution is taken as a criterion for resolution of two point sources, the distance between the two points must be at least 1.2 peculiar intervals if the sources are equal and more if they are not. Points scattered at random in two dimensions with a density n per unit area are separated on the average from their nearest neighbor by a distance $0.5\,n^{-\frac{1}{2}}$. Equating this distance to 1.2 peculiar intervals we find that the maximum number of resolvable sources in the whole sky is $\frac{D}{1.4\,\zeta}$, or approximately D. The number of sources per steradian is approximately $D/4\pi$ or one per effective solid angle Ω. In practice it is considered that there should be about 10 effective beamwidths per source in order to ensure that a substantial fraction of the apparrent sources are real, and perhaps 100 beamwidths per source if accurate flux density measurements are required.

68. Angular extent of a discrete source. The nulls in the response of an interferometer to a point source are filled in, if an extended source passes through the beam, by an amount depending on the width of the source. McCready, Pawsey and Payne-Scott used this phenomenon to determine the width of the sources of solar noise situated above sunspot groups, and Pawsey and Bracewell show that the width of a uniform one-dimensional source is given by

$$\frac{2\sqrt{3R}}{\pi S},$$

where S is the spacing in wavelengths and R is the ratio of the minimum to the maximum of the interference pattern.

The observation is essentially one of fringe visibility at a single aerial spacing since the visibility as defined in Sect. 56 is equal to

$$\frac{1-R}{1+R}.$$

Consequently the data are equivalent to one Fourier component of the source distribution T, normalized with respect to $\overline{T}(0, 0)$. Provided the aerials are not

[1] B.Y. Mills and A.B. Thomas: Austral. J. Sci. Res. A **4**, 158 (1951).
[2] F.G. Smith: Monthly Notices Roy. Astronom. Soc. London **112**, 497 (1952).

too widely spaced, this in turn is equivalent to a measure of the curvature of $\overline{T}$ at the origin, and by a theorem of Fourier transforms, this determines the standard deviation σ_x of T about its center of gravity, in the direction parallel to the base-line of the interferometer (taken as the x-direction). We find that

$$\sigma_x \approx \frac{\sqrt{\overline{R}}}{\pi \, S}$$

for small R. It is felt that this is a more apposite statement of source size than $2W$ the width of the equivalent uniform strip, or D the diameter of the equivalent uniform disc. A fourth measure also in use is σ_r the standard deviation of the equivalent two-dimensional Gaussian source. These various quantities are re-lated by

$$\sigma_x = \frac{2W}{2\sqrt{3}} = \frac{D}{4} = \frac{\sigma_r}{\sqrt{2}} .$$

If two width measurements at right angles were possible, $\sqrt{\sigma_x^2 + \sigma_y^2}$ would be a suitable size measure; but if only one is available and if the source possesses circular symmetry, or if circular symmetry forms a suitable basis, then σ_r or $\sqrt{2}\sigma_x$ is appropriate.

When R is small a single visibility observation thus measures the one-dimensional extent of a source; but when R is not small, the shape as well as the extent of the source are mixed together in the measurement.

69. Flux density of a discrete source. Suppose that a set of drift curves has been obtained at a set of properly spaced declinations and that they reveal a discrete source superimposed on a continuous background. The observer will often feel sufficiently confident to subtract the background by taking account of the surrounding continuum, which may appear to him to be relatively free of other discrete sources. Occasionally, helpful information on other frequencies will also be available. We shall assume that the background has been satis-factorily subtracted, leaving a remainder $T_a(x, y)$.

Now by Sect. 49 the flux density S of the source is given by

$$S = \frac{2k}{\lambda^2} \iint T \, dx \, dy,$$

where T is the true temperature distribution over the source. But we have only T_a. However

$$\iint T_a \, dx \, dy = \overline{T_a}\big|_{u=v=0}$$
$$= [\overline{A}\,\overline{T}]_{u=v=0}$$
$$= T\big|_{u=v=0},$$

provided $A(x, y)$ is normalized so that

$$\iint A \, dx \, dy = 1,$$

whereupon

$$\overline{A}\big|_{u=v=0} = 1.$$

But

$$\iint T \, dx \, dy = \overline{T}\big|_{u=v=0},$$

therefore

$$S = \frac{2k}{\lambda^2} \iint T_a \, dx \, dy,$$

and the flux density is correctly calculated from a knowledge of T_a only.

We now show that the integral can be evaluated merely by summing discrete values. If $\bar{T}_a = 0$ for $|u| \geq \frac{1}{2}$ and for $|v| \geq \frac{1}{2}$, then

$$\iint T_a \, dx \, dy = 4 \left[{}^2III(2u, 2v) * \bar{T}_a\right]_{u=v=0}$$

$$= \left[{}^2III(\tfrac{1}{2} x, \tfrac{1}{2} y) \, T_a\right]_{u=v=0}$$

$$= \iint {}^2III \left(\tfrac{1}{2} x, \tfrac{1}{2} y\right) T_a \, dx \, dy$$

$$= 4 \sum_{\mu=-\infty}^{\infty} \sum_{\nu=-\infty}^{\infty} T_a (x - \alpha \mu, y - \beta \nu)$$

provided α and β are both less than 2.

The essence of this proof is that the integral of T_a equals the height at the origin of the island representing $\bar{T}_a$ in the uv-plane. The distribution ${}^2III \, T_a$ (which contains information about T_a only at the sampling points where ${}^2III \neq 0$) has a transform consisting of an array of islands identical with $\bar{T}_a$ but centered on the points of a rectangular lattice. As before, the integral of ${}^2III \, T_a$ is equal to the height of the island system at the origin; but this is the same as before, provided the islands do not crowd too closely. When the sampling interval is so coarse that the islands just close in and touch each other, we have sampling at the peculiar interval (Sect. 54). However, since we only require the height of the central island at the origin to remain unchanged, we may allow overlapping— just so much that the edges of the nearest islands do not reach the origin. This allows the sampling interval to be twice as coarse as the peculiar interval.

Thus we correctly evaluate S by summing just one in four of the values of T_a necessary to define T_a. This result permits a time saving in observation and reduction or alternatively permits certain cross-checks when redundant data are available.

In the foregoing, T_a was restricted, as in Sect. 50, to a loss free aerial. Hence effective aerial temperatures actually observed will be less by a factor η, the aerial efficiency (Sect. 25). Furthermore, the procedure of background subtraction ignores stray contributions such as side reception by the feed horn of a paraboloid; hence the integral measured over the main beam is less than the integral over all directions by a factor $\mathfrak{B}$, the beam ratio (Sect. 38). Thus, in terms of the T_a actually observed, we have for the absolute flux density of an extended source

$$S = \frac{2k}{\beta \lambda^2} \iint_{\text{beam}} T_a \, dx \, dy$$

where the beam efficiency β is given by

$$\beta = \mathfrak{B} \eta.$$

For the flux density of a point source it is sufficient to measure T_m the maximum value of T_a actually observed. Then from the relation $k \, T_m \, \Delta f = \frac{1}{2} \, S A \, \Delta f$ it follows that

$$S = \frac{2k \, T_m}{A}$$

$$= \frac{2k \, T_m}{\alpha \, \mathfrak{A}}$$

where the aperture efficiency α was shown in Sect. 38 to be given by

$$\alpha = \beta \mathfrak{D}.$$

b) Extended sources.

70. Pencil beam surveys. Consider first a moderately extended field, not too near the poles, and not too extensive in declination. Then the lines of constant right ascension and declination form a rectangular network.

Drift curves will be taken at declinations spaced one peculiar interval apart until the whole field is covered. Other ways of covering the field are possible but an essential character of the data is always preserved, viz. that one independent variable varies continuously and the other assumes discrete values.

It is necessary to reduce the data to sets of isophotes without introducing spurious detail which occasionally afflicts published surveys and with due regard to aerial smoothing. Now a drift curve, which is a cross-section of a function $T_a(x, y)$ with a cut-off (two-dimensional) spectrum, has itself a cut-off (one-dimensional) spectrum, and hence should be free from high-frequency components. However such components may be present because of errors or noise, and should be removed since they are unwarranted. Noise is best smoothed out by the eye but accurate filtering can also be carried out simply by convolution with $\dfrac{\sin \pi X}{\pi X}$, if X is the coordinate along the drift curve measured in units of one peculiar interval. Since X varies continuously the convolution integral does not reduce to a summation; but to compute the integral one would evaluate a summation,

$$T_\varepsilon = T_a * III\left(\frac{X}{\varepsilon}\right)\frac{\sin \pi X}{\pi X}$$

which would approach the desired integral

$$T_F = T_a * \frac{\sin \pi X}{\pi X}$$

as ε approached zero. Just how small ε need be we now enquire. Beginning with the coarsest interval, $\varepsilon = 1$, we find $T_1 = T_a$, that is, no filtering has been achieved; but with $\varepsilon = \frac{1}{2}$ we find

$$T_{\frac{1}{2}} = T_a * III(2X)\frac{\sin \pi X}{\pi X}$$

whence

$$\overline{T}_{\frac{1}{2}} = \overline{T}\left\{\frac{1}{4\,s_c} \cdot III\left(\frac{s}{4\,s_c}\right) * II\left(\frac{s}{2\,s_c}\right)\right\}.$$

Hence $\overline{T}_{\frac{1}{2}}$ consists of a central part $II\left(\dfrac{s}{s_c}\right)\overline{T} = \overline{T}_F$ plus remoter parts. For many purposes this simple operation would be sufficient filtering, since the components to be rejected would often be chiefly just beyond the central region.

To recapitulate, T_a is filtered by reading off values at intervals of one half the peculiar interval and taking the convolution (more aptly serial product) with the series $\dfrac{\sin \pi X}{\pi X}$, where X runs through all half integers. This series is tabulated here, omitting the zeros and the value for $X = 0$.

X	$\dfrac{\sin \pi X}{\pi X}$	X	$\dfrac{\sin \pi X}{\pi X}$	X	$\dfrac{\sin \pi X}{\pi X}$	X	$\dfrac{\sin \pi X}{\pi X}$	X	$\dfrac{\sin \pi X}{\pi X}$
$\frac{1}{2}$	0.6366	$7\frac{1}{2}$	-0.0424	$14\frac{1}{2}$	0.0220	$21\frac{1}{2}$	-0.0148	$28\frac{1}{2}$	0.0112
$1\frac{1}{2}$	-0.2122	$8\frac{1}{2}$	0.0374	$15\frac{1}{2}$	-0.0205	$22\frac{1}{2}$	0.0141	$29\frac{1}{2}$	-0.0108
$2\frac{1}{2}$	0.1273	$9\frac{1}{2}$	-0.0335	$16\frac{1}{2}$	0.0193	$23\frac{1}{2}$	-0.0135	$30\frac{1}{2}$	0.0104
$3\frac{1}{2}$	-0.0909	$10\frac{1}{2}$	0.0303	$17\frac{1}{2}$	-0.0182	$24\frac{1}{2}$	0.0130	$31\frac{1}{2}$	-0.0101
$4\frac{1}{2}$	0.0707	$11\frac{1}{2}$	-0.0277	$18\frac{1}{2}$	0.0172	$25\frac{1}{2}$	-0.0125	$32\frac{1}{2}$	0.0098
$5\frac{1}{2}$	-0.0579	$12\frac{1}{2}$	0.0255	$19\frac{1}{2}$	-0.0163	$26\frac{1}{2}$	0.0120	$33\frac{1}{2}$	-0.0095
$6\frac{1}{2}$	0.0490	$13\frac{1}{2}$	-0.0236	$20\frac{1}{2}$	0.0155	$27\frac{1}{2}$	-0.0116	$34\frac{1}{2}$	0.0092

The test for adequacy of filtering is to compare T_a in the most unfavorable areas with the interpolated values discussed below. Where further filtering is required one repeats the first process with the same tabulated series but with $\varepsilon = \frac{1}{4}$.

There is an even simpler procedure which appears at first sight to be equivalent to filtering. If T_a is read off at discrete intervals corresponding to the desired cut-off, the set of values so obtained defines a function T' with cut-off spectrum. But this function is not the same as T_F since it is affected by high-frequency components of T_a. However, T' may often represent T_a adequately within the limits of accuracy of the observations.

The filtered data may be used to commence plotting points on isophotes. In areas where the isophotes are closely packed or highly curved it may be desirable to interpolate between the filtered drift curves. This is done by reading off a set of values along a transversal which intersects the drift lines. Convolution with the series tabulated above then gives an interpolated value.

When the field being surveyed is extended in declination the simplest procedure is to split it into smaller zones, and when the area is near the poles the drift curves may still be filtered and interpolation along hour circles is still possible. The apparent complication is considerably relieved by the use of a projection suited to the area being worked on.

Absolute values of sky brightness temperatures can be measured by taking account of losses as in Sect. 69. Suppose that a paraboloidal reflector points at an extended area of sky of brightness temperature T which fills the main beam (Sect. 38), thus eliminating considerations of aerial smoothing which will be returned to in Sect. 73. What will be the effective aerial temperature T_a actually observed? Power entering the aerial terminals distributes itself so that a fraction β is launched in the main beam (Sect. 38) and a fraction η all told is launched skywards (Sect. 25). Hence $1 - \eta$ is absorbed by the ground at temperature T_0. The remaining stray fraction $\eta - \beta$ proceeds skywards, but not in the main beam. By the principle of detailed balancing (Sect. 50) it follows that

$$T_a = \beta T + (1 - \eta) T_0 + (\eta - \beta) T_{\mathrm{av}},$$

where T_{av} is a weighted average of sky temperature over the directions taken by the stray radiation. Hence the temperature T, which it is desired to measure, is given in terms of the observed T_a by

$$T = \frac{T_a}{\beta} + \frac{1 - \eta}{\beta} T_0 + \frac{\eta - \beta}{\beta} T_{\mathrm{av}}.$$

The measurement of β and η has been discussed earlier. They are difficult to determine and may vary with aerial pointing. All the terms must, however, be estimated, and reference to the rich literature of sky surveys will reveal the wide variety of approaches that have been adopted.

71. Interferometric studies. From Sects. 55 and 56 it is clear that measurements of fringe visibility with an elementary interferometer at all spacings and in all azimuths give the Fourier transform of a source distribution and hence the distribution itself. The one dimensional form of this statement was given by McCready, Pawsey and Payne-Scott in their original paper on radio interferometric observations and the procedure has since been put into effect many times e.g. by Stanier[1] and Scheuer and Ryle[2] in one-dimension and by O'Brien[3] and Firor[4] in

---

[1] H. M. Stanier: Nature, Lond. **165**, 354 (1950).

[2] P. A. G. Scheuer and M. Ryle: Monthly Notices Roy. Astronom. Soc. London **113**, 3 (1953).

[3] P. A. O'Brien: Monthly Notices Roy. Astronom. Soc. London **113**, 597 (1953).

[4] J. Firor: Astrophys. Journ. **123**, 320 (1956).

two dimensions. The measurement of pha Γ is difficult and is not important in the case of symmetrical objects such as the quiet Sun except insofar as complete phase reversals occur. This has been investigated by allowing a small signal from a central aerial to leak into the transmission line in a way analogous to the determination of the phase in optical interference patterns by the admission of a little direct light.

In practice one can explore only that finite region of the transform plane corresponding to accessible interferometer spacings. Two-dimensional Fourier synthesis then yields a celestial distribution, which is not in general the true distribution, but the "principal solution" described below in Sect. 73. This procedure has been called aperture synthesis by RYLE[1]. When the elements of the interferometer are large the Fourier transform relation is modified (Sect. 55), but not greatly if, as is usually the case, the sources under study are compact relative to the beamwidth of a single element. When an interferometer has elements that are segmented or in any way complicated, it may be conveniently studied with the help of the two-dimensional spectral visibility island diagram (Sect. 65).

An important practical matter is to decide how many azimuths should be chosen and how many spacings. In the case of a bounded object such as the Sun, the two-dimensional sampling theorem shows that it suffices to sample the coherence at intervals of about 50 wavelengths as determined by the diameter of the radio Sun. It is not clear that radial symmetry in the observations is desirable, in fact it leads to redundancy at small spacings; it would be better to move the aerials along two perpendicular lines and arrange for the coherence to be sampled at points of a rectangular grid.

72. Fan beams and strip integration. It has not always been possible, during the rapidly expanding phase of observational radio astronomy, to bring to bear fine pencil beams or two-dimensional interferometric techniques, but important advances have been made with instruments having resolution in only one dimension. This was so with the early interferometric width measurements of solar noise storms and discrete sources. A long but narrow aperture is equivalent to an interferometer used at all spacings up to the maximum aperture but is speedier since it obtains the equivalent information in the one scan over the Sun. Nevertheless it is a highly deficient instrument since it receives on a knife-edge or fan beam and therefore confuses separate sources lying along the strip of sky lying in the beam. The purpose of this section is to show the limitations and possibilities of fan beams. They have been of undoubted importance and are likely to remain so, for at any stage of progress it is more feasible to improve resolution in one dimension than two, and experience has shown that new discoveries regarding detailed structure of emitting sources result from such partial improvement, though the effort of observation and reduction may then be too great to be suitable for the following phases of investigation.

Let $T(x, y)$ be a true distribution of brightness temperature and let a long and very narrow aerial receive radiation from a strip lying along the line

$$x \cos \vartheta + y \sin \vartheta - R = 0.$$

Then we distinguish between line integrating and strip integrating according as the strip is of infinitesimal width or is widened out with a profile A corresponding to the long though finite dimension of the aerial. The line integral of $T(x, y)$ is defined by

$$T_L(R, \vartheta) = \iint T(x, y)\, \delta(x \cos \vartheta + y \sin \vartheta - R)\, dx\, dy$$

[1] M. RYLE: Nature, Lond. **180**, 110 (1957).

and the strip integral by

$$T_S(R, \vartheta) = \iint T(x, y)\, A(x \cos \vartheta + y \sin \vartheta - R)\, dx\, dy.$$

If we have $T_L(R, \vartheta)$ for $\vartheta = \vartheta_1$ only, an important case in practice, the integral equation is not soluble; for if $T_1(x, y)$ satisfies the line integral equation when $\vartheta = \vartheta_1$, so also does $T_1(x, y) + T_2(x, y)$, where $T_2(x, y)$ is an invisible distribution, that is one such that

$$\iint T_2(x, y)\, \delta(x \cos \vartheta + y \sin \vartheta_1 - R)\, dx\, dy = 0.$$

If, however, in addition to $T_L(R, \vartheta_1)$ we have other information about the source distribution, then further progress may be made. For example if it is known that the source has circular symmetry then the problem may be fully solved, as explained below.

Now the strip integral equation will suffer from invisible distributions of the kind discussed in earlier sections, for $T_S(R, \vartheta_1)$ will differ from $T_L(R, \vartheta_1)$ by the absence of Fourier components beyond a cut-off set by the profile A. Hence $T_S(R, \vartheta)$, regarded as a two-dimensional distribution in the xy-plane, will have a Fourier transform in the uv-plane which does not extend beyond a central circular island, points on whose perimeter represent spatial frequencies equal to the cut-off value. It is thus sufficient to study the line integration problem here, i.e. the confusion of information, and to study the question of the loss of information and accompanying effects due to the cut-off in later sections on restoration.

In the case of a source with circular symmetry which has been scanned in one direction $\vartheta_1 = 0$ with a line beam, we have, writing $T_L(x)$ for $T_L(R, 0)$,

$$T_L(x) = \iint T(\sqrt{x^2 + y^2})\, \delta(x - R)\, dx\, dy$$

$$= 2 \int_{r=x}^{\infty} T(r)\, dy,$$

where

$$r^2 = x^2 + y^2,$$

and so

$$T_L(x) = 2 \int_x^{\infty} \frac{T(r)\, r\, dr}{\sqrt{r^2 - x^2}}.$$

Substituting $\xi = x^2$ and $\varrho = r^2$ and writing $T_L(x) = \hat{T}_L(x^2)$, $T(r) = \hat{T}(r^2)$,

$$\hat{T}_L(\xi) = \int_{-\infty}^{\infty} K(\xi - \varrho)\, \hat{T}(\varrho)\, d\varrho,$$

or

$$\hat{T}_L = K * \hat{T}$$

where

$$K(\xi) = \begin{cases} (-\xi)^{-\frac{1}{2}} & (\xi < 0), \\ 0 & (\xi \geq 0). \end{cases}$$

In this form as a convolution integral the equation is soluble, but the solution, $\pi \hat{T} = -K * \hat{T}_L'$, is not as convenient numerically as the direct inversion of the process $K * \hat{T}$, starting from the outlying parts where $\hat{T}$ is zero and working inwards.

In graphical terms (Fig. 54) one evaluates the line integral along AB over the shaded outer area where T is known, and deduces the next inner value from the correction necessary to give the observed full integral. A graphical generalization to the case of no symmetry is clearly possible, and has been discussed by BRACEWELL[1] in a paper which also presents a universal table of coefficients for the rapid inversion from $\hat{T}_L$ to $\hat{T}$.

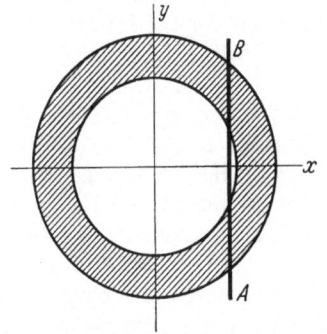

Fig. 54. Illustrating the inversion of line integration.

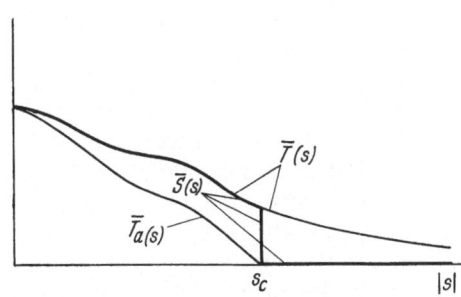

Fig. 55. The relation between $\overline{T}_a(s)$ and $\overline{T}(s)$.

c) Restoration.

73. The principal solution. It was shown in Sect. 53 that the normalized response of an aerial to a sinusoidal component of brightness falls with increasing spatial frequency s from unity, for a uniform distribution, to zero for a spatial frequency s_c cycles per radian, where the extent of the aerial is s_c wavelengths. Fig. 55 shows the relation between the transform of the observed distribution $\overline{T}_a(s)$ and that of the true distribution $\overline{T}(s)$. For the purposes of the figure, a one dimensional case has been illustrated in which $\overline{T}_a$ and $\overline{T}$ are real and even functions of s.

In Sect. 51 it was shown that there are an infinite number of solutions of the equation

$$T_a = A * T;$$

however, T_a itself is not one of them, and the question arises what to do to find a distribution which could have given rise to the observed T_a.

If there is no knowledge of $T(x)$ other than that contained in $T_a(x)$, the most that can be done is to restore to their full value those (low-frequency) components of T which, while present in T_a, have been reduced in amplitude. This gives a unique result, which is a solution of the equation and which is called the principal solution, $S(x)$. It may be defined as that solution whose transform is the same as the transform of the true distribution at all values of s for which $\overline{A}(s) \neq 0$, and zero elsewhere, that is,

$$\overline{S}(s) = \begin{cases} \dfrac{\overline{T}_a(s)}{\overline{A}(s)}, & (\overline{A}(s) \neq 0) \\ 0, & (\overline{A}(s) = 0). \end{cases}$$

The principal solution has the distinction that, among all the approximate solutions not containing components beyond s_c, it is the best least-mean-squares fit to the true distribution. For if $S(x) + H(x)$ is one of the said approximations

---

[1] R.N. BRACEWELL: Austral. J. Phys. **9**, 198 (1956).

then

$$\int_{-\infty}^{\infty} |T(x) - \{S(x) + H(x)\}|^2 \, dx = \int_{-\infty}^{\infty} |\overline{T} - (\overline{S} + \overline{H})|^2 \, ds = \int_{-s_c}^{s_c} |\overline{H}|^2 \, ds + \text{const},$$

which is a minimum when $\overline{H} = H = 0$.

When there are no errors, the information contained in S is exactly the same as that given by an interferometer consisting of two isotropic aerials used at all spacings from zero up to the full aerial width. However the interferometer gives the Fourier components of T directly at their full value, whereas an aerial consisting of a single aperture weights the components so that they must subsequently be restored. As components near the cut-off frequency receive little weight the error spectrum near the cut off is important and it has been felt that the application of a large compensating factor would be deleterious. Whilst this is true it does not follow that the variable-spacing interferometer gives a superior result, for the low weight assigned to the extreme components as received by the single aperture is that weight which is assigned evenly to the components observed interferometrically. The single aperture emphasizes the components of low spatial frequency, and restoration may be regarded as de-emphasis of the superabundant. This will be clear if a comparison is made between a Christiansen array of large paraboloids and a two element interferometer using a pair of the same paraboloids. But in this case the spectral sensitivity function does not fall uniformly away to zero, and the compensating factor does not become infinite. The case of a uniform aperture may appear to be different but here also a comparison may be made with the aerials proposed for the interferometer. Since the effective area of an aerial cannot fall below $\frac{1}{8}\lambda^2$, an apparently uniform aperture is equivalent to an array of elementary dipoles spaced 8 per square wavelength. Its spectral sensitivity function does not, strictly speaking, descend regularly to zero, and the compensating factor should never exceed the limit set by this consideration. Even so, full restoration may be deleterious, if a single dipole is insufficient on its own. It then leads to spurious detail to de-emphasize the components of low spatial frequency to the same state. The problem of optimum restoration in the presence of errors has been discussed[1]; the best spectral restoring factor, when there is no correlation between the error distribution and T_a, is

$$\frac{1}{\overline{A}\left[1 + \dfrac{\langle \overline{E}\,\overline{E}{}^* \rangle}{\overline{T}_o\, \overline{T}_a^*}\right]}$$

where $\langle \overline{E}\,\overline{E}{}^* \rangle$ is the ensemble average squared modulus of the error distribution.

The degree of approximation of the principal solution to the true distribution depends markedly on the form of T. Thus when T contains no spectral components at those frequencies where $\overline{A}(s) = 0$, S is identical with T, and when T has components only where $\overline{A}(s)$ is not very different from unity, restoration may be carried out with confidence. However, when the spectrum of T is still appreciable at the cut-off frequency s_c, the resulting discontinuity in the spectrum of the principal solution can cause spurious oscillations in S. This may result in S assuming negative values which in radio astronomy is impossible.

There are various ways of dealing with the discontinuity. In an analogous problem in X-ray crystallography[2] it is smoothed out arbitrarily. A little cautious extrapolation may be resorted to. A better approach seems to be to seek

[1] R. N. Bracewell: Proc. Inst. Radio Engrs. 46, 106 (1958).
[2] J. Waser and V. Schomaker: Rev. Mod. Phys. 25, 671 (1953).

some physical theory which predicts a form for T, or to adopt some multi-parameter representation which is physically acceptable, and then evaluate the parameters by use of the finite number of independent values of $\overline{T}$. This is an extension of the procedures for deducing the extent of a small source from a measurement of one value of $\overline{T}$, where a one-parameter family of shapes was fitted to the observation.

74. The method of successive substitutions. We may now consider actual methods of restoration in the light of the above. Suppose that $T_{\rm app}$ is some approximation to T. Then it can be tested by scanning with the aerial pattern and comparing with T_a. The discrepancy $T_a - A * T_{\rm app}$ is taken as a first estimate of the difference between $T_{\rm app}$ and T. This leads to a further approximation

$$T_{\rm app} + (T_a - A * T_{\rm app}).$$

If T_a itself is taken as the initial approximation to T, one obtains as the first approximate restoration

$$T_1 = T_a + (T_a - A * T_a).$$

By applying the same procedure to T_1 we have a second restoration

$$T_2 = T_1 + (T_a - A * T_1)$$

and so on. In practice the iteration is halted when smoothing the trial distribution with A gives a result agreeing with T_a within the experimental error. The method has been studied in detail by BRACEWELL and ROBERTS[1] who found the condition for convergence of the sequence to be that $|1 - \overline{A}(s)| < 1$ for all s such that $\overline{T}_a(s) \neq 0$, and showed that the limit of the sequence is the principal solution $S(x)$.

This method is feasible in practice and has been often used. However, it is laborious in two dimensions and probably will not be used in current high resolution studies except for special cases. A new approach to cases where there is much intricate detail is to admit the tentative character even of the principal solution and to seek methods of modest accuracy but which are simple to apply.

75. The chord contruction. An example of a simple method is afforded by the chord construction and its generalization to two dimensions[2]. In Fig. 56 the corrected value lies above the observed curve $T_a(x)$ by as much as the curve lies

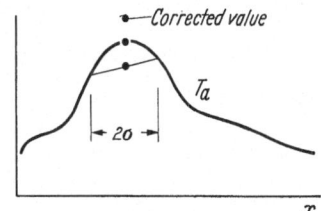

Fig. 56. The chord construction for restoration.

above the midpoint of a chord of constants pan 2σ, centered at the point in question. The quantity σ is the standard deviation of the instrumental profile in general, but in radio astronomy where σ is never finite, the span is determined by a matching technique. In two dimensions it is useful to imagine a plane corresponding to the chord, and the correction at any point, determined from the four neighbouring values, is

$$-\tfrac{1}{4}\,(\varDelta_{xx} + \varDelta_{yy})\,T_a,$$

where $\varDelta_{xx} T_a$ is the second difference of T_a when y is kept constant, and the interval over which the differencing is done is equal to $\sqrt{2}$ standard deviations for

[1] R.N. BRACEWELL and J.A. ROBERTS: Austral. J. Phys. **7**, 615 (1954).
[2] R.N. BRACEWELL: J. Opt. Soc. Amer. **45**, 873 (1955). — Austral. J. Phys. **8**, 54, 200 (1955).

a Gaussian beam. For beams other than Gaussian the correction becomes

$$- (\chi \Delta_{xx} + \psi \Delta_{yy}) \, T_a,$$

where χ, ψ and the differencing intervals α and β are fixed by matching $\chi \sin^2 \pi \alpha u + \psi \sin^2 \pi \beta v$ to $(\overline{A})^{-1} - 1$. The chord construction is equivalent to three stages of successive substitution.

Acknowledgment. This contribution contains material which was developed in connection with research in radio astronomy supported at Stanford University by the Office of Scientific Research of the United States Air Force.

Bibliography.

General.

Two books devoted entirely to radio astronomy, including techniques, are

Brown, R. H. and A. C. B. Lovell: The Exploration of Space by Radio. London 1951.
Pawsey, J. L., and R. N. Bracewell: Radio Astronomy, Oxford 1954.

The Proceedings of the Institute of Radio Engineers, Vol. 46, No. 1, 1958, ed. by F. T. Haddock is devoted to papers on radio astronomy which are very largely concerned with techniques.

Receivers.

Texts concerned with radar receivers are a source of information relative to receivers for radio astronomy. Two very detailed books are

Valley, G. E., and H. Wallman: Vacuum Tube Amplifiers, New York 1948,
Voorhis, S. N. van: Microwave Receivers, New York 1948,

and a more general book with valuable chapters on receiver practice is

Bowen, E. G.: A Textbook of Radar, 2nd ed., Cambridge 1954,

which also contains an excellent chapter on aerials. A rich variety of microwave techniques are presented by E. L. Ginzton, Microwave Measurements. New York 1958.

Aerials.

There is a wide range of literature and it is still increasing. There are no books on aerials for radio astronomy, but among books with chapters relevant to radio astronomy are the following:

Silver, S.: Microwave Antenna Theory and Design. New York 1949. (This book is useful for diffraction theory of apertures and detailed discussion of parabolic reflectors.)
Smith, R. A.: Aerials for Metre and Decimetre Wavelengths. Cambridge 1949.
Fry, D. W., and F. K. Goward: Aerials for Centimetre Wavelengths. Cambridge 1950.
Schelkunoff, S. A., and H. T. Friis: Antennas Theory and Practice. New York 1952.
(An undergraduate text concentrating on a logical development of relevant electromagnetic theory.)
Kraus, J. D.: Antennas. New York 1950. (An undergraduate text with much information on helical aerials.)

Aerial Smoothing.

Much of this subject will be found in volumes 7 to 9 of the Australian Journal of Physics. See **7**, 615 (1954); **8**, 54, 200 (1955); **9**, 198, 297 (1956) and Proc. Inst. Radio Engrs. **46**, 106 (1958).

Interferometers.

There are no books on radio interferometers, which so far have been discussed only by writers on radio astronomy. See Brown and Lovell, Pawsey and Bracewell, M. Ryle: Proc. Roy. Soc. Lond., Ser. A **211**, 351 (1952) and R. N. Bracewell: Proc. Inst. Radio Engrs. **46**, 97 (1958).

For early instrumental papers in the Russian language see V. V. Vitkevich: Dokl. Akad. Nauk. SSSR. **86**, 39 (1952); **91**, 1301 (1953); **102**, 469 (1955); Astronom. Zhurn. 29, 450 (1952); (with R. L. Sorochenko) **30**, 631 (1953); **34**, 217 (1957), and Transactions of the Fifth Conference on Questions of Cosmogony, Moscow 1956.

Diffraction theory.

A good deal of the theory underlying the observational techniques of radio astronomy applies also in other fields. For related reading in optics see

MARÉCHAL, A.: Handbuch der Physik, Bd. XXIV, S. 44, 1956.
FRANÇON, M.: Handbuch der Physik, Bd. XXIV, S. 171, 1956.
BORN, M., and E. WOLF: Principles of Optics, London 1958,
WOLF, E.: Proc. Roy. Soc. Lond., Ser. A **225**, 96 (1954); **230**, 246 (1955),
and in ionospheric physics,
RATCLIFFE, J.A.: Rep. Progr. Phys. **19**, 188 (1956).

Information theory, noise, and Fourier theory.

An extensive body of theory exists which continually arises in connection with all the subjects of this chapter. For an interesting combination of topics of this general kind see

WOODWARD, P.M.: Probability and Information Theory, with Applications to Radar. London 1953.
BRILLOUIN, L.: Science and Information Theory. New York 1956.

The classical papers on fluctuation theory from the standpoint of electric circuits are those of RICE, especially those from volumes 23 and 24 of the Bell System Technical Journal which may be found reprinted in

WAX, N.: Noise and Stochastic Processes. New York 1954.

Bibliography.

The literature of radio astronomy is widely scattered in journals devoted to physics, astronomy, and electrical engineering. Adequate abstracting is provided by Electronic and Radio Engineer Abstracts, which are reprinted monthly in the Proceedings of the Institute of Radio Engineers, especially in the section on Geophysics and Extraterrestrial Phenomena, but the sections devoted to aerials, wave propagation, reception, and others, are also highly relevant. The astronomy section of Science Abstracts is another source. A very competent bibliography prepared by MARTHA STAHR CARPENTER has also been issued at intervals by Cornell University.

Photographic Photometry.

By

HAROLD WEAVER.

With 29 Figures.

I. Preliminary remarks.

1. Objectives. The introduction of photography into astronomical observing techniques was, to the advance of astronomy, fully as important as the introduction of the telescope itself. Photography, as it has developed, has provided the astronomer with a means of recording permanently and accurately objects fainter than he can see and in colors to which his eyes are insensitive. It has permitted him effectively to increase his time at the telescope by recording simultaneously all features of the area observed. Later, in the laboratory, positions of the objects recorded on the photograph can be measured with relative ease; brightnesses of stars and extended objects can be derived over a wide range of intensity. It is to this latter photometric aspect of photography that the present article will be devoted. The discussion will cover certain problems of photographic determination of magnitudes and colors of stars from focal images. The point of view will be operational rather than historical. For a review of the historical aspects of the subject see WEAVER [1], where many additional references will be found.

The topics treated have been dictated in large part by the fact that, as photoelectric techniques have become more widely applied during the past decade, some aspects of photographic photometry have become less important than they were formerly. To these now less immediately useful aspects of the subject only brief attention will be given.

II. Observing materials.

2. Photographic materials. Presently available commercial photographic plates permit observations in a wide variety of spectral ranges and on a wide variety of emulsion types. As examples of the great choice of wavelength ranges available to the observer, spectral sensitivity curves of Eastman Spectroscopic Plates [2] are reproduced in Fig. 1. These spectral sensitivities (designated by letter) are available in various emulsion types (designated by number) which differ in their characteristics of contrast, granularity, resolving power, speed, and so forth (see Table 1).

The contrast of the various spectroscopic emulsions is illustrated by the collection of characteristic curves shown in Fig. 2. High contrast is not necessary for accurate photometry; fast emulsions of lower contrast may advantageously be used.

Granularity characteristics of several emulsion types are displayed in Fig. 3. The microphotometer tracings shown were made on plates uniformly darkened to density 0.3. An analyzing slit 0.5 mm long and 5 microns wide (as projected on the plate) was used in the tracing.

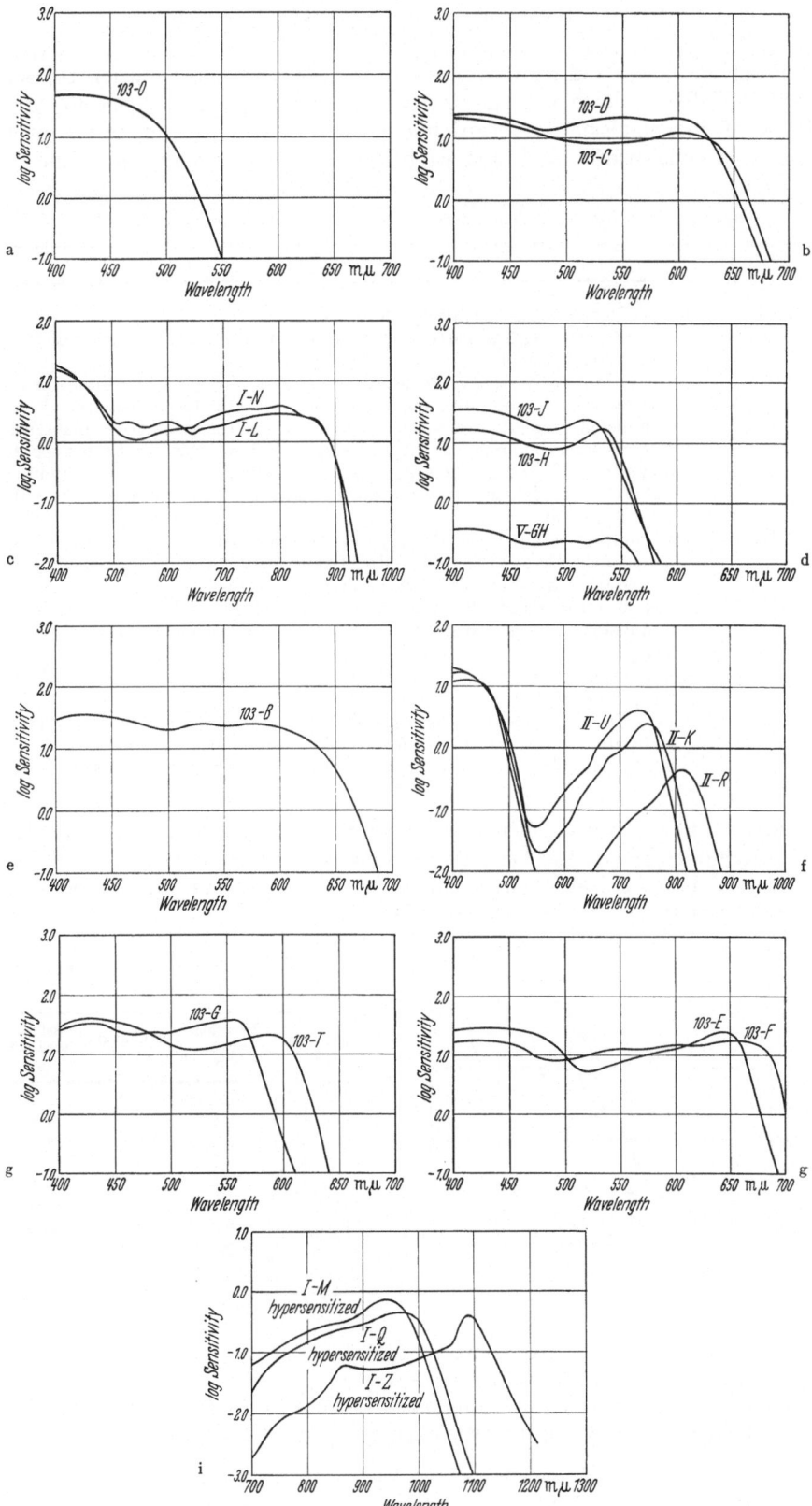

Fig. 1 a—i. Spectral sensitivity curves for Kodak Spectroscopic Plates (from Ref. 2).

Resolving power and relative sensitivity, S_A, for various emulsion types are listed in Table 2, which is from Ref. [2]. Resolving power, a measure of the ability of the emulsion to record fine detail, is specified in terms of the number of equal-width black and white (optical image contrast 20:1) lines per

Table 1. *Some Eastman Kodak emulsion types.* (Data compiled from Ref. [2].)

Emulsion type	Characteristics
I (visual)	An emulsion of high sensitivity and contrast.
I (infrared)	An emulsion of high sensitivity, high contrast, and slightly finer granularity than type I (visual).
103	An emulsion with sensitivity approaching that of type I emulsion, medium contrast, and granularity similar to that of type II emulsion.
103a	An emulsion of slightly lower sensitivity than type 103 to light of high intensity, but considerably more sensitive to light of low intensity than type 103. Type 103a is recommended for exposure times longer than two to five minutes.
II	An emulsion of high sensitivity and high contrast, with finer granularity than type I (visual).
IIa	An emulsion having considerably higher sensitivity than type II for long exposures to light of low intensity. Granularity is similar to that of type II emulsion. Available only with 0 and J sensitizings.
III	An emulsion of moderate sensitivity, high contrast, and low granularity.
IV	An emulsion of lower sensitivity than type III, high contrast, and low granularity.
V	An emulsion of lower sensitivity than type IV, very high contrast, and very low granularity.
33	An emulsion long used by astronomers; about 15% as sensitive as 103a−0, and half as sensitive as I−0 for astronomical exposures, sensitization equivalent to 0, high contrast, granularity similar to type II emulsion.

millimeter that are just resolvable visually under adequate magnification. Tabulated values represent the best results obtainable under optimum conditions. In general astronomical practice smaller resolving powers will be obtained. Relative sensitivity, S_A, is defined following a criterion suggested by DUNHAM[1].

Table 2. *Resolving power and relative sensitivity of some Eastman emulsions.* (Data compiled from Ref. [2].)

Emulsion type	Resolving power (lines/mm)	Emulsion type	Relative sensitivity (S_A)
103–E	60	103–0	5.0
		103a–0	11.0
I–0	60	I–0	3.2
II–C	75	II–0	2.4
		IIa–0	8.9
III–C	95	III–0	0.63
IV–C	120	IV–0	0.42
V–C	160	V–0	0.006
33	80	33	1.6

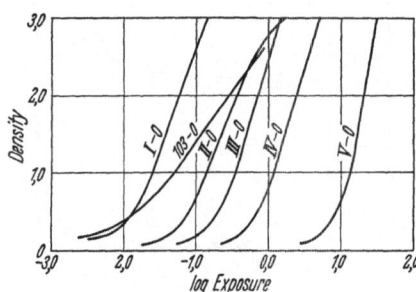

Fig. 2. Characteristic curves for Kodak Spectroscopic Plates (from Ref. 2).

It is the reciprocal of the exposure to tungsten radiation, expressed in meter-candleseconds, which will produce a density of 0.6 above gross fog when the emulsion is developed for the recommended time in Kodak Developer D-19. The tabulated relative sensitivities refer to the astronomical exposure time

[1] T. DUNHAM: Ber. VIII. Internat. Kongr. Photog., p. 287. Leipzig: Johann Ambrosius Barth 1932.

range, 10^4 sec; for another exposure range, for example, 10^{-2}sec, the values would differ greatly from those quoted.

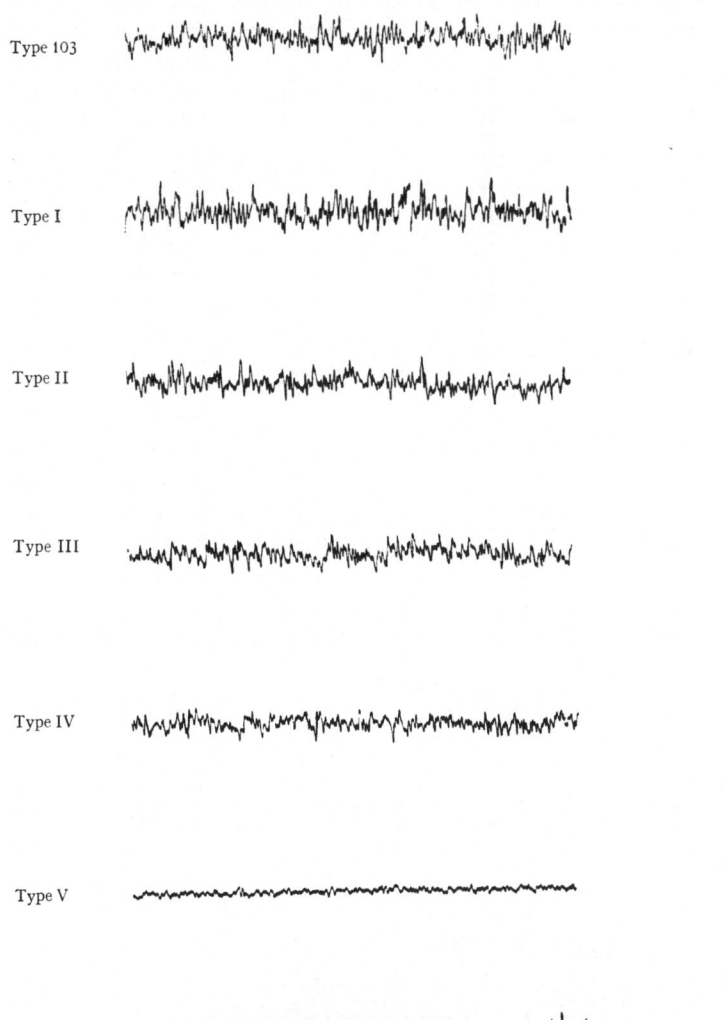

Fig. 3. Granularity characteristics of Kodak Spectroscopic Plates (from Ref. 2).

3. Photographic effects. The complex nature of the photographic process gives rise in the emulsion to a large variety of resultant physical properties or photographic "effects". Knowledge of some of these is basic to an understanding of the principles of photographic photometry.

Adjacency effects arise during development and exposure. In the region between the images of a close double star the developer becomes exhausted and reaction products accumulate to a greater degree than elsewhere. Development is retarded locally. The images develop asymmetrically and abnormally (Kostinsky effect). On the other hand, during the exposure, light from the two stars scattered by the emulsion adds in the region between the star images. This results in greater-than-normal blackening in the region between the images; the images appear drawn together (turbidity effect). These localized effects are difficult to allow for; they cause inaccuracies in the photometry of closely adjacent star images.

Particularly in extrafocal photometry (and in spectral photometry) adjacency effects are present at the image boundary. Developer reaction products diffuse from the dark image into the adjacent light area and retard development. Fresh developer from the light region diffuses into the dark image. It penetrates a short distance, gradually becoming chemically exhausted. The uniformly exposed area of the extrafocal image develops a high intensity rim (border effect) surrounded by an abnormally light area (fringe effect). Fig. 4, exhibiting microphotometer tracings of density variation

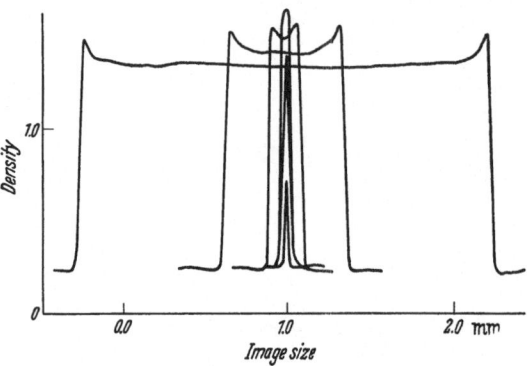

Fig. 4. Edge effects (from Ref. 2).

across narrow uniformly exposed images indicates that some caution is required in the interpretation of density measurements made in extrafocal images.

A photographic effect of prime importance is reciprocity failure, discovered by SCHEINER[1]. He observed that increasing exposure time by $\sqrt[5]{100}$ did not produce in the star images an increase corresponding to one magnitude. SCHWARZSCHILD[2], working in the range of stellar light intensities requiring exposures of a few minutes to a few hours, found that a constant photographic effect is produced so long as exposure time, t, and intensity, I, satisfy the relation

$$I t^p = \text{const.} \quad (3.1)$$

In this expression p, the Schwarzschild exponent, has a numerical value <1. KRON[3], in the first detailed study of the relation between intensity and exposure time, displayed his conclusions in the form of a graph,

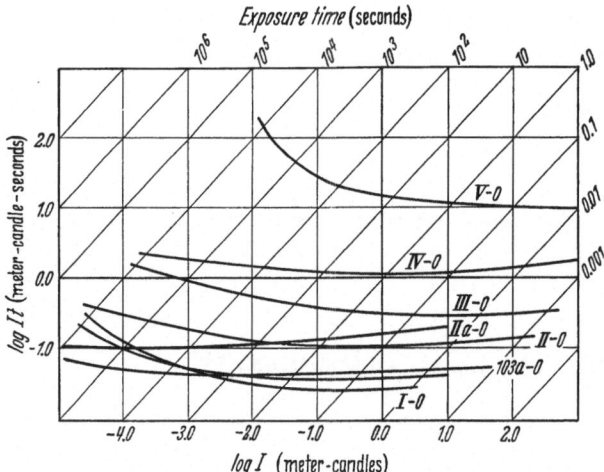

Fig. 5. Reciprocity effects for Kodak Spectroscopic Plates (from Ref. 2).

Fig. 5, showing the relation between $\log I$ and $\log It$ required to produce a constant specified photographic density. KRON's work demonstrated that there exists some optimum light intensity, I_0, which produces the specified photo-

[1] J. SCHEINER: Bull. du Comité Permanent International pour l'Exécution Photographique de la Carte du Ciel 1, 227 (1889).

[2] K. SCHWARZSCHILD: Astrophys. Journ. 11, 89 (1900). — Photogr. Korresp. 36, 109 (1899). — Publ. v. Kuffnersche Sternw., Bd. V, 1899.

[3] E. KRON: Publ. Astrophys. Obs. Potsdam 22, 1 (1913).

graphic density with the least expenditure of energy, that is, for which exposure time is a minimum, t_0. KRON found, empirically, that he could represent the relation between $\log I$ and $\log It$ by the hyperbolic equation

$$\log It = \text{const} + a \sqrt{\left(\log \frac{I}{I_0}\right)^2 + 1}, \tag{3.2}$$

where a is a shape factor for the curve. Later, HALM[1], applying KRON's results in stellar photometry, used the equation of a catenary, which KRON had also considered, to represent the relation between t and It:

$$It = \frac{I_0 t_0}{2} \left[\left(\frac{I}{I_0}\right)^a + \left(\frac{I}{I_0}\right)^{-a}\right]. \tag{3.3}$$

For low intensities (3.2) and (3.3) reduce to the simpler form

$$\log It = \text{const} - a \log I,$$

or

$$It^{\frac{1}{1+a}} = \text{const}. \tag{3.4}$$

If we put $1/(1+a) = p$, we recover the Schwarzschild equation, which is a special case of the more general Kron relation. That SCHWARZSCHILD found $p < 1$ is due

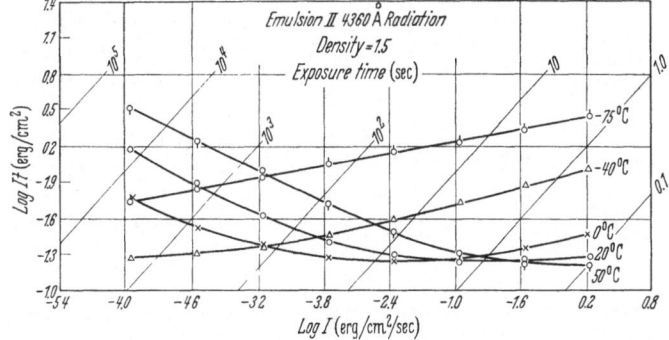

Fig. 6. Variation in reciprocity effect as a function of temperature. Results are for a high-speed, blue-sensitive, negative emulsion.

only to the fact that he worked with small values of I. If he had worked with large values of I, he would have found $p \geq 1$. Eqs. (3.2) or (3.3) are to be preferred to Eq. (3.1) in describing the I, It relation.

The form of the I, It relation expressing reciprocity failure for a specific emulsion can be made to vary in several ways. The values of I_0 and t_0, for example, are sensitive to the last stages of heat treatment in emulsion manufacture. An emulsion produced with a finishing treatment that places I_0 in the region of small I-values and gives t_0 a minimum value would be especially good for astronomical work. On the other hand, a finishing treatment placing I_0 among the higher I-values would be more satisfactory for short exposures to high intensities. Among the Eastman emulsions, those especially treated for astronomical applications (low light intensities) are designated by the letter "a".

The astronomical observer can, in some instances, vary the form of the I, It relation to his advantage by baking the plates, hypersensitizing the plates with mercury vapor, bathing them in a solution of ammonium hydroxide, and so forth. The I, It relation is affected by the temperature of the emulsion at the time of exposure. Different emulsion types show different degrees of temperature sensitivity. Fig. 6, taken from WEBB[2], illustrates the temperature effect,

[1] J. HALM: Monthly Notices Roy. Astronom. Soc. London **84**, 472 (1922).
[2] J. H. WEBB: J. Opt. Soc. Amer. **25**, 4 (1935).

which could have important astronomical applications. For the emulsion for which the data are illustrated in Fig. 6, an exposure requiring six hours if exposed at the summer temperature of $+20°$ C, would require less than 40 min if the emulsion were exposed at a temperature of $-40°$ C.

Wavelength of incident light appears to have little effect upon reciprocity failure. If the intensities of radiation in to different wavelengths are so adjusted that they produce equal photographic densities in equal times, the I, It curves for the two wavelengths will coincide.

WEBB and EVANS[1] have performed an experiment to investigate reciprocity failure at low intensities that is of direct importance to an understanding of the problems of photographic photometry. They gave an emulsion a series of exposures of equal energy, using light of two intensities. The high light intensity (abbreviated H.I.) was nearly optimal; the lower intensity (L.I.) was less by a factor of about one thousand. Each exposure was produced partly by H.I. light, partly by L.I. light. The exposure to the H.I. source could either precede or follow the exposure to the L.I. source; the fraction of the exposure, E, put on in the form if, say, H.I. light could be varied from 0 to 1. For any exposure

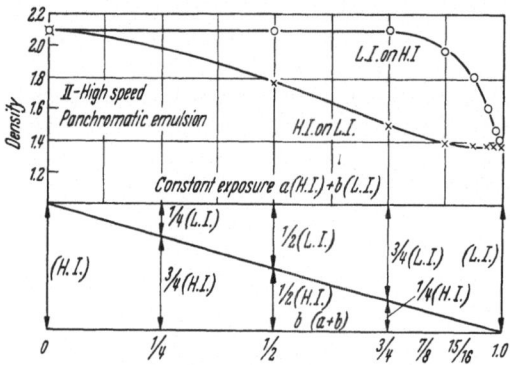

Fig. 7. Effect of equal energy, mixture exposures of part high intensity (H.I.) and part low intensity (L.I.) given in different time order.

$$E = a\,(\text{H.I.}) + b\,(\text{L.I.}); \left.\begin{array}{c} \\ a + b = 1, \end{array}\right\} \quad (3.5)$$

where a represents the fraction of E put on as H.I. light and b represents the fraction of E put on as L.I. light. The results of the experiment are exhibited in Fig. 7. The photographic density resulting from E is strongly dependent upon the order of exposure to the two sources. Exposure first to the H.I. source makes the subsequent exposure to the L.I. source much more effective than would have been the case if the L.I. source were used alone. In fact, for the emulsion tested, if as much as $\frac{1}{4}$ of the total exposure is given first as H.I. light, the $\frac{3}{4}$ of the exposure given as L.I. light is essentially as effective as though it were H.I. light. Reversal of the order of exposure, $\frac{3}{4}$ L.I. $+\frac{1}{4}$ H.I., produces a very pronounced decrease in the effectiveness of the L.I. light.

4. Physical explanation. The complex effects involving the action of light on the photographic emulsion can be explained in terms of the basic Gurney-Mott two-stage mechanism of photochemical change.

1. Light falling on a silver halide crystal frees electrons from the crystal lattice. The freed electrons and the resultant positive holes in the crystal lattice move within the crystal. The electrons and the positive holes tend separately to collect in the vicinity of certain traps within the crystal.

2. The redistribution of charge resulting from such preferential trapping initiates motion of silver ions from the vicinity of trapped holes. The ions move to the trapped electrons; silver atoms are formed. These react with the developer to produce the image.

---

[1] J. H. WEBB and C. H. EVANS J. Opt. Soc. Amer.: **28**, 431 (1938).

The efficiency of the radiation in forming the latent image is controled by the rate at which the electrons are freed, by the relative mobilities of the electrons, positive holes, and ions, by the efficiency of the trapping mechanisms, and by the stability of the clump of neutral silver atoms. For details of the process, see BERG and WEBB [1], or especially, MITCHELL [2].

The Webb and Evans experiment with H.I. and L.I. light can be interpreted as indicating that the clump of silver atoms formed by the migration of silver ions to the trapped electrons is unstable in its early stages of formation. Exposure of the emulsion to L.I. light produces few free electrons; the silver clump forms slowly. When the speck or clump exists only as a few silver atoms, it can easily be broken up. The mechism of break up of the speck may be assumed to be thermal ejection of electrons from the silver atoms and subsequent diffusion away of the reformed silver ions. Other mechanisms may be involved (see BERG and WEBB[1]).

L.I. light is ineffective in forming stable clumps of developable silver atoms; reciprocity failure for L.I. light is therefore large. On the other hand, if the emulsion is first given a short exposure to H.I. light so that the silver atoms form quickly enough to produce a stable clump, silver atoms later produced by L.I. light are added to, and remain in, this stable speck, which can grow into a development center. Feeding silver atoms into the stabilized silver speck formed by the H.I. light, the L.I. light attains its highest efficiency.

The process of thermal ejection of electrons from the silver atoms in the silver clump offered as the explanation of reciprocity failure at low intensities also provides an explanation of the suppression of low-intensity reciprocity failure as the temperature of the emulsion is decreased. Decrease in the emulsion temperature diminishes the rate of thermal ejection of electrons, which decreases the tendency of the slowly forming clump to disintegrate. Continued decrease in the emulsion temperature to very low values diminishes the mobility of the silver ions; this impedes the formation of silver atoms. Reciprocity failure is suppressed, but, at the same time, emulsion sensitivity is seriously decreased at extremely low temperatures as indicated in Fig. 6. For each emulsion there must exist a particular temperature which, for a given light intensity[3], will optimize the production of silver clumps.

Reciprocity failure at high intensity can be traced to failure in the second stage of the Gurney-Mott mechanism; that is, to insufficient ion mobility. The silver ions cannot move so fast as the electrons, hence cannot immediately neutralize the trapped electrons produced in very large numbers by high intensity radiation. A negative charge therefore tends to build up in the vicinity of the electron trap. The resultant repulsive force experienced by other approaching electrons reduces the electron trapping rate to that at which ionic mobility permits charge neutralization to take place. The energy in the high intensity radiation falling on the silver halide crystal and going into the production of free electrons is not so efficiently converted into latent image as it would be if the electrons were freed at a slower rate. Compared to the result obtained from radiation of intensity such that it produces electrons at the optimum rate there is, for such high intensity radiation, reciprocity failure.

[1] W. F. BERG and J. H. WEBB: In C. E. K. MEES, The Theory of the Photographic Process, Chap. 4. New York: Macmillan Company 1954.

[2] J. W. MITCHELL: Rep. Progr. Phys. **20**, 433 (1957).

[3] Specification of a light intensity on the emulsion implies specification of the rate of production of electrons.

III. Measuring instruments.

5. Image scales. To make visual photometric measurement of a focal image of a star, we may use a scale of graded focal images photographed with the same telescope as the plates to be measured. We assume that for the scale plate and the plates to be measured observing conditions (seeing, focus, guiding, ...) as well as photographic conditions (plate type, filter, development, ...) were identical. The scale is made by taking a series of exposures of duration t sec, kt sec, k^2t sec,..., the telescope being moved between exposures. The numerical value of the parameter k is generally in the range 1.4 to 2. From among the many image series photographed on the plate (one for each star in the field) we choose some particular series that is free of interference from other image scales and that covers the range of image sizes we wish to measure on the plates. We designate the largest image as 1, the next as 2, and so forth, and estimate where each image to be measured lies in this arbitrary but presumably uniform scale. To make the estimates we place the scale and the plate being measured emulsion to emulsion and view them simultaneously with a magnifying glass or we view the two by means of an optical system that permits us simultaneously to see the scale and the plate.

If accuracy of estimate is to be achieved,

(i) the images of the scale and plate must match in quality;

(ii) the images in the scale must be regularly spaced and in a straight line since irregularities in spacing produce irregularities in the estimates; and

(iii) the ratios of successive exposures must be accurately k. Non-uniformities in the scale must be determined and removed from the estimates before they are used. Such correction complicates the reduction process.

6. Other visual methods. Image diameters measured with a standard comparator or with a movable wire micrometer provide a photometric measure. A pair of converging lines of small angular separation ruled on a clear photographic plate provides a very simple means of estimating image size. With this device one estimates image size by specifying, with respect to a uniform linear scale marked along the lines, at what point the star image just fits between the converging lines.

7. Physical photometers. More frequently in present-day photometry, measurements are made with a physical photometer of the STETSON[1], SCHILT[2], or ROSS[3] type, which employs a diaphragm of fixed size, or of the SIEDENTOPF[4] type, which employs an iris diaphragm in the measuring beam.

The measurement principle of the fixed-aperture photometer is very direct. The star image is centered in a small-aperture light beam of fixed size. The percentage of light in the beam that passes around and through the image provides the photometric measure of the star image.

The iris-diaphragm photometer is a null-type instrument. The star image is centered in a light beam of variable aperture. The aperture of the beam is adjusted until a specified amount of light passes around and through the image. The aperture of the beam when so adjusted serves as the photometric measure. Effectively, the device measures image diameters, with image density changes playing a secondary role for the brighter stars, an increasingly important role for the very faint stars.

---

[1] H. T. STETSON: Popular Astron. **23**, 23 (1914). — Astrophys. Journ. **43**, 253 (1916).

[2] J. SCHILT: Bull. Astronom. Inst. Netherl. **1**, 51 (1922). — Groningen Astronom. Lab. Publ. **1924**, No. 32. — Bull. Astronom. Inst. Netherl. **2**, 135 (1934).

[3] F. E. ROSS: Astrophys. Journ. **84**, 241 (1936).

[4] H. SIEDENTOPF: Astronom. Nachr. **254**, 33 (1934).

8. The fixed-aperture photometer. The basic design of the fixed-aperture photometer is illustrated in Fig. 8. Light from a constant voltage lamp, L, passes through a collimating lens, C, through an aperture A, which is imaged on the plate, P, by the microscope objective, M_1. The light traverses the plate to the microscope objective, M_2, whence it passes through a field lens, F, to the photomultiplier, PM. The purpose of the field lens is to reduce the effects of any miscentering of the star image in the beam and to spread the light uniformly over a large area of the photocathode in order to minimize the effect of sensitivity irregularities over its surface. The photocell response is shown on a meter, M.

To use the photometer we insert a diaphragm (A) of an aperture such that when it is focused on the plate it just admits the brightest star to be measured. Illumination must be uniform over the area of the spot of light on the plate. A limited range of approximately four or five magnitudes can be measured with

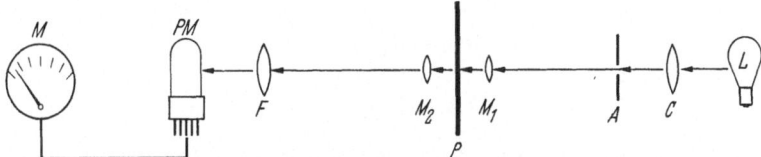

Fig. 8. Schematic fixed-aperture photometer.

one diaphragm. Often a diaphragm that projects into a spot that will just enclose the image of a star approximately five or six magnitudes above the plate limit is employed.

Making a photometric measurement involves the following steps:

1. Move the plate so that the star image to be measured is centered in the measuring beam.

2. Determine the meter deflection, d_*, when the star is so centered.

3. Make settings on the fogged plate immediately adjacent to either side of the star image and determine the average meter deflection, d_f, produced by the beam passing through the fogged plate. Care must be taken to see that the star image is definitely outside the measuring beam when these measurements are made.

The quotient

$$q = \frac{d_*}{d_f} \qquad (8.1)$$

is taken as the photometric measure; it ranges from 0 to 1. The former value corresponds to a completely dark image at least filling the measuring beam; the latter to a vanishingly faint image.

On a plate having no background fog we would obtain for the clear glass adjacent to the star image a meter deflection d_g, a constant depending only upon the intensity of the measuring beam and the sensitivity of the equipment. We would obtain for a star on this hypothetical fogless plate a deflection d_{*g}. The size and blackening of the star image is a function of the magnitude of the star, hence d_{*g}, which depends upon size and blackening of the image, is some function of the magnitude of the star:

$$d_{*g} = f(m). \qquad (8.2)$$

On any real plate background fog will be present; we shall observe a fog deflection $d_f = \varkappa d_g$, where $\varkappa$ is the factor by which the fog diminishes the reading that would have been obtained from a completely clear plate. Similarly, we may write $d_* = \varkappa d_{*g}$. This expression is correct to the extent

(i) that fog merely overlays the image, that is, to the extent that image and fog are additive and without interaction, and

(ii) that the photometer is linear in its response.

The second of these conditions is easily met; modern photometers employing photomultipliers are adequately linear even if no special meter is used. The older photometers that employed a thermopile in place of the photomultiplier were likewise adequately linear. The first condition is less easily met. Experience indicates that the assumption of no fog-image interaction is progressively less well obeyed as plates of increasing fog density are measured.

If we do assumed that the two conditions are fulfilled, then we may write

$$q = \frac{d_*}{d_f} = \frac{\varkappa d_{*g}}{\varkappa d_g} \tag{8.3}$$

from which

$$d_g q = d_{*g} = f(m). \tag{8.4}$$

Invert this to find

$$m = G(d_g q). \tag{8.5}$$

The quantity d_g is, presumably, a constant for the series of measurements made on one plate; for purposes at hand, d_g may be neglected. We shall consider only an m, q relation. Note, however, that we shall find a consistent m, q relation for a plate only so long as the light source and instrumental sensitivity remain constant throughout the measurements.

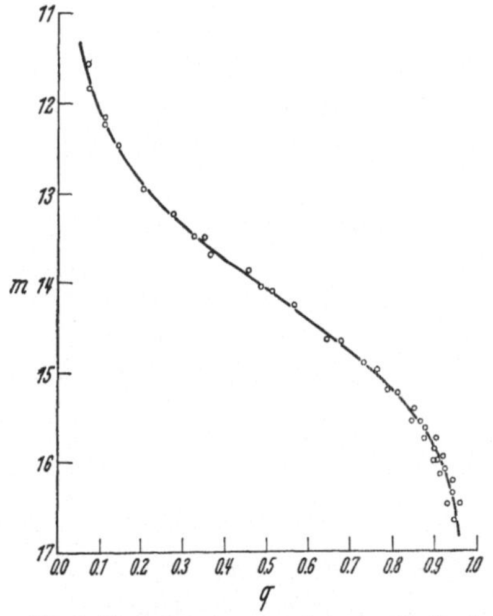

Fig. 9. Typical m, q curve derived with fixed-aperture photometer.

To the degree of approximation that there is no fog-image interaction, the effect of fog variation over the plate is removed through use of q as the photometric measure. If fog-image interaction takes place, the magnitude equation must be written in the more general form

$$m = F(d_g q, \varkappa). \tag{8.6}$$

The way in which $\varkappa$ and $d_g q$ combine cannot be specified. Provided $\varkappa$ is constant over the field in which measurements are made, an m, q relation may still be used. Difficulties are met when two photographs with different $\varkappa$ values are compared and the assumption is made that the m, q relations for the two photographs are the same. The m, q relations will, in fact, differ to the extent that the $\varkappa$'s are different and fog-image interaction has taken place.

Fig. 9 shows a typical m, q curve for a plate measured with a fixed-diaphragm photometer. Very little fog was present on the plate. The curve is approximately linear over a range of two or three magnitudes. In practice, the curve can be used with good results over the q-range 0.1 to 0.95, and can be used even beyond these limits. Uncertainty in the inferred magnitude increases as the q-value is increasingly distant from the linear portion of the curve.

For consistency in a set of measurements, it is important that the plate emulsion be maintained accurately in the focal plane of M_1 (see Fig. 8) as the plate moved is to place different star images in position for measurement. Movement of the plate in and out of the focal plane will change the diameter of the projected spot on the plate and hence the derived q-value. It is readily shown that

$$\Delta q = 2(1 - q)\frac{\Delta f}{f} \tag{8.7}$$

where f is the focal length of M_1, Δf is the distance the plate moves out of focus, q is the q-value for an in-focus measurement, and Δq is the error in q resulting from the defocusing.

9. The iris photometer. The basic principles of the iris photometer are shown in Fig. 10. Light from the constant voltage lamp, L, passes through the collimat-

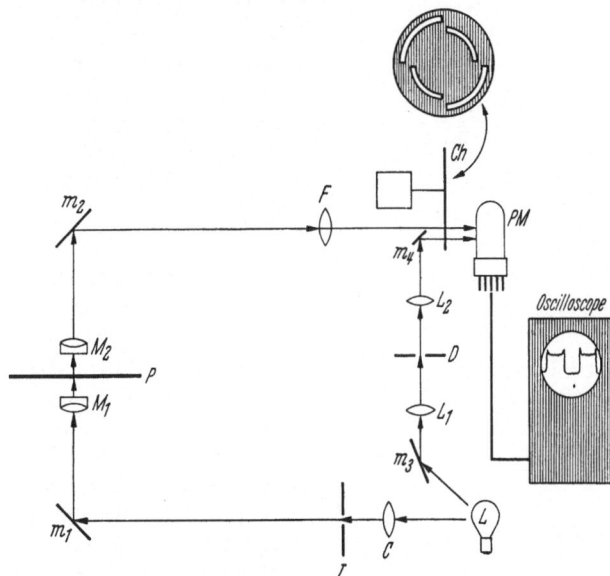

Fig. 10. Schematic iris photometer.

ing lens, C, through the iris diaphragm, I, and is reflected from the mirror, m_1. The iris is imaged on the plate by the objective, M_1. The light passes through the plate, P, to a second objective, M_2, is reflected from the mirror, m_2, and passes through the field lens, F, to the photomultiplier, PM. The field lens serves the same purpose as in the fixed-aperture photometer.

Light from the lamp L_1 is also utilized in a comparison beam, reflecting first from mirror m_3, then passing through lenses L_1 and L_2, and through a diaphragm, D. It is reflected from mirror m_4 into the photomultiplier, PM. In the path of the two beams, one that passes through the iris diaphragm and the plate, the other that comes direct from the lamp, is a chopper, Ch, that permits first one beam then the other to enter the photomultiplier. The output of the photomultiplier as first one beam then the other falls on it is viewed finally on an oscilloscope screen, which permits rapid visual comparison of the intensities in the two beams.

Iris-photometers have been described by EICHNER, HETT, SCHILT, SCHWARZSCHILD, and STERLING[1], HAFFNER[2], CUFFEY[3], and BECKER and BEBER[4].

[1] L. C. EICHNER, J. H. HETT, J. SCHILT, M. SCHWARZSCHILD and H. T. STERLING: Astronom. J. **53**, 25 (1947).
[2] H. HAFFNER: Veröff. Univ.-Sternwarte Göttingen, **1953**, No. 106.
[3] J. CUFFEY: Sky and Telescope **15**, 258 (1956).
[4] W. BECKER and C. BEBER: Z. Astrophys. **41**, 52 (1956).

To make a photometric measurement, the observer first adjusts the intensity of the comparison beam by means of diaphragm D or its equivalent in order to produce the most satisfactory form of reduction curve. The adjustment must be made on the basis of experience with the photometer. Examples of form of reduction curve for different comparison beam intensities have been given by CAMERON[1]. A star is centered in the measuring beam; the size of the iris diaphragm is adjusted so that the measuring beam and comparison beam are of equal intensity. The resultant reading on the iris, which we shall still term q, though it no longer is a quotient of two quantities, is taken as the photometric measure.

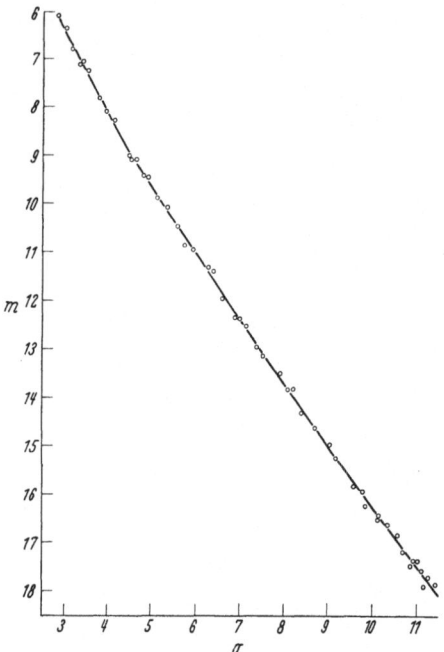

Fig. 11. Typical m, q curve derived with fixed-aperture photometer.

A typical m, q-curve for an iris-diaphragm photometer is shown in Fig. 11. The remarkable feature of the iris photometer is that it produces essentially straight-line reduction curves over a range of very great extent, approximately ten magnitudes in the figure. Such a range is impossible with the fixed-diaphragm photometer.

As the iris photometer is ordinarily used (in the manner described) no allowance is made for fog variation over the plate as it is with the fixed-aperture photometer. This constitutes no serious objection to the process. In a measurement with an iris photometer the measuring beam, when adjusted to proper diameter, cuts into, or approaches very close to, the star image. This is a situation very different from that encountered with a fixed-aperture photometer, where the star image may fill only a small fraction of the area of the measuring beam. Because of the image-diameter-beam-diameter relation for the iris photometer, fog plays an unimportant role except, possibly, for the very faintest images. For such faint images the hypothesis of a superposed fog is rather unrealistic. Fog-image interaction effects are more likely to be encountered. Such interaction effects are not removed by procedures of the type used to determine q with the fixed-aperture photometer. The interaction effects constitute a source of additional error, particularly in the magnitudes of the faintest stars, and cannot be removed.

However, fog corrections can be made to the same extent they were for the fixed-aperture photometer. Let

d represent the diameter of the iris that would be obtained if the star image measured were seen against a "standard" fog background;

D represent the diameter of the iris obtained in the actual measurement of the star, a value that differs from d since the star is seen against a fog background different from the standard;

d_{sf} represent the iris diameter when adjusted to balance the comparison beam, the "standard" background fog being centered in the measuring beam:

---

[1] D. M. CAMERON: Astronom. J. **56**, 92 (1951).

d_I represent the iris diameter when adjusted to balance the comparison beam, the particular background fog (average on either side of the image being measured) being centered in the measuring beam.

One then derives

$$d^2 = D^2 - (d_f^2 - d_{sf}^2). \tag{9.1}$$

If we know d_f, the fog reading next to the image being measured, and d_{sf}, which may be taken as the average for the whole plate, we may then, through Eq. (9.1) reduce the reading D to that value, d, that would have been obtained if there were everyhwere on the plate a uniform background fog.

Eq. (9.1) provides some interesting numerical results on the importance of variations in the background fog. For an image of a very bright star, $D \gg d_I$, while, for a vanishingly faint star that is just visible on the plate, $D \approx d_f$. If we assume a fluctuation of 5% in the plate fog, we find that, for a very bright star, the numerical value of D will differ insignificantly from that of d. On the other hand, for a very faint star whose image is just visible, the numerical value of D will differ from that of d by approximately 5.3%. Even for very faint stars, the error caused by fog variations is small. An error of this size is, nevertheless, detectable. It can be corrected, or at least diminished, through use of Eq. (9.1) if such correction seems desirable.

10. Accuracy of measurement. In principle, one should be able to approach the accuracy of reproducibility of a photographic plate with either of the photometers. For a star image well above the plate limit, a mean error for a single measurement of a star image may be expected to lie in the range 0.03 to 0.04 mag for good quality, large reflector photographs and approximately 0.06 mag for Schmidt films. The size of error increases for fainter stars and may become quite large (≈ 0.1 mag or greater) for the very faintest stars which show as only a few grains of developed silver. In all instances the plate error is much greater than the measuring error. CUFFEY[1] states that, with his iris photometer, for round, sharp, well-exposed images, the mean error of a photometer setting is 0.007 to 0.015 mag; for a very poor image it may rise to 0.024 mag. BECKER and BEBER[2] find similar or slightly smaller mean setting error for their iris photometer.

The greater ease of making reduction curves for the iris photometer and the longer magnitude range over which it permits measurements to be made make it the more desirable type of instrument at the present time.

IV. Some properties of photographic star images.

11. Image factors. Factors contributing to the properties of a star image can be divided into two broad classes: (i) image-form factors; (ii) intensity-chromatic factors.

Examples of image-form factors are: physical properties of the photographic emulsion (aside from spectral sensitivity), focal properties of the camera, mechanical quality and adjustment of the telescope, seeing conditions. These are factors that determine the sharpness and appearance of the image of a specific star at a particular time.

In discussing image-form factors in photometry we shall ignore purely photographic factors since, for the purpose at hand, they are of minor importance. They are readily under the control of the observer. Plates of different emulsion

[1] J. CUFFEY: Sky and Telescope **15**, 258 (1956).
[2] W. BECKER and C. BEBER: Z. Astrophys. **41**, 52 (1956).

properties need not be reduced together in a comparison. Only photographs taken on identical emulsion types developed together will be considered. Generally only one plate, which may contain two exposures, will be employed in such a discussion. Photometric plates should be developed with a high degree of uniformity. If any nonuniformity or irregularity of development is detectable, the plate should not be used for photometric purposes.

Intensity-chromatic factors include spectral sensitivity of the emulsion, the wavelength-dependent transmission of lenses and filters or reflectivity of aluminized or silvered mirrors in the optical system, the extinction of starlight in the Earth's atmosphere. This latter subject will be treated in a separate section in considerable detail.

Intensity-chromatic factors are of very great importance in photometry. Generally, their effects are more subtle than those of the image-form factors. They must be carefully investigated and thoroughly understood by the observer of accurate photometry is to be achieved. In particular, if the observational results are to be in any way compared or combined with other photometric data, accurate matching of the color systems of the two sets of data is important.

In general, change in any image-form or intensity-chromatic factor will be detectable in the properties of the images and will produce a photometric effect.

V. Image-form factors.

12. The optimum image point. Under optimum conditions a camera concentrates the light of a star within a certain minimum area on the photographic emulsion. If conditions deteriorate, the starlight will be less well concentrated than originally; the star images will increase in size. A bright star that produces a large dense image on the emulsion under optimum conditions will produce a still larger dense image on a plate exposed during less-than-optimum conditions. The light of such a bright star is sufficiently intense to affect the emulsion by direct or indirect processes even if the light is attenuated by being spread over a larger-than-normal area. For a faint star the effect is just the opposite. The feeble light of the faint star must remain concentrated on a small area of the emulsion for a long period of time in order to produce a developable image. If the light of a faint star is spread over a larger-than-normal area by image-quality deterioration, the photographic density of the image will be less than it would have been if it had been obtained under optimum conditions. The very faintest stars recorded on a plate taken under optimum conditions will not appear on a plate taken under less-than-optimum conditions. Deterioration of image quality causes the bright stars recorded on a plate to appear brighter, and the faint stars to appear fainter than would have been the case for images obtained under better conditions. There thus exists some image of intermediate brightness that is not affected by deterioration of image quality.

The existence of such an "optimum image point" on the m, q-curve has been recognized by TRUMPLER[1] for many years and has been mentioned by ZUG[2] and others who worked with TRUMPLER. BLAAUW[3] made it the basis of a plate reduction method that was later generalized by WEAVER[4].

Experience has shown that, for moderate changes in image quality of the nature encountered in practice through changes in seeing, focus, and guiding,

[1] R. J. TRUMPLER: Lick Obs. Bull. **18**, 176 (1938).
[2] R. S. ZUG: Lick Obs. Bull. **16**, 121 (1933).
[3] A. BLAAUW: Bull. Astronom. Inst. Netherl. **9**, 141 (1940).
[4] H. F. WEAVER: Astrophys. Journ. **106**, 366 (1947).

the optimum image point occurs about 1.5 mag. above the plate limit, at a constant q-value as measured with a fixed-aperture photometer of specified characteristics. The same should be true for an iris photometer, but this assertion has not yet been tested observationally. From a series of exposures taken with the 8-inch Draper Telescope (focal length 1.26 meters) at the Harvard College Observatory, BLAAUW[1] found for the optimum image point $q_0 = 0.850$. From a series of plates taken with the Mount Wilson 60-inch reflector (focal length 7.60 meters) diaphragmed to 40-inches, WEAVER[2] derived the value $q_0 = 0.822$.

To a first approximation, any change in image quality causes a rotation of the m, q-curve about the optimum image point, q_0. Change in curvature of the m, q-curve traceable to change in the image-form factors appears primarily in the region of large and small q-values as measured with a fixed-aperture photometer.

13. Optical image-form factors. Optical aberrations cause image quality to vary over the field of the telescope. Such aberrations are usually circularly symmetric about the optic axis; image quality depends upon distance from the point at which the optic axis pierces the plate, but not upon position angle. Coma provides an example; it changes regularly from the optic axis outward. A reflector produces images on the optic axis that are smaller and sharper than those at some distance from the optic axis. The m, q-curve for the star images will also be a function of distance, ϱ, from the center of the plate. To a first approximation the change in the m, q-curve will consist of a rotation about q_0, the angle of rotation increasing with distance, ϱ.

In practice, a single m, q-curve will be employed in the reduction of the plate, that is, in the transformation from measured q for any star to the magnitude of the star. The magnitude so derived is in general incorrect since in general an incorrect m, q-curve will have been used. To the derived magnitude of each star there will be a distance-from-center correction, Δm. The correction is given, for a particular q-value, by the difference between the m, q-curve that should have been used on the basis of the star's distance ϱ from the plate center and the single mean m, q-curve that was used. Δm is thus a function of ϱ and q.

Determination of the distance-from-center correction, $\Delta m(\varrho, q)$, may not always be easy in practice. If many stars of known magnitude are available in the field, it is advantageous to divide the field into annuli centered on the plate center (which is presumably identical with the point at which the optic axis pierces the plate), determine the m, q-relation for each annulus, and discuss the variation of the m, q-relation as a function of ϱ in the form $\Delta m(\varrho, q)$, which may be displayed as a family of curves, $\Delta m(\varrho)$, each curve referring to a different q-value.

Generally, the large number of stars necessary for such a discussion will not be available. The family of curves $\Delta m(\varrho)$ is therefore derived by discussing magnitude residuals of the stars when a single m, q-curve is used for the entire plate. Effectively, such a single m, q-curve refers to the mean distance, $\bar{\varrho}$, of the standard stars from the optic axis; $\bar{\varrho}$ is fixed by the distribution of the particular set of standard stars used. The family of $\Delta m(\varrho)$ curves derived should thus not be used directly to correct the magnitudes of stars found from an m, q-curve referring to a different mean distance, $\bar{\varrho}'$, from the optic axis.

A star field in which there is correlation between m and ϱ for the stars of known magnitude should be avoided for the determination of the $\Delta m(\varrho, q)$ rela-

[1] A. BLAAUW: Bull. Astronom. Inst. Netherl. **9**, 141 (1940).
[2] H. F. WEAVER: Astrophys. Journ. **106**, 366 (1947).

tion. Care should be taken to investigate the effect of combining $\Delta m(\varrho, q)$ results derived from star fields measured with diaphragms of different size in a fixed-aperture photometer of with comparison beams of different intensity in an iris photometer.

Often, the quality of images produced by an optical system is not symmetrical about the plate center; it is a function of position angle, ϑ, as well as ϱ. In such a case the photometric errors will also be a function of both ϱ and ϑ.

Position-angle variation of image quality can be produced by maladjustment of the components of the optical system, displacement of the optic axis from the plate center, flexure of the telescope, poor guiding combined with optical aberrations, plate tilt, and other defects. Photometric errors from such causes can best be discovered and allowed for by patient recording of residuals on a chart of each plate and drawing of contours of equal error size. Often the source of the error can be identified from such a chart and the defect can be corrected. Displacement of the plate center from the optic axis will result in a distribution of residuals symmetrical or essentially symmetrical about the displaced center. Plate tilt will result in a distribution of residuals symmetrical with respect to the line of tilt. A somewhat similar distribution of residuals can result from a combination of effects such as guiding error and coma.

Precision photographic photometry demands telescopes of high quality in excellent adjustment both optically and mechanically.

VI. Chromatic factors.

14. Color effects. Color plays a dominant role in stellar photometry. Chromatic aberration in a camera, for example, profoundly influences the photometric results obtained with the instrument. Consider telescopes T_1 and T_2 identical except for their color curves. The minimum of the color curve of T_1 lies in the blue at λ_1; that of T_2 at λ_2 is located somewhat to the longward of λ_1; $\lambda_2 > \lambda_1$. Identical filters and photographic emulsions are used in the two telescopes to photograph a pair of stars, one blue, star B, the other red, star R. The magnitude difference mag_B-mag_R observed with T_1 will differ from the magnitude difference observed with T_2 since the most intense portion of the spectrum of star B will be concentrated on the emulsion more effectively in T_1 than in T_2, while for star R the more intense portion of the spectrum will be concentrated more effectively in T_2. In view of the optimum image principle, it may be anticipated that the measured magnitude difference may depend upon star brightness or, more specifically, upon the photometric measure, q. Such an effect is equivalent to the Purkinje effect in the human eye.

Magnitudes, $m(T_1)$, derived with T_1 for a sequence of stars of different colors will differ from magnitudes $m(T_2)$ found with T_2 for the same stars. For any particular star the observed magnitude difference $\Delta m = m(T_1) - m(T_2)$ will be a function of the color of the star,

$$m(T_1) - m(T_2) = f(C), \tag{14.1}$$

or, in the more general case, of the color and magnitude (preferably q-value) of the star:

$$m(T_1) - m(T_2) = f(C, m). \tag{14.2}$$

Eqs. (14.1) and (14.2) are termed color equations; they permit transformation of magnitudes from one color system to another color system. While the equations are written for magnitudes, it is evident that similar equations could be used to transform colors from one system to another.

The principal photometric effects found for two telescopes with different color curves will also be found if one telescope is used with two slightly different plate-filters to observe a magnitude sequence. One plate-filter combination will respond more to blue stars, the other to the red stars. A color equation will exist between the two sets of derived magnitudes. In general photometric practice an analogous situation is invariably encountered: we employ a particular telescope-plate-filter combination to observe a magnitude sequence for which published magnitudes, m_1, were derived with a somewhat different telescope-plate-filter combination. A color equation will exist between the observed and published magnitudes. To derive the color equation, we establish the mean m_1, q-curve from the measured q's, read off m_2's with the measured q's and discuss the differences $m_1 - m_2$ according to Eqs. (14.1) or (14.2). Once we have determined the color equation we can, in principle, then transform from one color system to the other.

In the great majority of past photometric investigations, $f(C)$ has been taken to be a linear function of C or, if m is included, a function involving at most the product of m and C:

$$m_1 - m_2 = a + bC, \qquad (14.3\,\text{a})$$

$$m_1 - m_2 = a + (b + cm)\,C. \qquad (14.3\,\text{b})$$

The magnitude scales of m_1 and m_2 are taken to be identical. The zero-point parameter a insures that the mean color of the stars employed in the discussion does not influence the value of the parameter b, the color coefficient; c is the Purkinje coefficient.

The modern trend is to utilize Eq. (14.1) and to treat the problem by purely graphical means, not imposing any functional form on the curve.

Correlation between the colors and magnitudes of stars employed in derivation of the m, q-curve causes difficulties in separation of the color equation and the true m, q-relation. If correlation between m and C exists and a color equation is present, the correlation produces a distortion of the derived m, q-curve. When residuals from this distorted curve are used to find the color equation, an incorrect $\varDelta m$, C-relation will be derived inasmuch as a part of the color equation has already been absorbed into the m, q-curve. Correlation between m and C can result in the injection of a spurious Purkinje effect into the color equation.

The existence of correlation between m and C among the stars of the North Polar Sequence and of other standard magnitude sequences has introduced difficulties in photographic photometry. Ideally, one wishes a full range of color in every magnitude interval in any standard sequence used to derive the m, q-curve and to investigate the properties of the telescope-plate-filter combination employed for the observations. The small number of faint blue stars makes this ideal situation difficult to achieve. The difficulty is compounded by the fact that coloration of a star occurs either because of the intrinsic form of the star's energy distribution curve or because of the effects of interstellar extinction on that curve. The form of the color equation is different in the two cases. Transformation from one color system to another in general introduces inaccuracies and reduces the precision of the observations.

If the magnitude scales of the two magnitude series being compared are not identical, that is to say, if the light ratio between magnitude intervals is not the same for the two series, an additional term of the form dm must be added to Eq. (14.3 b). The parameter d is termed the scale coefficient. Considerable care must be taken to avoid confusion of scale and color coefficients when correlation between m and C is present in the observed magnitudes.

10*

In modern photographic observations, which frequently are derived from a photoelectric magnitude sequence, scale errors are not likely to be significant. In older photographic work scale errors were common. However, judgment must be exercised in interpreting published discussions of scale divergences, particularly in those instances in which the observational errors in the magnitudes are large. Generally, in determination of the scale error the difference $m_1 - m_2$ was discussed as a function of m_1 or m_2. This practice introduces a fictitious scale error since a selection effect is involved in the ordering of the residuals. The principles and

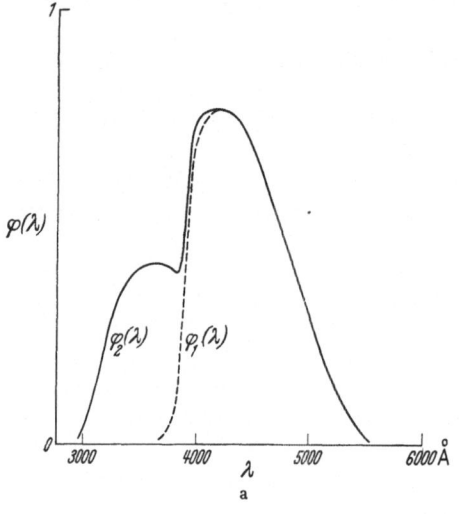

the removal of such effects have been treated extensively by TRUMPLER and WEAVER[1]; an example of elimination of fictitious scale error has been given by STOY[2].

Recent work, particularly by JOHNSON[3] and JOHNSON and MORGAN [3] has shown that $f(C)$ is complicated in form (it may be multivalued) unless precautions are taken in the choice of the telescope-plate-filter combination. For blue magnitudes in particular, the form of $f(C)$ depends strongly upon the ultraviolet admittance of the equipmental response function[4] and upon the energy distribution in the stellar spectrum. The character of the Balmer lines of

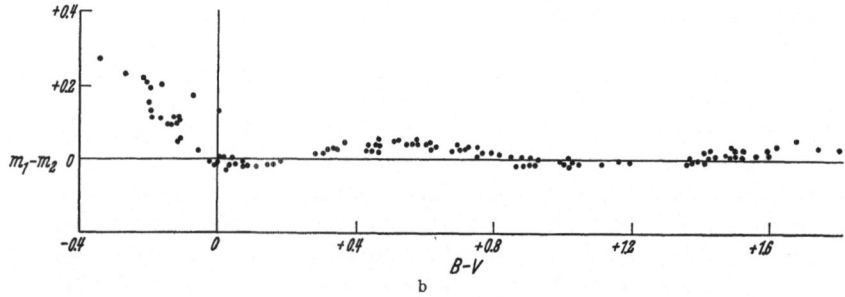

Fig. 12. (a) Response functions $\varphi_1(\lambda)$, $\varphi_2(\lambda)$. (b) Differences $m_1 - m_2$ for luminosity class V stars observed with response functions φ_1, φ_2.

hydrogen and the amount of the Balmer discontinuity at 3646 Å profoundly affect the observed magnitudes. If one set of blue magnitudes is determined with a response function that includes the wavelength region 3600 to 3900 Å and the second set of magnitudes is determined with a response function that excludes the wavelength range shortward of 3900 Å, the transformation between the two systems is very complicated. For some stars the distribution of energy in the wavelength range 3600 to 3900 Å may differ rather considerably, yet be rather similar for the wavelength

[1] R. J. TRUMPLER and H. F. WEAVER: Statistical Astronomy, Chap. 5. Berkely: University of California Press 1953.

[2] R. H. STOY: Month. Notes, Astronom. Soc. of South Africa **8**, 41 (1949).

[3] H. L. JOHNSON: Astrophys. Journ. **116**, 272 (1952).

[4] The equipmental response function, $\varphi(\lambda)$, describes for a specified telescope-plate-filter combination the relative effectiveness of radiation in the wavelength range λ to $\lambda + d\lambda$ in the production of the photographic image.

range to the longward of 3900 Å. Stars recorded as essentially identical in brightness by the response function terminating at $\lambda\,3900$ may be indicated as having rather different brightnesses by the response function with extension into the wavelength range 3600 to 3900 Å. The effect will be systematic depending upon the spectral characteristics of the stars. Presumably, the intrinsic spectral energy distribution in a star can be specified adequately for purposes of photometric transformation by means of the MK spectral type and luminosity class (see Ref. [3]). Luminosity class might be considered as an additional parameter of the transformation. However, the spectral energy distribution is modified by any interstellar extinction suffered by the starlight. This introduces a major complication in the achievement of accurate transformation from one magnitude system to another.

An example of a color equation (or transformation from one color system to another) is shown in Fig. 12. Magnitudes, m_1, were observed with the response function designated (1) in Fig. 12a; magnitudes m_2 were observed with the response function designated (2) in Fig. 12a. The differences $m_1 - m_2$ are plotted as a function of color in Fig. 12b. To avoid complications, only stars of MK luminosity class V have been utilized in the construction of the diagram. The observations utilized were made photoelectrically; they are taken from the work of JOHNSON[1]. The principle that they illustrate is as important to photographic photometry as it is to photoelectric photometry: the complicated nature of transformation from one magnitude system to another demands that careful consideration be given to the choice of the response function used in making the observations. The necessity of making transformations should be avoided whenever possible; if they must be made, the corrections to transform from one system to another should be as small as possible.

The question of choice of magnitude system and related problems will be discussed in Division VIII.

VII. Atmospheric extinction.

15. Factors involved in extinction. In traversing the Earth's atmosphere, light is diminished in intensity. For a given telescope-plate-filter combination the amount of such extinction depends upon

(i) the properties of the atmosphere along the light path,

(ii) the path length traversed by the light, and

(iii) the wavelength distribution of stellar radiation.

16. The extinction equation. Let $I(\lambda)$ represent the wavelength distribution of intensity of stellar radiation as it would be measured outside the Earth's atmosphere. Let $\varphi(\lambda)$ represent the response function of the camera, photographic emulsion, and optical devices in the optical train, all taken together as one unit. For convenience, we assume $\varphi(\lambda)$ to be normalized so that $\int_0^\infty \varphi(\lambda)\,d\lambda = 1$. The effective total intensity of radiation, L, that would be recorded by the telescope-plate-filter combination if it were located outside the Earth's atmosphere and pointed at a specific star is

$$L = K \int_0^\infty I(\lambda)\,\varphi(\lambda)\,d\lambda. \qquad (16.1)$$

The parameter K matches the units, takes into account the effective area of the telescope lens or mirror, and so forth. The numerical value of K will never concern

---

[1] H. L. JOHNSON: Astrophys. Journ. **116**, 272 (1952).

us unless we attempt to measure stellar energy in absolute units; K will vanish from the equations since we shall always use the telescope as a comparison device.

Generally, the wavelength range of $\varphi(\lambda)$ is not great, and variation of $I(\lambda)$ over this wavelength range is small. We here ignore the very rapid variation of $I(\lambda)$ caused by the spectral lines and concentrate interest on the envelope of the distribution. Following STRÖMGREN[1] we expand $I(\lambda)$ about a point λ_0:

$$I(\lambda) = I_0 + (\lambda - \lambda_0) I_0' + \frac{(\lambda - \lambda_0)^2}{2!} I_0'' + \cdots \tag{16.2}$$

whence

$$
\left.
\begin{aligned}
L = K & \left[I_0 \int_0^\infty \varphi(\lambda)\, d\lambda + I_0' \int_0^\infty (\lambda - \lambda_0)\, \varphi(\lambda)\, d\lambda + \right. \\
& \left. + \frac{1}{2!} I_0'' \int_0^\infty (\lambda - \lambda_0)\, \varphi(\lambda)\, d\lambda + \cdots \right] \\
= K & \left[I_0 + \frac{1}{2} \mu_2 I_0'' + \cdots \right].
\end{aligned}
\right\} \tag{16.3}
$$

Here

$$\lambda_0 = \int_0^\infty \lambda \varphi(\lambda)\, d\lambda; \tag{16.4}$$

μ_i represents the i-th order central moment of $\varphi(\lambda)$.

Since atmospheric extinction is wavelength dependent, it modifies the distribution $I(\lambda)$ as the radiation traverses the atmosphere. As modified by extinction, $I(\lambda)$ becomes $I^*(\lambda, z)$. The asterisk signifies the modified distribution and z, representing zenith distance, is an index to the path traversed by the radiation, hence to the degree of modification that has taken place in $I(\lambda)$. For the star under consideration we write

$$I^*(\lambda, z) = I(\lambda)\, T(\lambda, z), \tag{16.5}$$

where $T(\lambda, z)$ specifies the transmittance of the atmosphere for light of wavelength λ, along a path which is described by the star's apparent zenith distance, z. The Earth's atmosphere is presumed to be homogeneous in horizontal planes; extinction does not depend upon azimuth. Such a simple plane-layer model of the atmosphere has been found adequate for all ordinary photometric purposes. We neglect the effects of refraction, which is significant only at zenith distances at which photometry becomes impracticable.

With the modified function $I^*(\lambda, z)$ we find

$$L^*(z) = K \int_0^\infty I^*(\lambda, z)\, \varphi(\lambda)\, d\lambda = K \left[I_0^*(z) + \tfrac{1}{2} \mu_2 I_0^{*\prime\prime}(z) + \cdots \right]. \tag{16.6}$$

The symbol $L^*(z)$ represents the effective total intensity of radiation traversing the atmosphere and recorded by the telescope. The atmospheric extinction, $\delta m(z)$, measured in magnitudes and representing the amount of light lost in traversing the atmosphere, is, by the definition of magnitude,

$$\delta m(z) = 2.5 \log \frac{L^*(z)}{L}. \tag{16.7}$$

For the ratio on the right-hand side of the equation we find, to terms of the second order,

$$\frac{L^*(z)}{L} = \frac{I_0^*(z) + \tfrac{1}{2} \mu_2 I_0^{*\prime\prime}(z)}{I_0 + \tfrac{1}{2} \mu_2 I_0''}. \tag{16.8}$$

---

[1] B. STRÖMGREN: Handbuch der Experimentalphysik, Vol. 26, p. 392. Leipzig: Akademische Verlagsgesellschaft 1937.

We compute the derivatives of $I^*(z)$ from its definition, Eq. (16.5), and insert these in Eq. (16.8), to find, after some rearrangement of terms,

$$\frac{L^*(z)}{L} = \frac{T_0(z) + \frac{\mu_2}{2 I_0}[I_0 T_0''(z) + 2 I_0' T_0'(z) + I_0'' T_0(z)]}{1 + \frac{\mu_2}{2 I_0} I_0''} \; ; \qquad (16.9)$$

here subscript 0 indicates that the quantity so designated has been evaluated at $\lambda = \lambda_0$; primes indicate differentiation with respect to λ.

If we expand the denominator and operate upon the expansion, keeping only first order terms in the result, we find

$$\frac{L^*(z)}{L} = T_0(z)\left[1 + \frac{\mu_2}{2} \cdot \frac{T_0''(z)}{T_0(z)} + \mu_2 \frac{I_0'}{I_0} \frac{T_0'(z)}{T_0(z)}\right]. \qquad (16.10)$$

To proceed with the problem, we require knowledge of the form of $T(\lambda, z)$. We assume that the atmosphere is plane, horizontal-layered and that its properties are constant in any plane. We neglect refraction. The adequacy of this simple model may be judged from the difference between air mass, $\sec z$, on the plane model, and air mass, $F(z)$, on the basis of BEMPORAD's much more complete model. The data have been tabulated by SCHOENBERG[1].

In traversing a path of length dh perpendicular to the direction of layering at height h above the observer, light of wavelength λ has its intensity, originally unity in a beam of unit cross section, diminished to the value

Table 3. *Comparison of air-mass values computed from plane-layer model and Bemporad model.*

z	$\sec z$	$F(z)$	$\sec z - F(z)$
65°	2.366	2.357	0.009
70	2.924	2.904	0.020
75	3.864	3.816	0.048

$$T(\lambda, 0 \mid h) = \exp[-\varkappa_\lambda \varrho(h)\, dh]. \qquad (16.11)$$

Here $\varkappa_\lambda$ is the mass extinction coefficient at wavelength λ, $\varrho(h)$ is the atmospheric density at height h, and dh is the path length in the layer. The parameter $\varkappa_\lambda$ may also be a function of h; this will not influence the form of the final result. For rectilinear propagation of light not perpendicular to the direction of layering, the path length may be written as $dh \sec z$, where z is the zenith angle made by the path. Thus

$$T(\lambda, z \mid h) = \exp[-\varkappa_\lambda \varrho(h) \sec z\, dh]. \qquad (16.12)$$

If we now consider the whole atmosphere lying above the observer rather than a layer of thickness dh at height h, it is evident that the function $T(\lambda, z)$ of interest to us is

$$T(\lambda, z) = \exp\left[-\varkappa_\lambda \sec z \int_0^\infty \varrho(h)\, dh\right] = \exp[-\alpha_\lambda \sec z] \qquad (16.13)$$

where

$$\alpha_\lambda = \varkappa_\lambda \int_0^\infty \varrho(h)\, dh,$$

a quantity that, in principle, can be calculated. Calculation requires that we know the structure and properties of the atmosphere over the observing station.

From Eq. (16.13) we compute the derivatives of $T(\lambda, z)$ and insert them in Eq. (16.10) to find, to terms of the first order,

$$\frac{L^*(z)}{L} = T_0\left[1 - \frac{\mu_2}{2} \alpha_{\lambda_0}'' \sec z - \mu_2 \frac{I_0'}{I_0} \alpha_{\lambda_0}' \sec z\right] \qquad (16.14)$$

where α_{λ_0}' and α_{λ_0}'' indicate derivatives of α_λ with respect to λ, evaluated at $\lambda = \lambda_0$.

[1] E. SCHOENBERG: Handbuch der Astrophysik, Bd. II/1, p. 183. Berlin: Springer 1929.

The expression for $L^*(z)/L$ consists of three terms. The first term on the right represents the extinction we would find if we were observing monochromatically at λ_0. The second term corrects for variation of atmospheric extinction within the wavelength range in which the observations are made. It does not depend upon $I(\lambda)$ and involves the response function $\varphi(\lambda)$ through μ_2. This term would exist even if all the stars were equienergy sources. The third term allows for variation in the distribution of intensity in the star's spectrum; it also involves the variation of α_λ and the response function through μ_2.

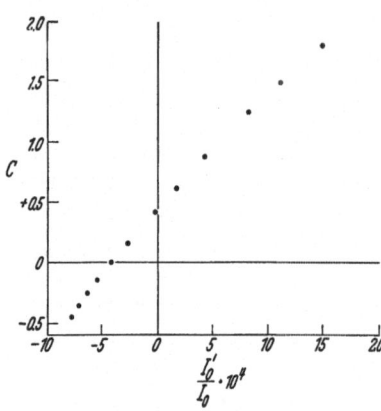

Fig. 13. Illustration of the linearity of the relation between I_0'/I_0 and C.

To the first order, the quantity I_0'/I_0 is proportional to the color index of a star. We can demonstrate this most easily for black body radiation. From the Planck function one computes

$$\frac{I'}{I} = \frac{1}{\lambda}\left[\frac{x\,e^x}{e^x - 1} - 5\right], \qquad (16.15)$$

where $x = c_2/(\lambda T)$. The reflectivity of two silvered mirrors can be derived from the measurements made by PETTIT[1]; the sensitivity of an E 40 plate has been tabulated by MEES[2]. These permit the response function $\varphi(\lambda)$ for the International Photographic System[3] to be found. From $\varphi(\lambda)$ one computes $\lambda_0 = 4238$ Å, which must be used in Eq. (16.15) to find I_0'/I_0 as a function of T. SEARES and Miss JOYNER [4] have calculated black body color indices (on the International System) as a function of T; these have been matched with the equivalent I_0'/I_0-values and used in the construction of Fig. 13, which illustrates the high degree of linearity in the I_0'/I_0, C relation.

We take the logarithm of Eq. (16.14) to express $\delta m(z)$ in terms of stellar magnitudes, and expand the term in square brackets [Eq. (16.14)] to find to first order terms

$$\delta m(z) = 2.5 \log e \left(-\alpha_{\lambda_0} - \frac{\mu^2}{2}\alpha_{\lambda_0}'' - \mu_2 \frac{I_0'}{I_0}\alpha_{\lambda_0}'\right)\sec z. \qquad (16.16)$$

If we make use of the fact that $I_0'/I_0 = a + bC$, where a and b are parameters, we may also write Eq. (16.16) as

$$\delta m(z) = (\beta_0 + \beta_1 C)\sec z, \qquad (16.17)$$

where

and

$$\left.\begin{array}{l} \beta_0 = 2.5 \log e \left(-\alpha_{\lambda_0} - \dfrac{\mu_2}{2}\alpha_{\lambda_0}'' - \mu_2 \alpha_{\lambda_0}' a\right) \\[2mm] \beta_1 = -\,2.5 \log e\, \mu_2 \alpha_{\lambda_0}'\, b. \end{array}\right\} \qquad (16.18)$$

A star's zenith-extinction coefficient $(\beta_0 + \beta_1 C)$ is a linear function of the zero-atmosphere color, C, of the star. Zero atmosphere signifies the color that would have been observed if there had been no atmospheric extinction. The parameters β_0 and β_1 are functions of the response function of the instrument through μ_2, a, and b, and of the properties of the atmosphere through α_λ and its derivatives, all evaluated at λ_0.

[1] E. PETTIT: Publ. Astronom. Soc. Pacific **46**, 27 (1934).
[2] C. E. K. MEES: J. Opt. Soc. Amer. **21**, 767 (1931).
[3] Described in Division VIII, p. 155.

17. Factors influencing the zenith-extinction coefficient. α) *Color-index effect.* Table 4 displays the values of the zenith-extinction coefficient as a function of color, C. The values have been computed by Eq. (16.16) for the International Photographic response function, portrayed in Fig. 16, and with the atmospheric transmission coefficients tabulated by PETTIT[1] for the average air mass above Mt. Wilson.

The color indices, Col. 2 of Table 4, were computed by SEARES and Miss JOYNER [4] for stars radiating as black bodies and are on the International System; they are zero-atmosphere colors. Over the color range tabulated, −0.441 mag to +1.797 mag, the extinction coefficient varies by 0.100 mag, from 0.319 to 0.219 mag.

β) *Response-function effects.* SEARES and Miss JOYNER [4] have computed the zenith-extinction coefficients for the International System Photovisual response function, Fig. 16. They find the value (0.135 ± 0.001) mag over the entire color range −0.441 to +1.797 mag. For the International Photovisual System, there is essentially no variation of $\delta m(0)$ with C; $\delta m(0)$ is very different for the International Photographic and Photovisual Systems. Extinction values are sensitive functions of $\varphi(\lambda)$.

γ) *Atmospheric effects; night-to-night variations.* WEAVER[2] has published ac-

Table 4. *Zenith extinction on the International Photographic System tabulated as a function of zero-atmosphere color.*

T (degrees)	C (zero atm.)	Zenith extinction coefficient in mag. [Eq. (16.16)]
30×10^3	− 0.441	0.319
22	− 0.347	0.316
17	− 0.249	0.313
13.7	− 0.141	0.309
11	0	0.303
9	+ 0.165	0.297
7	+ 0.421	0.286
6	+ 0.612	0.278
5	+ 0.873	0.266
4	+ 1.241	0.249
3.5	+ 1.489	0.237
3	+ 1.797	0.219

curate photoelectrically determined zenith-extinction values for five nights in an interval of 30 days. The coefficients refer to a star of color zero on the instrumental system and to a response function more compact and slightly to the longward of the International Photographic System.

Values of zenith extinction on five nights

0.270 mag
0.355 mag
0.320 mag
0.257 mag
0.337 mag

0.308 mag (average)

Errors up to 0.05 mag at the zenith would have resulted from use of a direct mean value. Night-to-night variation of rather large amount may be expected.

18. Check on the first-order theory. The zenith-extinction values computed by Eq. (16.16) can be checked by direct integration:

$$\delta m(0) = 2.5 \log \left[\int_0^\infty \varphi(\lambda)\, I(\lambda)\, T(\lambda)\, d\lambda \Big/ \int_0^\infty \varphi(\lambda)\, I(\lambda)\, d\lambda \right]$$

where $T(\lambda)$ represents the atmospheric transmission tabulated by PETTIT[1]. The difference $\delta m(0)$ in the sense first order theory *minus* integration result is system-

---

[1] E. PETTIT: Astrophys. Journ. **91**, 159 (1940).
[2] H. F. WEAVER: Astrophys. Journ. **116**, 612 (1952).

atic; plotted as a function of color index, it defines a second order curve, indicating a maximum difference of 0.008 mag in the two ways of calculation.

19. Some extinction traditions in photographic photometry. If accurate extinction corrections are to be made for a particular star, one must know the color of the star outside the Earth's atmosphere. Numerical values of β_0 and β_1 must be available for the particular night. Determination of β_0, β_1, and C presents no difficulties in photoelectric photometry. In photographic photometry this is not the case. It has therefore been customary in photographic photometry to adopt some mean color, $\bar{C}$, for all stars and some mean values $\bar{\beta}_0$ and $\bar{\beta}_1$ for all nights to find

$$\delta m (z) = (\bar{\beta}_0 + \bar{\beta}_1 \bar{C}) \sec z = \bar{\beta} \sec z. \tag{19.1}$$

This expression is used for extinction correction of all stars regardless of color; errors are thus introduced in the derived magnitudes.

To the extent that mean values $\bar{C}, \bar{\beta}_0, \bar{\beta}_1$ are applicable, Eq. (19.1) permits us to remove from the observed magnitude of a star the effects of extinction; to "reduce the star's magnitude to outside the Earth's atmosphere". In present-day photoelectric photometry such reduction to outside the atmosphere is customary. In photographic photometry (particularly the older work) custom dictated that correction for extinction should reduce a star's magnitude to what it would have been if the star had been seen at the zenith of the observing station. In photographic photometry the extinction is thus traditionally written

$$\Delta m (z) = \bar{\beta} (\sec z - 1). \tag{19.2}$$

Errors presently exist in many photographic magnitudes, including the International Standards of the North Polar Sequence, because of these traditions. Seares and Miss Joyner [4] first pointed out the errors in the North Polar Sequence Stars from these sources.

At Mt. Wilson the practice was to adopt for all "photographic" (blue) magnitudes the value $\bar{\beta} = 0.295$ mag. This corresponds to a very blue star. An M star, $C \approx 1.4$ mag, observed at the pole and reduced to the Mt. Wilson zenith with this extinction coefficient would be too bright by the amount $(0.295 - 0.228)$ $(1.71 - 1) = 0.054$ mag. Such errors exist in the International Standards as a kind of fictitious color equation caused by the reduction process.

Errors were also introduced by the custom of reducing to the station zenith. When results were obtained at different observatories, residual effects always remained in the magnitudes because of different station altitudes. The effect has been illustrated by Velghe[1].

Because of the very much smaller variation in the photovisual extinction with color, the error in photovisual magnitudes through use of a mean extinction coefficient for all stars is insignificant provided a proper mean value is used for the station. Some altitude effects remain.

In the field of precision photometry there is encountered no problem more delicate than that of properly removing the effects of atmospheric extinction. Careful thought must be given to each step in the process of making such corrections. It is highly desirable to make use of photoelectrically determined β_0-, β_1-values in precision photographic photometry. This requires careful matching of the response functions of the photographic and photoelectric observations. The problem will be discussed in Division XI below (p. 172).

---

[1] A. Velghe: Comm. Obs. Roy. Belg., No. 9, (1949).

VIII. Photometric systems.

20. System and response function. In view of the complications and ambiguities that can arise in making transformations from one photometric system to another, the choice of system on which the observations are made is of great importance. Choice of a system is equivalent to choice of a response function, $\varphi(\lambda)$.

21. Basis for choice of system. The purpose for which the observations are to be used provides the basis for choice of the system on which they are to be made. If the observations are to be combined or compared with other photometric data, they must be on a system identical with that of the other data. Examples of programs involving comparison and combination of photometric data are numerous in present-day research.

(i) Color-magnitude diagrams of galactic and globular clusters. The purpose of such observations is to provide data on clusters as a class. The results will have the greatest value if all investigators observe on the same photometric system so that all results are immediately and precisely comparable.

(ii) Photometric data for galactic structure studies involving, say, determinations of photometric distances. The observations must be on the photometric system on which absolute magnitudes and intrinsic colors are available. The value of χ, the ratio of total photographic extinction to color excess, must also be known on the system. Mismatching among the photometric systems of any of the quantities leads to incorrect distances and erroneous conclusions.

(iii) Determination and comparison of luminosity functions in the solar neighborhood, in clusters, in associations, in galaxies, and so forth. A strictly meaningful comparison will result only if identical photometric systems are employed for all observations. If dissimilar photometric systems are used and only the functions $L_\odot(M_1)$, $L_{cl}(M_2)$, $L_{assoc}(M_3)$, $L_{gal}(M_4)$, ... are available, no transformation is possible and the comparison loses sharpness and significance.

These are examples of a broad class of research problems which are most satisfactorily carried through to completion if all photometry is done on a "standard" system, that is, with standard response functions. If the observations are on a standard system, data from various sources and of various kinds can be combined with a minimum of difficulty and with the introduction of a minimum of uncertainty in the final results.

Another class of problems, often astrophysical in nature, require for their solution photometric observations in specific wavelength regions that are not related to any present standard response functions. Examples of such problems are:

(i) the study of the wavelength dependence of interstellar extinction;

(ii) the determination of the spectral energy distributions in stars of different galactic clusters and the general field;

(iii) investigation of the wavelength dependence of limb darkening in the components of an eclipsing variable star;

(iv) classification of stellar spectra by photometry.

For the solution of such problems one might make observations with a series of narrow response functions. The wide choice of filters now available makes such programs completely feasible though, admittedly, in the present stage of development of the photometric art, one would make the observations photoelectrically rather than photographically.

22. Differences in the types of programs. Basic differences exist in the two types of programs. The photometric requirements of the galactic astronomer, of the investigator comparing general properties of the galaxy and extragalactic systems, of the investigator using for statistical purposes the photometric data in a catalog will best be met if the observations are made on a standard system. Such procedure will simplify the investigator's work; it will permit him to take full advantage of an already existing body of photometric data. The greater this body of data becomes, the greater are the opportunities for him to make significant discoveries. Difficulties and frustrations arise when the data used are on mixed systems; transformations may be poor or impossible; they can only detract from the accuracy of the results.

Present-day photometric observations attain a high degree of precision. The greater the precision of the measurements, the greater the care that must be exercised in standardization if the full potentialities of the measurements are to be realized.

The photometric needs of the investigator interested in physical properties of stars or of the interstellar medium will generally not be adequately satisfied by observations made on standard systems. Often the investigator in these fields wishes to determine how some quantity varies with wavelength. He makes observations in various narrow wavelength ranges; there is no need for him to use any standard system. In fact, it might be disadvantageous for him to use standard systems. His interest focuses on the curve expressing variation of some quantity with wavelength and it is this curve, not the individual observations, that provides the information for his discussion and that he compares and combines with the results of other observers.

Likewise, the photometric observers interested in exploring new fields of research for their equipment will not, in all probability, be adequately served by any present standard systems. These observers may require in their research many response functions of varied properties. The classification of stellar spectra by photometry mentioned above is a case of this type. Exploratory results are now being obtained[1]; experimentation with various filters (photometric systems) is being carried on. Once a completely satisfactory set of filters (response functions) is found for the solution to the problem, other observers will tend to adopt the same set of response functions for their own work. The particular set of response functions (photometric systems) will tend to become standard. A particular photometric system can pass from the experimental and nonstandard into the standard category once the experimental part of the program is concluded and a body of data becomes available in the literature.

The central point of the discussion can be summarized as follows: the requirements of the problem under investigation must specify the photometric system (response function) to be employed. In many instances great advantages accrue from observing strictly on a standard system; in other instances specific, nonstandard systems (response functions) are required. If there is no compelling reason for an observer to use a nonstandard system, he will generally find a standard system the most satisfactory for his work.

23. Operational procedures for observing on a standard system. 1. In the traditional approach to the problem of matching observations to some specified standard system, the investigator required no precise knowledge of the physical properties of either his own instrumental system or the standard system. He matched the response function of his equipment roughly to that of the standard system and observed the magnitude sequence that defined the standard system.

---

[1] B. STRÖMGREN: Vistas in Astronomy, Vol. 1. London and New York: Pergamon Press 1955. — B. STRÖMGREN and K. GYLDENKERNE: Astrophys. Journ. **121**, 43 (1955).

Through a transformation equation he adjusted the photometric values he observed to reproduce the photometric values of the standard system.

The assumption is fundamental in this traditional approach that an accurate transformation can be found even when the instrumental and standard response functions are only approximately matched. In practice, the resulting transformation was not always of high precision by modern standards. Mismatches of system were frequent when magnitudes derived by such a process were combined with other photometric data.

A practical deficiency in many older photometric observations was that, because of the amount of work involved in obtaining them, colors were generally not available for individual field stars studied. Thus, while the observer might give a transformation equation to put the instrumental system on the standard system, the transformation could not be made for the field stars because the necessary data were lacking.

2. A second approach to the problem is based upon the concept of matching the instrumental and standard systems through knowledge of their physical properties. The matching is accomplished before any observations are made; laboratory investigations may be required. By proper choice of observing materials, the observer tailors the response function of the instrumental system to reproduce the response function of the standard system. From this point of view it is primarily the response function that defines the standard system, not a magnitude sequence in the sky. The sequence in the sky is secondary; a photograph of it provides the direct observational test of the accuracy of the matching. Because of the closely matched response functions, there should be essentially no differences in the observed and standard photometric values. Transformation, if required at all, should involve only very small corrections to the observed values.

This second approach can provide a more satisfactory result than the traditional approach and is greatly to be preferred to it.

24. The standard system. Traditionally, the International System, which developed empirically by use and which exists primarily in the form of photographic and photovisual magnitudes of stars in the North Polar Sequence (NPS), has been the standard system to which, presumably, all observers referred their results. The early history of the NPS has been summarized by WEAVER [1]. The first acceptable magnitudes and colors of the stars comprising the NPS were given by SEARES[1]. From time to time following the publication of the 1922 values, SEARES made small revisions in the individual magnitudes and colors of the NPS stars; the latest was published[2] in 1945.

The physical characteristics of the International System have never been well established. SEARES and Miss JOYNER [4] have only recently briefly discussed the physical basis of the system. A detailed discussion cannot be made since, in particular, precise sensitivity data for the photographic plates and yellow filter used in the establishment of the system are lacking and cannot now be obtained.

25. Critical examination of the NPS and the International System. Inadequacies in the NPS were evident early in its history. SEARES[3], the principal investigator of the NPS, summarized many years of effort in the words:

---

[1] F. H. SEARES: Trans. Internat. Astronom. Union **1**, 69 (1922).

[2] F. H. SEARES and MARY C. JOYNER: Astrophys. Journ. **101**, 15 (1945).

[3] F. H. SEARES, F. E. ROSS and M. C. JOYNER: Carnegie Inst. Wash., Publ. No. 532, 1941.

"Attempts to transfer the photometric system of the NPS to other parts of the sky have met with serious difficulty. The standards are few and scattered, B-type stars are lacking, and the red stars do not accurately define the color system. As a group, these standards can be observed only with large-field cameras which require complicated corrections depending upon color, brightness, positions of the stars, and frequently also on focus and other factors such that each plate must be treated by itself. The bright polar standards thus failed of the purpose for which they were intended."

Similarly, complaints can be made about the faint stars. With large instruments the establishment of a scale of magnitudes by comparison with the NPS is very difficult. The pole is poorly located; for most observatories it is unnecessarily low in the sky; it is generally to far away from the area being observed and changes in seeing owing to great change of azimuth are likely; guiding at the pole is likely to be done differently from that at another region of the sky. These factors cause systematic errors in scales established; large telescopes are especially susceptible to such errors. Some of the difficulties encountered in the bright stars are also found among the faint stars: "blue stars are lacking and the red stars do not define the color system". There is correlation between color and magnitude. The faintest stars depend only upon a few Mt. Wilson observations and have not been checked by other observatories.

Definition of a photometric system solely or primarily through observed magnitudes and colors of a sequence of stars is, in view of present trends in photometry, undesirable. Specification of the physical properties of the photometric system is a present-day necessity. Even if one overlooks the lack of adequate physical specification of the International System, now defined by the NPS, there is no assurance that the published values for the stars in the sequence are self-consistent. The necessity of making frequent small adjustments in individual values indicated the existence of inconsistencies.

26. Observational inconsistencies; the P, V system. Contradictions began to appear when accurate photoelectric observations of the NPS stars were made. STEBBINS and WHITFORD[1] found a divergence between photographic and photoelectric magnitudes of the brighter stars in the sequence. However, the response functions of the two measuring devices were very different. The photographic response function was much broader and extended farther into the ultraviolet than did the photoelectric response function (see Fig. 16). SEARES[2] attributed the apparent scale divergence to the existence of hydrogen lines and the Balmer discontinuity in the ultraviolet that affected the photographic results but not the photoelectric measurements. This was the first indication of the seriousness of the transformation problems so clearly brought to light by JOHNSON[3]. Somewhat unheeded at the time, it pointed out the need for understanding the physical properties of any photometric system under discussion.

Acute difficulties were encountered by the photoelectric observers in the late 1940's when they attempted to transform their observations to the International System. In part, these difficulties arose because identical response functions were not used by all observers. The observations of some investigators were more affected by the part of the spectrum to the shortward of $\lambda 3900$ than were those of other investigators. This severely complicated transformations between systems. The importance of the ultraviolet admittance of the response function was not clearly realized at the time.

---

[1] J. STEBBINS and A. E. WHITFORD: Astrophys. Journ. **87**, 237 (1938).

[2] F. H. SEARES: Astrophys. Journ. **87**, 257 (1938).

[3] H. L. JOHNSON: Astrophys. Journ. **116**, 272 (1952).

Difficulties also arose because of apparent inconsistencies in the NPS stand-ards. The results obtained by an observer transforming his instrumental system to the International System depended upon the specific NPS stars he used for making the transformation. To eliminate this source of trouble, a group of active photoelectric observers agreed to adopt a perticular set of nine NPS stars (Nos. 6, $2r$, 10, $4r$, 13, $8r$, 16, 19, $12r$) as defining the International System for photoelectric work[1]. This group of stars was selected because it had been well observed photoelectrically by STEBBINS, WHITFORD, and JOHNSON[2]. The values of the photographic magnitudes and colors given by them were adopted for the nine stars. Photographic magnitudes on this "nine-star system" were indicated by the symbol P; colors were indicated by the symbol P-V.

27. The U, B, V system. As a still more satisfactory substitute for the Inter-national System, JOHNSON and MORGAN [3] and later JOHNSON[3] proposed the three-color U, B, V system. Its physi-cal properties are well defined; response functions are shown in Fig. 14. U, B, V values for approximately 400 stars of a wide variety of types have been publi-shed by JOHNSON[3] and serve as obser-vational standards for the system.

JOHNSON and MORGAN [3] have summarized the principles on which they have based their three-color photo-metric system as follows:

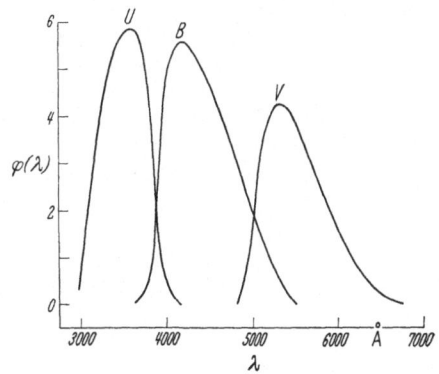

Fig. 14. Response functions for the U, B, V system.

"Because of the nonlinear (and sometimes multivalued) relationships between various systems of color indices, the definition of a fundamental system of magnitudes and colors becomes difficult; in particular, for regions like that of the North Polar Sequence, where no early-type stars are available, a system for stars bluer than class $A0$ (and for reddened B stars in general) can hardly be considered to exist. The principal difficulty here is that it is not justifiable to extrapolate a color equation determined from A-K stars to those of class B; in addition, the spectrum of a reddened B star does not have the same energy distri-bution as does that of a later type star; therefore, a different color equation may be necessary for reddened and unreddened stars.

A fundamental photometric system for stars should therefore include:

1. Magnitudes and color indices for reddened and unreddened stars from all parts of the H-R diagram; these should include white dwarfs and subdwarfs, as well as supergiants, giants, and main sequence stars.

2. The same photometric data for stars having interstellar reddening and of known spectral type and luminosity class.

3. A series of color indices extending from the altraviolet to the infrared, so that reduc-tions to the standard color system can be made by a process of interpolation rather than extrapolation; the six-color photometry of STEBBINS and WHITFORD satisfies this condition extremely well.

4. A determination of the zero point of the color indices in terms of a certain *kind* of star which can be accurately defined spectroscopically; that is in terms of a kind of star whose spectral energy distribution can be predicted accurately from its spectral type and luminosity class; in addition, the stars used for the zero point should be plentiful.

The above requirements cannot be satisfied by using small selected regions of the sky; the standard stars are, of necessity, scattered over the sky, and many of them must be very bright if condition 1 is to be satisfied.

---

[1] O. J. EGGEN: Astrophys. Journ. **114**, 144 (1951). — H. L. JOHNSON and W. W. MORGAN: Astrophys. Journ. **114**, 522 (1951). — H. F. WEAVER: Astrophys. Journ. **116**, 617 (1952).
[2] J. STEBBINS, A. E. WHITFORD and H. L. JOHNSON: Astrophys. Journ. **112**, 469 (1950).
[3] H. L. JOHNSON: Ann. d'Astrophys. **18**, 292 (1955).

It is, therefore, of importance to supplement the scattered standards with regional secondary standards which fulfill some of the conditions and which are located in a small region. The most useful of the regional standards would be accurately observed mainsequence stars in open clusters, together with some yellow giants.

The scale of the U, B, V system is accurately POGSON; it derives directly from the photoelectric photometer with which the observations were made.

The magnitude zero point of the U, B, V system is fixed in such a way that the V magnitudes of the nine NPS stars specifically mentioned in Sect. 26 above in connection with the P, V system agree with the International photovisual (IPv) magnitudes derived for those stars by STEBBINS, WHITFORD, and JOHNSON[1]. Thus V in the U, B, V system is identical with V in the P, V system, and with the International Photovisual System as that system is defined by the STEBBINS, WHITFORD, JOHNSON magnitudes for the nine NPS stars. For these nine NPS stars JOHNSON and MORGAN [3] find

$$V = \text{IPv} + 0.000 + 0.002\,(B-V) \quad \text{(probable error)}.$$
$$\pm\, 0.006 \pm 0.005 \tag{27.1}$$

To define the zero point of colors (or, alternatively, of the U and B magnitudes) JOHNSON and MORGAN [3] have followed their own suggestion that the zero point should be defined "in terms of a kind of star whose spectral energy distribution can be predicted from its spectral type and luminosity class". They have also followed the principles of the procedure employed in establishing the zero point of colors in the International System. For six stars classified as $A\,0\,V$ in the MK system, JOHNSON and MORGAN have imposed the condition that

Table 5. *Zero point observations for U and B in the U, B, V system.*

Star	$B-V$ (mag)	$U-V$ (mag)
H-R 3314 . .	−0.02	−0.02
γ UMa − . .	0.00	+0.01
109 Vir − . .	0.00	−0.03
α CrB	−0.02	−0.02
γ Oph	+0.04	+0.04
α Lyr	0.00	−0.01

$$\overline{U-B} = \overline{B-V} = 0. \tag{27.2}$$

Observational results for the six stars are shown in Table 5.

The departures from zero are larger than can be explained purely as observational errors. The spectral energy distributions of the six stars are not identical; polarization measurements serving as criteria for the existence or nonexistence of material in the line of sight would be of interest. From the criterion stated by MORGAN and JOHNSON to govern the choice of the stars to be used for the zero-point determination ("the spectral energy distribution can be predicted from the spectral type and luminosity class") one infers that unreddened stars are to be utilized in the definition of the zero point. The fact that the entries in Table 5 are not all zero indicates that while for the six stars $\overline{U-B} = \overline{B-V} = 0$, for $A\,0\,V$ stars seen through perfectly transparent space $U-B$ and $B-V$ may not be zero. Hence the zero point of intrinsic colors in the U, B, V system and the zero point of the observational scale in the same system may correspond to different spectral types.

With the zero point specified by the six $A\,0\,V$ stars it is found that for $K\,0\,\text{III}$ stars $B-V = 1.01 \pm 0.007$ (p.e.) mag. This is consistent with the original definition of the zero point of the International System, which was originally overspecified.

The U, B, V system is based on the transformation concept; it was devised to permit transfers from one system to another by interpolation rather than by

---

[1] J. STEBBINS, A. E. WHITFORD and H. L. JOHNSON: Astrophys. Journ. **112**, 469 (1950).

extrapolation. One can, in principle, weight the various U, B, V values to reproduce any magnitude system in the wavelength range covered by U, B, V and thus achieve very satisfactory transformation. In discussing the choice of response function for measuring on the U, B, V system, JOHNSON[1] states (note that the first remarks apply to photoelectric photometry with a $1 P 21$ photomultiplier tube):

1. If possible, use filters having the same numbers as the ones that were used in defining the U, B, V system ($U =$ Corning 9863, $B =$ Corning $5030 + 2$ mm Schott GG 13, $V =$ Corning 3384). With such filters, the deviations of the transformations from linearity will be slight. If it is not possible to use these filters, satisfactory substitutes are: $U = 2$ mm Schott UG 2, $B = 1$ mm Schott BG 12 $+ 2$ mm Schott GG 13, $V = 12$ mm Schott GG 11.

2. Aluminized reflecting telescopes should be used if possible. Silvered mirrors or refracting telescopes will not affect the transformations to B and V but will introduce non-linear and multi-valued transformations for U. Some photo-multipliers, such as EMI, have somewhat less response in the ultraviolet than do RCA $1 P 21$'s and may also affect the U transformations.

3. Regardless of the filters, optics, or photomultipliers, it is recommended to observe a large number of stars of all kinds selected from Table 3 (in JOHNSON's paper). These observations will allow one to check on the quality of transformations. Once the character of the transformations has been established, continued observation of a much smaller number of stars will serve to standardize the observations.

4. Although several regions have been supplied for the use of photographic observers, it is always best to use photoelectric transfers to the photographed region and to use a photoelectrically set up standard sequence in the region. One thereby avoids many of the systematic errors that can creep into photographic photometry.

5. Photographic photometry on the U, B, V system if the proper filters are used. The recommended filters are: $U =$ Corning 9863 or 2 mm Schott UG 2 with a blue-sensitive plate; $B = 2$ mm Schott GG 13 with a blue-sensitive plate; $V = 2$ mm Schott GG 11 with a yellow-sensitive plate.

Experience has shown that if these simple precautions are observed, particularly if one observes enough stars, it is possible to transform the observed magnitudes and colors to the U, B, V system without a significant loss in precision. Such transformed values are, then, directly comparable with all the values for the standard stars and with derived relations such as the intrinsic colors and color versus spectral-type relation.

For photoelectric work Johnson stresses the need for precise reproduction of the response function of the equipment, but for photographic work he leaves the specification quite loose. If the photographic observer is to avoid unnecessary complications, he must carefully match the response functions of his equipment to those of the U, B, V system. Such matching is of especial importance in those instances in which colors may not be determined for each star, hence transformation from the instrumental system to the U, B, V system cannot be made. Initial matching of the response function is, in such a case, of crucial importance. The observer must know the properties of his camera thoroughly and must choose the filters and plates with care if accurate reproduction of the U, B, V response function is to be achieved. Each instrument presents a separate problem; specification of a filter and a general plate type as in the recommendation above is inadequate for accurate photometry.

Fig. 15 illustrates some examples of matching achieved with certain plate-filter combinations used with aluminized mirrors. Note that filter thickness as well as type must be considered.

At present the U, B, V system is the most satisfactory one available for general photometry. It is well defined physically; an ever-increasing body of observations is being made upon it; photometrically matched auxiliary data and auxiliary parameters are becoming available. It is equivalent to the International

---

[1] H. L. JOHNSON: Ann. d'Astrophys. **18**, 292 (1955).

Photovisual System and is not vastly different from the International Photographic System. The older observational data are therefore not useless for qualitative or semi-quantitative comparison with newer material.

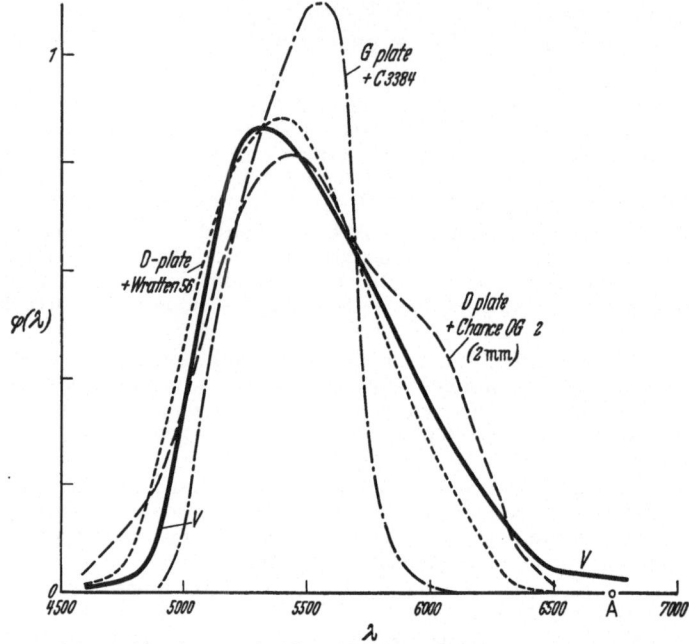

Fig. 15. Illustration of the matching of response functions of V system achieved with various plate-filter at the Newtonian focus of an aluminized reflector.

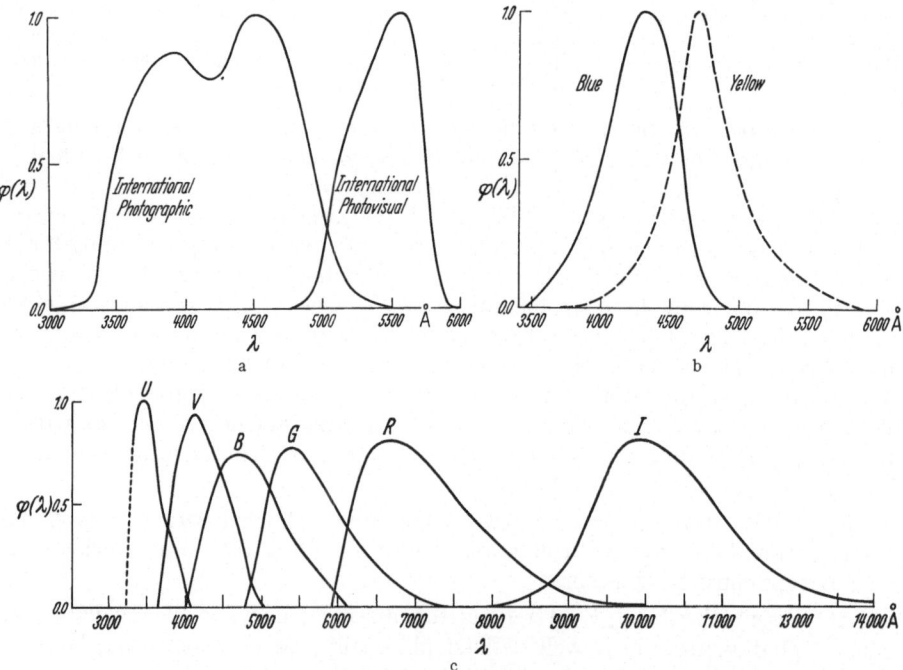

Fig. 16 a—c. Response functions of various photometric systems found in the literature. (a) International photographic and photovisual. F.H. SEARES and M.C. JOYNER: Astrophys. Journ. 98, 302 (1943). (b) Response functions for C$_g$ colors. J. STEBBINS and C.M. HUFFER: Publ. Washburn Obs. 15, 217 (1934). (c) Six-color photometry. J. STEBBINS and A.E. WHITFORD: Astrophys. Journ. 98, 20 (1943). — Cf. Fig. 16 d on next page.

28. Other systems. Other important photometric systems are displayed in Fig. 16 by means of their response functions.

Designation of a photometric system as standard logically should signify that the response function that defines it covers a wavelength range that permits a number of investigators to solve problems in which they are interested; that observations have been or are being actively made on the system. It is advantageous for an investigator to observe on such a system since he can compare and combine his results with existing observations. In his discussions he can employ matched data derived by other investigators. Each set of observations made with a specified response function strengthens the system and more firmly establishes it as standard. In practice, use is an important factor in establishing a system as standard.

The response functions of the International Systems are largely the result of historical accident; they were not tailord to any particular astronomical problems. The response functions of the U, B, V system are conditioned in part by those of the International System, in part by the spectral characteristics of stars in the ultraviolet, in part by the ready availability of filter materials and photomultipliers. If additional photometric systems are in future adopted as

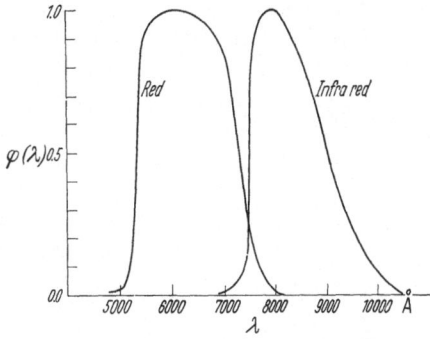

Fig. 16 d. Kron red, infra-red magnitudes. (The curve for the red magnitudes is slightly smoothed.) G.E. KRON and J.L. SMITH: Astrophys. Journ. **113**, 324 (1951). — Publ. Astronom. Soc. Pacific **63**, 89 (1951).

standard by a number of observers, careful consideration should be given to optimizing the response function or functions defining the system to provide the maximum information to solve some particular problem or group of problems. Once the standard system is established and its physical properties are well established, observers should make every effort to reproduce the standard response function with their own equipment in order to gain maximum value from their observations.

IX. Establishment of a fundamental magnitude scale by photography.

29. Observational data required. To establish a fundamental magnitude scale we require at least two photographs of a star field in which corresponding images are identical in every respect except that they differ by a constant known magnitude. For practical reasons we assume, additionally, that the number of stars in the field is not too small and that the camera with which the plates were taken produces images of good photometric quality. An elegant analytic solution to the problem of establishing a scale from such a plate pair was devised long ago by K. SCHWARZSCHILD[1]. However, the fact that the solution is analytic and has the danger that it may distort portions of the scale by imposing a particular mathematical form on the reduction curve is a serious drawback from the contemporary point of view. We shall consider only a purely graphical method that in no way forces a particular form on the reduction curve.

30. Attenuation devices. Numerous methods of taking a plate pair so that the images on one photograph differ from those on the other by a known magnitude,

---

[1] K. SCHWARZSCHILD: Astronom. Nachr. **172**, 65 (1906).

k, have been suggested. For a review see, for example, WEAVER [1]. We mention here only two methods that offer possibilities of accurate results, and discuss only one of these in detail.

Use of a parallel wire grating in front of the objective produces for each star a series of images that differ in brightness by calculable amounts. The grating method has the advantage over many others that the necessary pair of images is obtained on one plate at one time; atmospheric, telescopic mechanical, and fog effects are identical for the two images. Unfortunately, however, the first and higher order images are short spectra and hence are not strictly comparable to the zero order central image. Shortening the spectra through design of the grating also moves the images closer together; photographic adjacency effects become more pronounced. One cannot predict, a priori, how such photographic effects and how photometric properties of images that are short spectra will affect the results. In particular, there exists the possibility that magnitude, color, and the effective grating magnitude constant may be related. RIGGS[1] has presented a long series of observations showing that the grating "constant" varies with the q-value of the star images. Gratings appear to offer less promise than neutral half-filters for the establishment of a fundamental magnitude scale by photography. However, the reduction methods to be described can be applied to the plates obtained with either device.

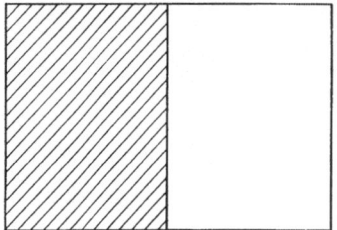

Fig. 17. Neutral half-filter.

A neutral half-filter consists of a glass optical filter of such properties that it produces the required response function. On one half of this filter has been deposited by sputtering or evaporating a uniform neutral metal film (see Fig. 17). Neutrality is of prime importance; non-neutral filters diminish the light of stars of different colors by different amounts. Platinum[2] or Chromel A films[3] have proved useful. Platinum appears to be most nearly neutral when deposited in such thickness that it diminishes the light intensity by approximately one magnitude. For references see WEAVER [1]. Fortunately, such a filter constant is very convenient. Uniformity of the filter over the usable area of the photographic field is also of prime importance. Lack of such uniformity introduces complications in the reductions and may introduce significant errors in the final results. Every effort should be made to construct a uniform filter. A thorough laboratory investigation of each filter for uniformity and filter constant, k, measured in mag. on a Pogson Scale, should be made before the filter is used at the telescope.

31. Use of the half filter. In use, the half filter is placed directly in front of the photographic plate (usually with metal film away from the emulsion) and a photograph is taken of the star field in which the scale is to be established. Under such circumstances, one half of the field is photographed through the clear glass ("unfiltered images"), the other half through the glass and neutral filter ("filtered images"). Since we require filtered and unfiltered images for each star, this exposure by itself does not suffice. Therefore, the filter is rotated 180° in the plate holder so that clear glass and filter portions are interchanged, the plate is shifted slightly in position, and a second exposure, of the same duration as the

[1] P. S. RIGGS: Publ. Astronom. Soc. Pacific **60**, 307 (1948).

[2] H. KIENLE: Handbuch der Experimentalphysik, Vol. 26, p. 768. Leipzig: Akademische Verlagsgesellschaft 1937.

[3] MARY BANNING: J. Opt. Soc. Amer. **37**, 686 (1947). — H. F. WEAVER: Publ. Astronom. Soc. Pacific **62**, 167 (1950).

first, is made on the star field. The first photograph was taken with the filter in what we shall designate as position 1, the second with the filter in position 2. On a photograph taken in the manner described we obtain two images of every star in the field (except, of course, that the faintest stars photographed through the clear glass produce no images when photographed through the filter), one through the filter, the other through the glass. Note, however, that we cannot assume that the corresponding images are, except for magnitude reduction, identical in their photometric properties.

(i) They were taken at different times and we cannot assume that seeing, focus, guiding, and other image-form factors were identical during the two exposures. Failure to allow for this possibility of change between exposures has tended to introduce errors into the results obtained by earlier reduction procedures.

(ii) The two exposures have received different treatment, in that exposure 1 was post-exposed to sky light; exposure 2 was pre-exposed to sky light.

Point (ii) is the principal objection to the neutral half-filter method. There appears to be no satisfactory way to allow for differences in pre-exposure and post-exposure effects if they are found to exist. It is particularly unfortunate that in pre-exposure we have high intensity on low intensity; in post-exposure we have low intensity on high intensity. The exposure effects in the two instances are very different. See the results of WEBB and EVANS, Sect. 3. Experience has shown that the half-filter method is practical only if sky fog is so slight that pre-exposure and post-exposure effects are negligible. Only experimentation with the method can establish a valid upper limit on the amount of sky fog that is tolerable; different emulsions behave differently. Treatment of the emulsion before and after exposure also plays a role in fixing the allowable upper limit of fogging arising from pre-exposure and post-exposure to sky light. Fortunately, the process of photometric reduction provides for each plate a check on the importance of pre-exposure and post-exposure effects on that photograph. Each half-filter exposure provides a magnitude scale. If the scales derived from the two exposures do not agree, it is probable that pre-exposure and post-exposure effects have reached a level where they are objectionable. The plate should be discarded.

The danger warned against in point (i) is easily removed. To remove it, we require a second photograph of the field through a glass filter identical to the one on which the metallic film forming the neutral filter was deposited. This second photograph, termed the reference plate, need not be taken on the same night as the half-filter photographs, nor does it need to be of precisely the same exposure time as the half-filter photographs. But we must take care to assure that the reference plate, like the half-filter plate, is photometrically homogeneous. There must be no significant variation of photometric properties or quality over the measured protion of the field. It is advantageous to center the half-filter and reference plate exposures about the time of meridian passage of the field in order to minimize extinction color-effects.

32. The reduction process. Consider first only those images on the half-filter plate taken with the filter in position 1. For star images in the half of the field not covered by the neutral filter and taken through the glass we have measurements q_g; for star images in the half of the field covered by the half-filter we have measurements q_f. For all stars we have measurements q_r on the reference plate. For each star having a q_g, we plot q_g as a function of the corresponding q_f (Fig. 18) and draw the best smooth curve through the plotted points; this is the transformation curve—for brevity, the T-curve—that permits as to infer the q_g-value for any star for which we have a q_r-value.

For the filtered images for which we have q_f-values, we also have q_r-values. For each of these q_r's we read off from the T-curve the corresponding q_g. The T-curve and the reference plate permit us to infer for each filtered image what

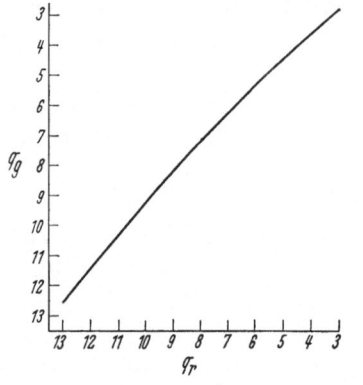

Fig. 18. The T-curve.

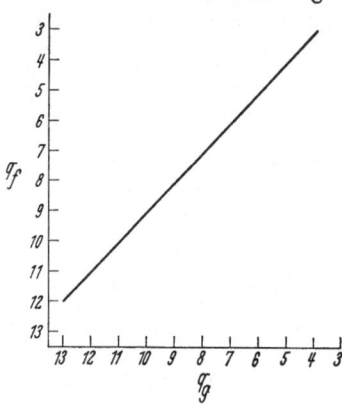

Fig. 19. The χ-curve

q-value would have been measured if no neutral filter had been present. We thus obtain a consistent set of q_f, q_g values which differ (aside from observational errors) only because of the diminution of brightness caused by the neutral half-filter.

From this set of q_f, q_g values we plot q_f as a function of q_g. The best smooth curve drawn through this set of points gives what SCHWARZSCHILD designated as the χ-curve, Fig. 19. The relation $q_f = \chi(q_g)$, which specifies for any q_g-value what q-value would be measured for a star precisely k magnitudes fainter, provides the solution to the problem. From it we derive the reduction curve, which permits us to find magnitudes for all the stars measured.

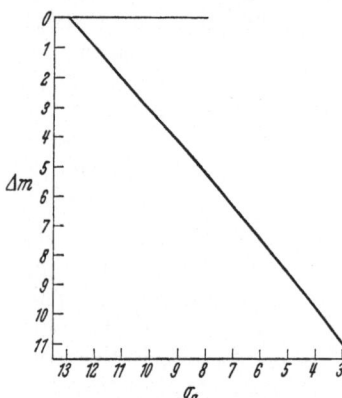

Fig. 20. The reduction or $\varDelta m$, q_g curve.

To draw the reduction curve we assign to an arbitrary point on the χ-curve the magnitude zero. For convenience we take $q_g = 13.00$ as corresponding to $\varDelta m = 0.00$. A star of magnitude $\varDelta m = k$ would then correspond to the value q_{f1} which is read from the χ-curve for $q_g = 13.00$; $q_{f1} = \chi(13.00)$. A star of magnitude $2k$ would then, on the q_g scale, correspond to the value $q_{f2} = \chi(q_{f1})$; a star of magnitude $3k$ would then, on the q_g scale, correspond to the value $q_{f3} = \chi(q_{f2})$; and so forth. In this way we find a series of points at $\varDelta m = 0$, k, $2k$, $3k$, ... on the reduction curve. Through these points we draw the smooth curve representing magnitude, $\varDelta m$, as a function of q_g (Fig. 20).

The self-consistency of this curve can be tested. At some point on the reduction curve, say at $\varDelta m = 2.5k$, we read off the q_g value and then from the χ-curve find corresponding values at $0.5k$, $1.5k$, $3.5k$, ... and plot these newly determined points to see if they lie on the reduction curve previously drawn through the points k, $2k$, ... If they do not fall on the original curve, adjustments in the form of the curve must be made until it is self-consistent.

The exposure with the filter in position 1 thus provides a magnitude scale with arbitrary zero-point. The entire procedure of deriving a T-curve, χ-curve, and reduction curve must then be repeated for the photograph taken with the filter in position 2. We then obtain a second magnitude scale, independent of the first except that the same reference plate was used in deriving each of them.

33. Tests for pre-exposure, post-exposure effects. To determine whether pre-exposure, post-exposure effects have in any way vitiated the scales derived, we test one scale against the other. Magnitudes on the scale derived from the filter 1 position will be designated $\Delta m^{(1)}$; those on the scale derived from the filter 2 position, $\Delta m^{(2)}$. Table 6 illustrates a simple way to make the test. Col. (1) lists an arbitrary set of q_r-values, say 3.0, 4.0, 5.0, ... Col. (2) and (3) give the equivalents of the q_r-values as found from the $T^{(1)}$-curve and the $T^{(2)}$-curve.

Table 6. *Test for pre-exposure, post-exposure effects.*

q_r (Col. 1)	$q_g^{(1)}$ (Col. 2)	$q_g^{(2)}$ (Col. 3)	$\Delta m^{(1)}$ (Col. 4)	$\Delta m^{(2)}$ (Col. 5)	$\delta \Delta m$ (Col. 6)
q_{r1}	$q_{g1}^{(1)}$	$q_{g1}^{(2)}$	$\Delta m_1^{(1)}$	$\Delta m_1^{(2)}$	$\Delta m_1^{(1)} - \Delta m_1^{(2)}$
q_{r2}	$q_{g2}^{(1)}$	$q_{g2}^{(2)}$	$\Delta m_2^{(1)}$	$\Delta m_2^{(2)}$	$\Delta m_2^{(1)} - \Delta m_2^{(2)}$
q_{r3}	$q_{g3}^{(1)}$	$q_{g3}^{(2)}$	$\Delta m_3^{(1)}$	$\Delta m_3^{(2)}$	$\Delta m_3^{(1)} - \Delta m_3^{(2)}$
q_{r4}	$q_{g4}^{(1)}$	$q_{g4}^{(2)}$	$\Delta m_4^{(1)}$	$\Delta m_4^{(2)}$	$\Delta m_4^{(1)} - \Delta m_4^{(2)}$
$\vdots$	$\vdots$	$\vdots$	$\vdots$	$\vdots$	$\vdots$

In Col. (4) is the $\Delta m^{(1)}$-value corresponding to $q_{gi}^{(1)}$; this is found from the reduction curve for position (1). In Col. (5) is the $\Delta m^{(2)}$-value corresponding to $q_{gi}^{(2)}$. In Col. (6) is the difference $\delta \Delta m_i = \Delta m_i^{(1)} - \Delta m_i^{(2)}$. The quantity $\delta \Delta m$ should be essentially a constant. If it varies systematically by a significant amount—a few hundredths or a few tenths of a magnitude, depending upon the level of accuracy the investigator requires—then it is clear that photographic effects, primarily pre-exposure, post-exposure effects, are present and have affected the magnitude scales derived. In this case the plate and the results derived from it should be discarded.

34. Reduction to a homogeneous scale. If the differences $\delta \Delta m$ are found to lie within the tolerance set by the investigator, then two further steps can be taken to put all the data on a homogeneous system.

(i) The mean system $\Delta m_i = \frac{1}{2}(\Delta m_i^{(1)} + \Delta m_i^{(2)})$ is formed and used in drawing new reduction curves. This is easily accomplished by plotting $q_{gi}^{(:)}$, Col. (2), against Δm_i and $q_{gi}^{(2)}$, Col. (3), against Δm_i. From these new reduction curves Δm-values are read off for all the $q_g^{(1)}$- and $q_g^{(2)}$-values. These Δm's represent the final data derivable from the measurements of the unfiltered images.

If the neutral filter is of excellent quality, Δm-values derived from the measured q_f's and corrected by the filter factor, k, may be included in the discussion. Generally, the Δm's derived from the filtered images will be of distinctly lower weight than those derived from the unfiltered images.

(ii) The reference plate can be used to improve individual magnitude values. Two ways to do this may be suggested. (a) We plot Δm_i as a function of q_{ri}, Col. (1), draw the reduction curve thus defined, and read off Δm's for all measured q_r-values. These Δm's are then combined directly with those derived from $q_g^{(1)}$ and $q_g^{(2)}$ and possibly $q_f^{(1)}$ and $q_f^{(2)}$. (b) It may be undesirable to go through the transformation curves involved in Cols. (1), (2), (3). In this case, we deal with individual stars. A preliminary mean Δm is formed for each star, Δm's from $q_g^{(1)}$, $q_g^{(2)}$, $q_f^{(1)}$, $q_f^{(2)}$ being included, weighted if necessary. We plot the Δm of each star as a function of the star's q_r-value and draw through the many plotted points the smooth best curve of Δm as function of q_r. This reduction curve is then used to derive Δm's from the measured q_r's; the Δm's so derived are then combined with Δm's found from $q_g^{(1)}$, $q_g^{(2)}$, $q_f^{(1)}$, $q_f^{(2)}$ to form the final values.

In either method, weights can be applied to the Δm-values found from q_r, q_g, q_f if such are thought necessary in deriving the final results.

The reduction process just described provides magnitudes on a Pogson scale for all stars for which q-values were measured. The zero-point of this scale is arbitrary. It remains to determine the constant which, when added to the Δm's, will convert them to some standard magnitude system. Determination of this constant must be made by a comparison procedure.

X. Scale and zero-point transfers.

35. Transfers. A photometric scale and zero point can be transferred by comparing focal images of the group of stars under study (the "field") with focal images of a group of stars having accurately known magnitudes on some specified photometric system (the "standard field"). The procedure involves various assumptions in regard to similarity of image properties on the two exposures. Successful application requires that the observer pay careful attention to a number of details.

(i) The exposures on the field and the standard field must be of identical duration.

(ii) Times of exposures should be so arranged that the zenith distances of the field and standard field should be equal at their respective midexposure times. In this case extinction is identical for the two fields; no differential extinction correction is necessary. If this ideal situation cannot be achieved, the difference in zenith distances at midexposure times should be as small as possible so that the correction for difference in extinction is minimized.

(iii) The exposures should not be superimposed on one plate. Such a procedure causes pre-exposure for one field, post-exposure for the other. The camera should be arranged so that each field is photographed on one half of a plate that is offset in such a way that the center of the half plate being exposed is on the optic axis of the camera. If an arrangement for such offsetting and exposing of one half plate at a time is not possible, then each field should be photographed on a separate plate, the plates used for the photographs being two that were packed face to face in the box. Such a pair of plates must be developed together.

It is possible that in the case of a Schmidt telescope of large unvignetted field and of excellent image quality, this precaution can be ignored. A portion of the Schmidt plate is uncovered for each exposure; the fields are photographed off the optic axis. Before this procedure is followed, the investigator should assure himself that there are no corrections depending on position angle and distance or that, if there are such, he can determine and apply them.

(iv) The exposures must be made under such conditions and the development must be so handled that background fog for the two exposures will be light; it must be identical for the two exposures. Unequal fog on the two exposures, particularly if it is dense, is an indication that poor results will be obtained.

If such precautions as these are observed and reliable plates of uniform surface sensitivity have been used, it is generally assumed that the photometric properties of images on the two exposures are identical. One then derives the m, q-curve from the stars in the standard field, and, from this curve, infers the magnitudes of field stars for which q-values have been measured. This simple and frequently-used procedure can, at times, yield excellent results and, at other times, lead to serious systematic errors in the magnitudes. Such systematic errors are caused by those image-form factors over which the observer has little or no control as, for example, changes in seeing, focus, and guiding from one exposure to the next. The effect of such changes has been discussed earlier. Primarily, they cause the m, q-curve to rotate about the optimum image point, q_0 (see Sect. 12).

36. Optimum-image-point methods of reduction. TRUMPLER[1], BLAAUW[2], and WEAVER[3] have devised reduction methods based on the optimum-image-point concept. These methods have as their purpose the elimination of m, q-curve rotation caused by changes in image-form factors. We here follow the basic procedure suggested by WEAVER; it is the most widely applicable of the proposals.

The optimum-image-point method has as its physical basis two empirically observed facts.

(i) To a first approximation a change in the image-form factors causes a rotation of the m, q-curve around the optimum image point, q_0.

(ii) Empirically, q_0 is found to have a fixed numerical value, at least for a given camera and for measurements made with a photometer of specified characteristics. It appears to be at most only slightly dependent upon camera properties; it is probably more sensitive to photometer characteristics. TRUMPLER, specifying the optimum image point in a different manner, stated that it is located approximately 1.5 magnitudes above the plate limit.

To employ the optimum-image-point method of reduction, we require several comparisons of a star field with a standard field. Preferably, each comparison is of a different exposure time, $e_1 > e_2 > e_3 \ldots$. The longest and shortest of these should differ by a factor of 10 or more. It is not necessary that the comparisons all be made with the same standard field, but it is imperative that the magnitudes in the standard fields be on one homogeneous system.

For each comparison the optimum image point occurs at the same q_0-value. Because of the variety of exposure times, this corresponds in each comparison to a different apparent magnitude, namely, a value 1.5 magnitudes above the plate limit. If we transform all of the measurements to one q-scale and plot all of the m, q-curves together, marking on each curve where its optimum image point occurs, these points will spread along the magnitude scale. This family of plotted curves illustrates directly how the scale determinations made from the various comparisons by the customary transfer procedure would have agreed. To derive the best curve from this family, we rotate each m, q-curve around its q_0 point so that each rotated curve coincides as nearly as possible with all the q_0 points. When the rotations have been performed, a single mean curve is drawn to represent all the adjusted m, q-curves. This master curve (transferred to the q-system of each plate) is then used for all conversions of measured q-values to magnitudes.

An example will illustrate the steps required.

37. Example. We have three comparisons between a field and a standard field. The three plates, each of which contains two exposures, one on the field, the other on the standard field, will be designated α, β, γ; $e_\gamma > e_\beta > e_\alpha$. We choose

[1] R. J. TRUMPLER: Lick Obs. Bull. **18**, 176 (1938).
[2] A. BLAAUW: Bull. Astronom. Inst. Netherl. **9**, 141 (1940).
[3] H. F. WEAVER: Astrophys. Journ. **106**, 366 (1947).

the q-system of the field of plate γ as the one to which we will transform. For the transformation, we consider only the field exposures. Subscript F designates that the quantity so marked refers to the field stars. Fig. 21 a, b illustrates the transformation curves. We designate $q_{\alpha F}$ transformed to the $q_{\gamma F}$-system by q_{α}^{*F}; $q_{\beta F}$ transformed to the $q_{\gamma F}$-system by $q_{\beta F}^{*}$.

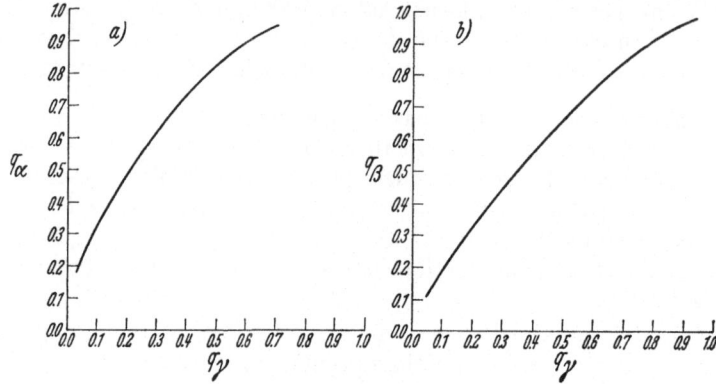

Fig. 21 a and b. Transformation curves.

We next derive the three m, q-curves for the standard field as shown in Fig.22 a to c. Subscript S designates that the quantity so marked refers to the standard field.

The m, q-curves are next placed together on one diagram. Care is here necessary because of the extinction differences between the various exposures. How-

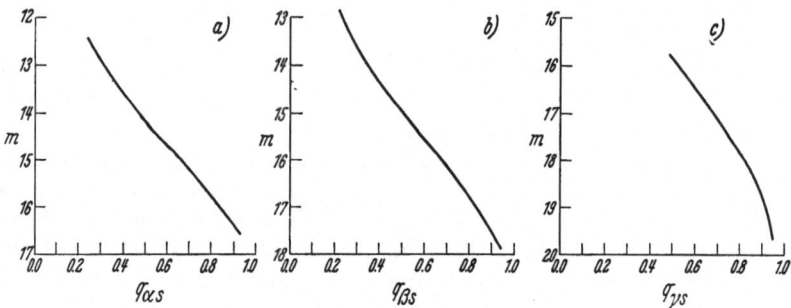

Fig. 22 a—c. The m, q curves for the three exposures.

ever, we adopt the customary method of photographic photometry, ignore differences in colors of the stars and employ a mean extinction coefficient. The quantities $q_{\alpha F}^{*}$ and $q_{\beta F}^{*}$ are put on the extinction system of $q_{\gamma F}$ by the transformation curves. For homogeneity, we must also reduce the standard fields to the extinction system of the field on plate γ. We therefore add the necessary extinction corrections and obtain the curves

$$(m + \delta m_{\alpha}, q_{\alpha S}), \quad (m + \delta m_{\beta}, q_{\beta S}), \quad (m + \delta m_{\gamma}, q_{\gamma S})$$

where, in terms of Eq. (19.2)

$$\begin{aligned}
\delta m_{\alpha} &= \bar{\beta}(\sec z_{\alpha S} - \sec z_{\gamma F}), \\
\delta m_{\beta} &= \bar{\beta}(\sec z_{\beta S} - \sec z_{\gamma F}), \\
\delta m_{\gamma} &= \bar{\beta}(\sec z_{\gamma S} - \sec z_{\gamma F}).
\end{aligned} \right\} \tag{37.1}$$

Remark on the extinction correction. We have followed tradition in this treatment of the extinction correction; it is an adequate treatment for the purpose at hand. More important is the question of the earlier treatment of the magnitudes in the standard field. If, as is customary in photographic photometry, they are magnitudes observed at zenith distance z and reduced to the station zenith with a constant coefficient, they then contain a fictitious color equation as do the NPS stars discussed earlier. If it is possible to make the correction, the magnitudes in the standard field should be converted to observed magnitudes at the zenith distance at which the photograph was taken. For this one must know the color of each star, the extinction coefficients β_0 and β_1, and precisely how the magnitudes tabulated for the standard field were originally corrected for extinction.

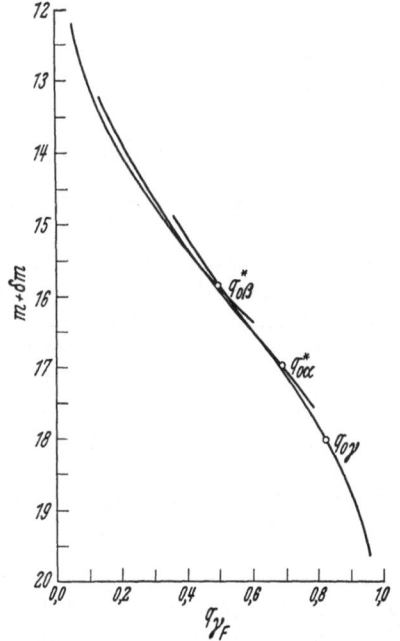

Fig. 23. The m, q curves transformed and plotted together.   Fig. 24. Final mean $\overline{m}, q_{\gamma F}$ curve.

We assume that a q_{iF}-system is equivalent to the corresponding q_{iS}-system and, through the transformation curves, form and plot together the curves

$$(m + \delta m_\alpha, q^*_{\alpha S}), \quad (m + \delta m_\beta, q^*_{\beta S}), \quad (m + \delta m_\gamma, q_{\gamma S})$$

as shown in Fig. 23. The three curves are not coincident because of the incorrectness of the assumption of equivalence of q_{iF} and q_{iS}. The diagram gives an immediate visual impression of the disagreement we would have encountered in the magnitudes that would have been derived by traditional procedures. Note further that if the customary procedure had been employed and we had formed for each star the mean of all observed magnitudes, we would have obtained a final magnitude system with discontinuities in it since each exposure does not cover the entire range of magnitudes observed.

Adjustment by the optimum-image-point method is accomplished by rotating each m, q-curve (traced on transparent paper) about its q_0 point until it is in the most satisfactory agreement with the other q_0 points. A mean curve, $(\overline{m}, q_{\gamma F})$, Fig. 24, is then drawn. This adjustment leaves us with a smooth magnitude system.

To derive magnitudes for the field stars, we may employ the final $\overline{m}, q_{\gamma F}$-curve for plate γ. For plates α and β, we employ the transformation curves to form

$$(\overline{m} + \delta\overline{m}_\alpha, q_{\alpha F}), \qquad (\overline{m} + \delta\overline{m}_\beta, q_{\beta F})$$

where

$$\left.\begin{array}{l} \delta\overline{m}_\alpha = \overline{\beta}\,(\sec z_{\alpha F} - \sec z_{\gamma F}), \\[2mm] \delta\overline{m}_\beta = \overline{\beta}\,(\sec z_{\beta F} - \sec z_{\gamma F}). \end{array}\right\} \tag{37.2}$$

These curves may be used to find magnitudes from the $q_{\alpha F}$, $q_{\beta F}$-values. The final magnitude for any star is then simply the average of all the observed values for that star.

The basic optimum-image-point method of reduction is capable of many variations. Some of these have been discussed by WEAVER[1].

XI. Photography as a method of photometric interpolation.

38. Remarks on photographic scale transfers. Experience has shown that use of the optimum-image-point method (or of one of its many variations) produces the most satisfactory photometric transfers. However, scale transfer by photography is, at best, a difficult, tedious, and uncertain procedure. It is possible to establish fairly reliable photographic scales over the range of magnitudes lying between the optimum image points of the longest and shortest exposures. But even in this range all m, q-curves will not coincide even when rotated; the mean curve is therefore still a compromise. Outside the optimum-image-point range, particularly among the brighter stars, variations in the image-form factors during the exposures on the field and standard field to some extent change the shape of the m, q-curve as well as rotate it. The amount of this change of shape we cannot determine; this leaves the final transferred scale uncertain.

39. Present-day trends. In recent years there has been an effort to raise the general level of accuracy of photographic photometry. To a large extent this has been made necessary by the increasingly general use of photoelectric photometers. Observers are no longer satisfied with the accuracy previously achieved by photographic procedures. Many problems of modern astronomy cannot be solved with the inaccurate photometric results of the past.

There has been a strong trend to combine photoelectric and photographic methods in photometry to take advantage of certain properties of each. Photography is a poor device for transferring photometric scales and zero-points; it is very sensitive to image-form factors. On the other hand, it can provide a means of interpolating magnitudes with very satisfactory accuracy if a sequence of stars of known magnitudes is available on the plate. It provides a saving of telescope time since it records simultaneously all the stars in the field. Photoelectric methods are excellent both for transferring scales and zero-points and for interpolating magnitudes in a sequence of known values. Extremely high accuracy can be achieved in either procedure. But photoelectric photometry is a process that requires much telescope time. For those problems in which magnitudes must be measured for many stars in one area, the trend has therefore been to use photoelectric techniques to establish a magnitude sequence over the magnitude range in which measurements are to be made. The scale can be very adequately established if three or four stars per magnitude interval are measured photoelectrically. A photograph of the region is then taken and measured photometrically. The stars with photoelectrically determined magnitudes

---

[1] H. F. WEAVER: Astrophys. Journ. **106**, 366 (1947).

are used to derive an m, q-curve for the plate. From this curve accurate magnitudes of all other stars measured in the area can be derived. This process appears at present to produce the best results for the least expenditure of time and effort. It is to be recommended above all others if accurate results are required.

40. Observational precautions. Two precautions must be taken by the observer using such a photographic interpolation technique.

1. He must match the response function of his telescope-plate-filter combination to that of the photoelectric photometer. Alternatively, he might match the response function of the photoelectric photometer to that of the camera. At any rate, ideally, photometer and camera should have identical response functions which, presumably, have been chosen as the best to solve the particular problem at hand.

In photoelectric photometry observations are always made in at least two colors. The photographic observer should likewise determine magnitudes in at least two colors. Then if, in practice, the photoelectric and photographic response functions are not precisely matched, the observed magnitudes can be accurately transformed from one system to another provided the required corrections are small.

2. Photoelectric magnitudes are generally given for outside the Earth's atmosphere. The photographic plate of the field in which magnitudes are to be derived will be taken at zenith distance, z. Extinction affects the magnitude of each star in the photoelectric sequence; the photoelectric magnitude of each sequence star should therefore be corrected for extinction in accord with the formula

$$\delta m(z) = [\beta_0 + \beta_1 (B\text{-}V)] \sec z \qquad (40.1)$$

before being used in construction of the m, q-curve. The zero-atmosphere color, $B\text{-}V$, is known for each sequence star; the parameters β_0 and β_1 should preferably by known from photoelectric observations on the night for which the photograph was taken, otherwise mean values can be used. Failure to allow for extinction in the manner suggested can add appreciably to the scatter of points used in drawing the m, q-curve. The procedure advocated reduces the magnitudes of the sequence stars strictly to their observed values at zenith distance z; we derive for the field stars magnitudes that refer to zenith distance z. If we derive from plate i magnitudes on the B system, designated by $m_{\text{obs}\,(i)}^{(B)}$ when affected by extinction, from n plates, taken at zenith distances z_i, we have

$$B = m_{\text{obs}\,(1)}^{(B)} - [\beta_{0\,(1)}^{(B)} + \beta_{1\,(1)}^{(B)} (B\text{-}V)] \sec z_1$$

$$B = m_{\text{obs}\,(2)}^{(B)} - [\beta_{0\,(2)}^{(B)} + \beta_{1\,(1)}^{(B)} (B\text{-}V)] \sec z_2$$

$$\vdots$$

$$B = m_{\text{obs}\,(n)}^{(B)} - [\beta_{0\,(n)}^{(B)} + \beta_{1\,(n)}^{(B)} (B\text{-}V)] \sec z_n$$

or, from all the observations

$$B = \overline{m_{\text{obs}}^{(B)}} - \overline{\beta_0^{(B)} \sec z} - \overline{\beta_1^{(B)} \sec z} (B\text{-}V) \qquad (40.2)$$

where the averages of $\beta_0^{(B)}$ and $\beta_1^{(B)}$ are weighted by $\sec z$.

If we have also made observations on the V-system we can write

$$V = \overline{m_{\text{obs}}^{(V)}} - \overline{\beta_0^{(V)} \sec z} - \overline{\beta_0^{(V)} \sec z} (B\text{-}V). \qquad (40.3)$$

Combining the two expressions,

$$B\text{-}V = \frac{\overline{m_{obs}^{(B)}} - \overline{m_{obs}^{(V)}} - \overline{\beta_0^{(B)} \sec z} + \overline{\beta_0^{(V)} \sec z}}{1 - (\overline{\beta_1^{(V)} \sec z} - \overline{\beta_1^{(B)} \sec z})}. \tag{40.4}$$

This last formula is very similar to the analogous one employed in photoelectric photometry.

If we have a photoelectric sequence in a field and if we know the extinction coefficients β_0 and β_1 in both B and V (or whatever other response functions are used) we can photographically derive magnitudes and colors for outside the atmosphere. The equations involved are very similar to those of photoelectric photometry. The technique is the same for the two methods; we find first the color, then the magnitude. The advantage of having matched response functions for the camera and the photoelectric photometer is evident; one can, in such a case, employ the photoelectrically determined extinction coefficients for the photographic work and no later transformation of systems is required.

XII. Determination of color indices.

41. Color indices. The problem of determining color indices by photographic methods has been solved in a variety of ways. In the most direct approach to the problem photographic and photovisual scales are set up by transfer from some standard field (the NPS) and photographic and photovisual magnitudes derived for each field star of interest. The difference of the two magnitudes is, by definition, the required color index or, for short, color. This straightforward procedure appears to be the most satisfactory to use in modern work with the difference, however, that scales are now established photoelectrically in the area in which photometry is to be carried on. Photography serves only as an interpolation device for refering the field stars to the photoelectric sequence.

The difficulty and uncertainty of photographic scale transfer made earlier investigators seek methods of determining color indices directly. For the astronomer setting up fundamental scales, such a direct determination of color was important; it provided a check on the parallelism of the photographic and photovisual scales. Various methods of obtaining yellow and blue exposures on one plate and procedures for deriving color indices from them were devised[1]. However, in view of the present-day understanding of transformation problems and the subtleties of photographic effects, these methods do not appear now to offer important possibilities for precise quantitative development. If they have applications at the present time, it is, rather, in some of the qualitative photometric classification or discovery methods derived from them or related to them.

XIII. Quantitative classification photometry.

42. Multicolor photometry. Wide-band and narrow-band spectrophotometry, giving information about the spectral energy distribution in stellar spectra and other sources, provides important physical data about both the stars and the interstellar medium. With certain notable exceptions, such multicolor photometry has been carried on photoelectrically. Since photoelectric methods permit observation of only one star at the time, observations accumulate slowly. If

---

[1] F. H. SEARES: Proc. Wash. Nat. Acad. Sci. **2**, 521 (1916). — G. A. TICHOFF: Bull. Acad. Russe Sci. **5**, No. 8 (1916). — Astronom. Nachr. **218**, 145 (1923). — R. J. TRUMPLER: Lick Obs. Bull. **14**, 89 (1929). — H. O. ROSENBERG: Astrophys. Journ. **83**, 67 (1936). — Many additional references are given by WEAVER [1].

data are required for many objects in a single field, photographic interpolation among photoelectrically established standards offers promising possibilities if plate-filter combinations can be found to provide the proper response functions. No new photometric problems arise; the photographic interpolation techniques required for such problems are identical with those described in Sect. 40.

Two types of investigations where such methods might be found useful are (1) STRÖMGREN's[1] photometric spectral classification procedure which makes use of very narrow band interference filters, and (2) the BECKER[2] two-color method of establishing the spectral type and the amount of interstellar reddening. The U, B, V system has been effectively utilized by JOHNSON and MORGAN [3] in a slight modification of the Becker method.

Particular care will be required in the photographic application of the Strömgren technique; magnitudes must be established with an error of less than 0.02 mag.

XIV. Qualitative classification photometry.

43. General remarks. Quantitative classification photometry provides the classification of an object on a precise numerical scale. It provides, for example, the spectral type and the absolute magnitude of a star. Qualitative classification photometry provides the answer to a simpler question: Is an object of a certain specified type or not? In qualitative classification photometry we might use photographic means to select from among all objects in a certain field and down to some specifiable magnitude limit, all those objects of a specific type as, for example, all planetary nebulae or all OB stars. To specify a method of making such a selection, we examine the distribution of energy in the spectra of objects of the type we wish to detect and then devise comparisons of spectral regions that will permit us to distinguish the specified object from all others we photograph. Choice of practical spectral regions may not always be easy; practical solution to the problem may not always be possible with the observational materials available. A practical solution is one that involves no measurements. The observer should be able to determine whether an object is or is not of the required type by visual inspection of the plate or plates.

44. A photometric sieve to detect planetary nebulae. The spectrum of a typical planetary nebula is shown in Fig. 25. It consists primarily of isolated emission

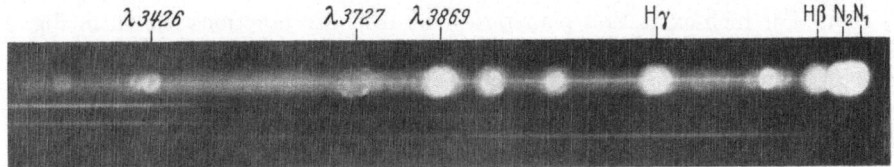

λ3426   λ3727 λ3869   Hγ   Hβ N₂N₁

Fig. 25. Spectrum of a typical planetary nebula (Lick Observatory photograph).

lines arising from the nebular shell. Clearly, we can detect all such nebulae if we take two exposures, the first with a plate-filter combination having a response function centered on an intense emission line, the second with a plate-filter combination having a response function very near to the first, but just to one side of the emission line. Comparing the two plates in a blink microscope, we shall find the planetary nebulae recorded on one plate but not on the other.

[1] B. STRÖMGREN: Vistas in Astronomy, Vol. 1. London and New York: Pergamon Press 1955. — B. STRÖMGREN and K. GYLDENBKERNE: Astrophys. Journ. **121**, 43 (1955).

[2] W. BECKER: Astrophys. Journ. **107**, 278 (1948).

We shall find the stars recorded on both plates with little change in intensity, provided we properly adjust the exposure times.

From practical considerations, we might choose Hα, λ 6563, and [NII], $\lambda\lambda$ 6548 and 6584, as the group of emission lines on which we center one response function, and adopt the spectral region just to the longward as the companion region. Fig. 26 shows computed response functions of this type. The plate-filter combinations were chosen by inspection of the plate and filter catalogs, with subsequent computation of the response functions for promising plate-filter combinations.

The measurements of line intensities made for planetary nebulae by MIN-KOWSKI and ALLER[1] show that for the high-excitation planetaries the energy contained in the N_1, N_2 lines of (OIII] at $\lambda\lambda$ 5007 and 4959 plus Hβ at λ 4861 is greater than that contained in Hα and [NII] in the ratio of approximately

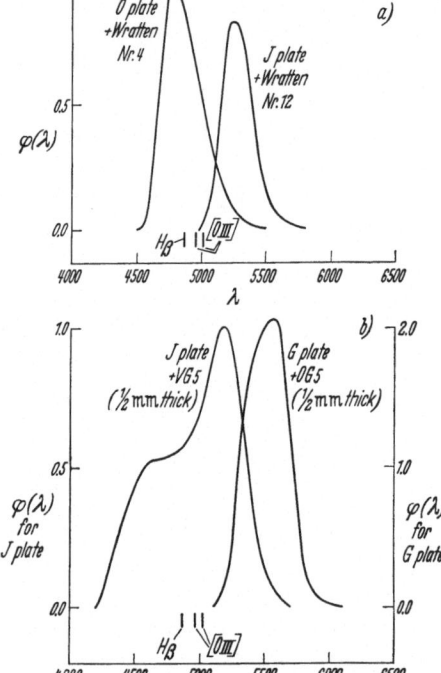

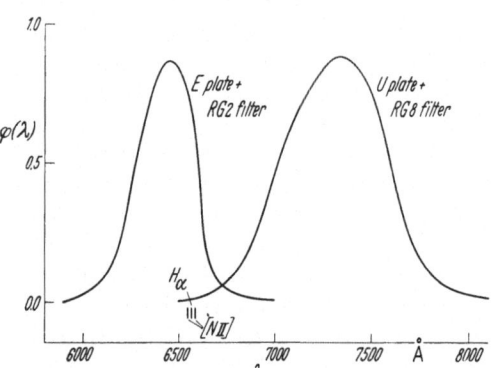

Fig. 26. Plate-filter combination to permit discovery of planetary nebulae from Hα images.

Fig. 27 a and b. Two different plate-filter combinations to permit discovery of planetary nebulae from Hβ, N_1, N_2 images.

5 to 1. For high-excitation planetaries the response functions shown in Fig. 27 should be more satisfactory than those illustrated in Fig. 26. In Fig. 27, the response function that permits photography of the planetary nebulae includes Hβ, $N_1 N_2$; the response function that is blind to the planetaries lies just to the longward of Hβ.

Both combinations have worked well in practice. They are convenient in that the rather narrow response functions diminish the prominence of the stars on the plate. Moreover, photography can be carried on with them in bright moonlight with little fogging of the plates.

The Hα region is less affected by interstellar extinction than the Hβ, N_1, N_2 region. When the interstellar extinction becomes as great as six magnitudes for the Hβ, N_1, N_2 region, the Hα region is the better of the two even for the high-excitation planetaries. Ordinary emission-line stars should not cause confusion with either combination, though one might find that for such objects the images on the Hα or Hβ photograph are slightly stronger than normal for their counter-

[1] R. MINKOWSKI and L. ALLER: Astrophys. Journ. **124**, 93 (1956).

parts on the companion plate. Some hydrogen emission nebulae may be found, but generally their forms will reveal their nature. Some checks on the nature

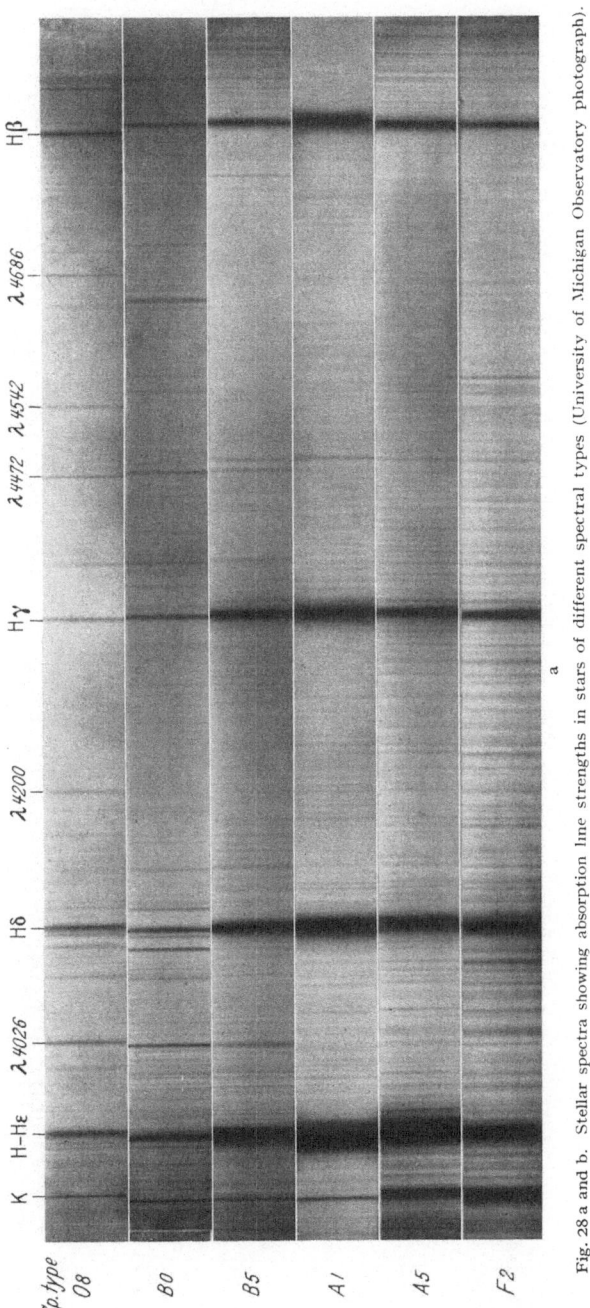

Fig. 28 a and b. Stellar spectra showing absorption line strengths in stars of different spectral types (University of Michigan Observatory photograph).

of the objects discovered can be made if both the Hα and Hβ techniques are utilized in the search.

45. A photometric sieve for *OB* stars. Fig. 28 exhibits a series of spectra ranging in type from *O* to *M*. In the wavelength range covered in the figure,

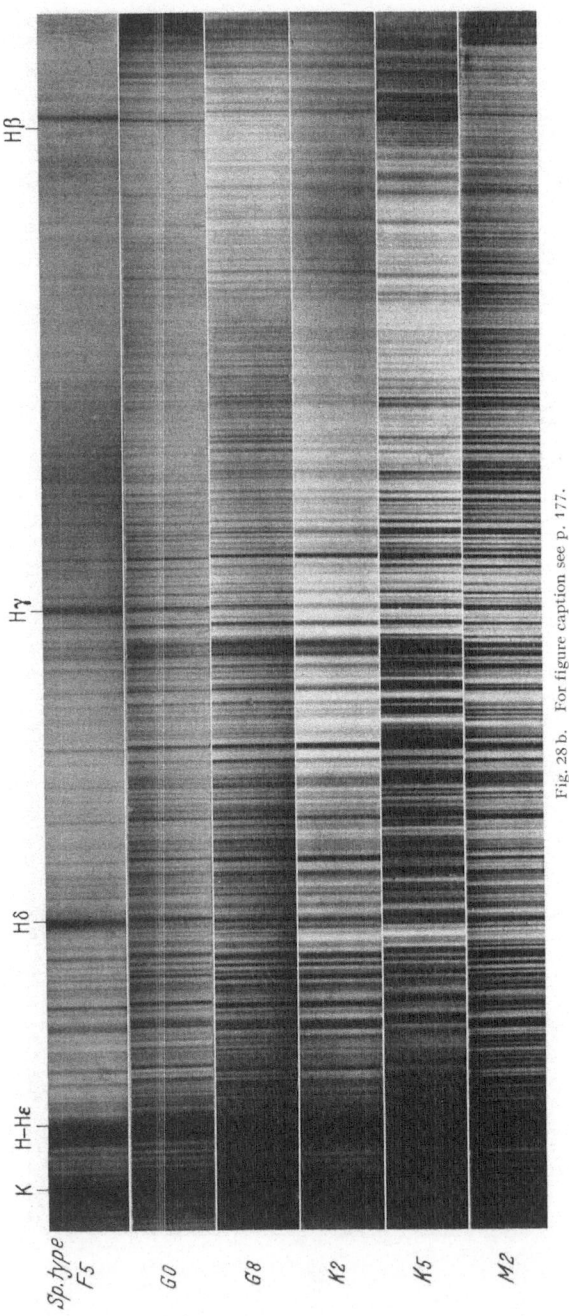

Fig. 28 b. For figure caption see p. 177.

the outstanding features that distinguish the *O* and early *B* stars from the later types are:

1. The K line of CaII at $\lambda\,3934$ is very weak in the early-type stars. In these early-type the line is, in fact, interstellar. It increases enormously in strength starting with the A stars.

2. The H_ε line of hydrogen at $\lambda\,3970$ is weak in the O and early B stars. It increases in strength as we proceed to the later B stars and reaches a maximum of intensity at $A\,0$. For stars $A\,0$ and later the strength of the H line of CaII is added to that of H_ε.

The early-type high-luminosity stars are similar in spectral properties to those of the low-luminosity stars illustrated in the figure, though the lines of hydrogen and calcium are less broad. Features described in (1) and (2) are observed in both high- and low-luminosity stars.

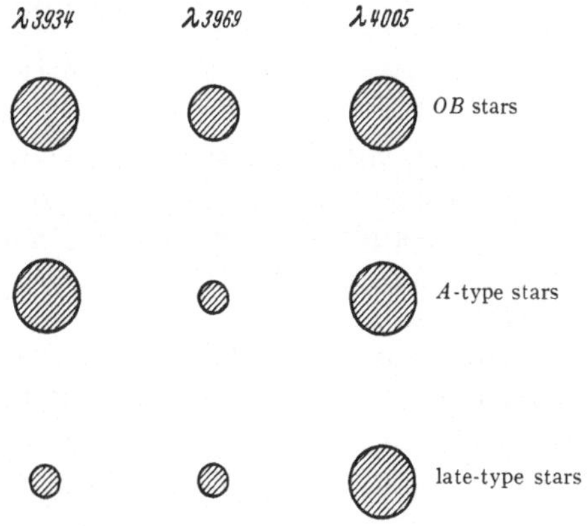

We can unambiguously distinguish OB stars from all other types if we employ interference filters to compare light intensities in three narrow wavelength ranges of, say, 15 to 30 angstroms half-width, centered at $\lambda\lambda\,3934$, 3969, 4005. We make three exposures, one through each filter, to obtain on the plate three images of each star. The plate is moved slightly between exposures. The appearance of image sets for different spectral types is shown schematically in Fig. 29. The use of a narrow band-pass filter involves, for the OB stars, a loss of limiting magnitude of approximately 4 mag compared to B magnitudes in the U, B, V system.

Fig. 29. Image sets for OB stars, A stars, and later-type stars.

Various other OB star sieves can be devised in which the Balmer jump and Balmer continuous absorption are utilized in the process.

Bibliography.

[1] WEAVER, H. F.: Popular Astron. **54**, Nos. 5, 6, 7, 8, 9, 10 (1946).
[2] EASTMAN Kodak Company, Kodak Photographic Plates for Scientific and Technical Use, 7th ed. Rochester 4, New York, 1953.
[3] JOHNSON, H. L., and W. W. MORGAN: Astrophys. Journ. **117**, 313 (1953).
[4] SEARES, F. H., and MARY C. JOYNER: Astrophys. Journ. **89**, 302 (1943).

La nature de la surface des planètes et de la Lune.

Par

AUDOUIN DOLLFUS.

Avec 44 Figures.

Certains astres du système solaire sont entourés d'une atmosphère opaque qui dérobe leurs sols à nos moyens d'observations.

D'autres planètes au contraire montrent plus ou moins complètement leurs surfaces. Ce sont Mercure, la Lune, Mars, certains astéroïdes, les principaux satellites de Jupiter.

Nous examinerons les conditions physiques particulières qui caractérisent les surfaces de ces différents corps célestes.

I. La Lune.

La grande proximité de la Lune permet une connaissance élaborée de sa surface. Privée d'atmosphère, elle montre son sol sans l'interposition de voiles.

Nous connaissons les propriétés de la surface lunaire selon deux échelles. D'une part, l'analyse des propriétés physiques de la lumière permet de préciser des structures comprises entre quelques millimètres et quelques dizaines de microns. D'autre part, l'observation télescopique révèle des formations topographiques ou orographiques comprises entre quelques centaines de mètres et plusieurs centaines de kilomètres.

a) Structure microscopique de la surface lunaire.

Nous examinerons, avant de les comparer, les conclusions auxquelles conduisent les différentes techniques d'observation.

1. Photométrie de la surface de la Lune. Des mesures photométriques et colorimétriques de la surface de la Lune ont été entreprises par de nombreux observateurs[1-6], et résumées dans un travail d'ensemble par J. van DIGGELEN[7].

La brillance d'une région lunaire varie fortement selon l'angle d'incidence du faisceau qui l'éclaire. L'éclat maximum est observé non pas sous l'incidence normale, c'est-à-dire au point de la surface lunaire ayant le soleil au zénith, mais à la pleine Lune quand le rayon incident revient sur lui-même.

Les mers, les régions claires, les fonds des cratères, les remparts et les chaines montagneuses ont des indicatrices de diffusions du même type et de même aspect. Cette propriété caractérise une structure du sol semblable à elle-même en tous

[1] A.L. BENNETT: Astrophys. J. **88**, 1 (1938).
[2] K. GRAFF: Mitt. Wiener Sternw. **4**, 79 (1949).
[3] N.P. BARABASHEV: Astronom. Nachr. **217**, No. 5207; **221**, No. 5296; **226**, No. 5409; **229**, No. 5473. — Astrophys. J. **3**, No. 1 (1926). — Publ. Obs. Kharkov de 1946 à 1959.
[4] V.V. CHARONOV: Publ. Obs. Léningrad de 1941 à 1959.
[5] N.N. SYTINSKAYA: Publ. Obs. Léningrad **17**, 76 (1959).
[6] N.S. ORLOVA: Astronom. J. URSS. **33**, 93 (1956).
[7] J. van DIGGELEN: Recherches Astron. Obs. Utrecht **14**, No. 2.

les points de la surface, ceci malgré des variations locales importantes de l'éclat et de la topographie.

De nombreuses courbes photométriques sont données dans le mémoire de J. van Diggelen[1] (exemple Fig. 1).

Les courbes photométriques caractérisent l'état de rugosité du sol lunaire. La structure du sol peut être précisé par ces courbes, soit à l'aide du calcul, soit par l'étude comparative au laboratoire d'échantillons divers.

Le calcul des ombres portés par les cavités et les rainures du sol de différentes formes et sous différents éclairements a été entrepris par N. P. Bara-bashev (1922), E. Schoenberg (1925), A. Bennett (1938) et finalement d'une façon plus complète par J. van Diggelen (1959).

Selon ce dernier travail, les observations semblent s'interpréter convenablement en supposant que les deux tiers de la surface du sol lunaire sont constituées par des cavités, en moyenne semi-ellipsoïdales, le grand axe étant perpendiculaire à la surface.

La photométrie ne donne pas d'indications sur les dimensions de ces cavités. Mais l'étude comparative d'échantillons divers, mesurés au laboratoire, permet de préciser encore dans une certaine mesure la structure de la surface.

Selon de nombreuses mesures données par N. S. Orlova (1956)[2], seules conviennent les surfaces très rugueuses, poreuses et présentant de nombreuses cavités. Des granits, des basaltes, des tufs de cendres volcaniques, des scories volcaniques ont été étudiées. Les caractéristiques photométriques de la surface lunaire se retrouvent à peu près sur les scories bulleuses très aérées, criblées de cavités jointives irrégulières de toutes dimensions.

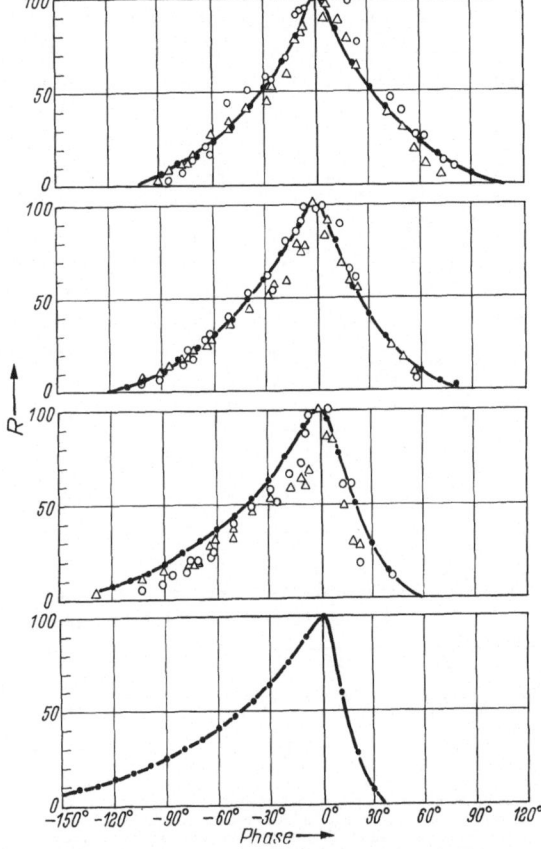

Fig. 1. Courbes photométriques des fonds des cratères, sous différentes longitudes sélénographiques.
$\lambda =$ de 0 à 10°,
$\lambda =$ de 10 à 30°,
$\lambda =$ de 30 à 50°,
$\lambda =$ de 50 à 90°.
D'après J. van Diggelen[1].

J. van Diggelen a mesuré des cendres volcaniques poudreuses, des grains polis vitreux de tektites, et des plaques métalliques artificiellement perforées de cavités diverses. Ces mesures confirment également une structure constituée de cavités de toutes dimensions, généralement plus profondes que larges. Il semble possible que ces cavités soient recouvertes d'une couche mince de grains très petits, analogue à nos cendres volcaniques.

---

[1] Voir note 7, p. 180.
[2] Voir note 6, p. 180.

Cette structure très tourmentée est probablement due à la chute et à l'explosion au sol des très nombreux petits météorites dont les cavités se sont accumulées pendant de très nombreux millénaires.

Les mesures colorimétriques de la surface lunaire donnent quelques indications sur la composition chimique superficielle. Les brillances et les indices des couleurs ont été simultanément relevées sur une centaine de régions à la pleine Lune par N. N. Sytinskaya[1]. Les résultats sont présentés sous forme d'une diagramme donnant en abscisse le pouvoir réflecteur et en ordonnée l'indice de couleur rapporté à celui du Soleil. Les points correspondants aux mesures se groupent tous dans une même région du plan figuratif, et définissent un domaine caractéristique (Fig. 2).

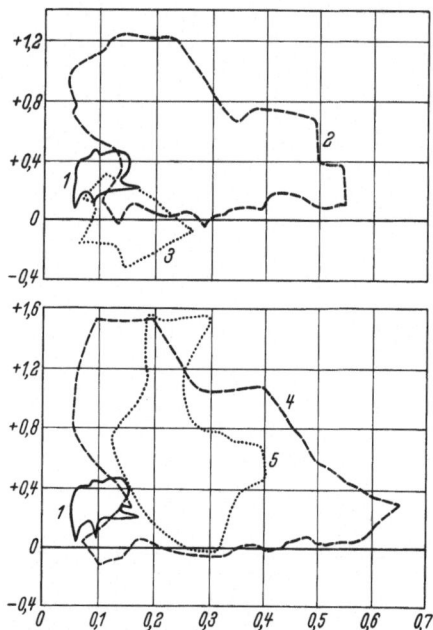

Fig. 2. Photométrie et colorimétrie de la surface de la Lune. D'après N.N. Sytinskaya[1]. Mesures à la pleine Lune. En abscisse: Pouvoir réflecteur du sol lunaire. En ordonnée: Indice de couleur du sol lunaire, rapporté à celui du Soleil. Courbe 1: contour du domaine enveloppant les mesures sur la Lune; Courbe 2: contour du domaine enveloppant les mesures sur les calcaires; Courbe 3: contour du domaine enveloppant les mesures sur les basaltes; Courbe 4: contour du domaine enveloppant les mesures sur les granits; Courbe 5: contour du domaine enveloppant les mesures sur les grès.

Les mesures effectuées au laboratoire sur de très nombreux échantillons minéraux dessinent d'autres domaines du plan figuratif. Sur la Fig. 2, on reconnaît les domaines des calcaires, des grès, des granits, des basaltes. Les gneiss ont également été mesurés, ainsi que des météorites. Aucun type de substances ne reproduit exactement les propriétés colorimétriques du sol lunaire. N. N. Sytinskaya pense que la surface lunaire doit sa couleur à l'influence des chocs de très nombreux petits météorites; ceux-ci produisent localement de grands dégagements de chaleur. Ils peuvent décomposer les silicates en oxydes de fer, très sombres. Les olivines du type fayalites, sous l'effet de la chaleur, donnent en effet de la silice et des oxydes très absorbants.

2. Polarisation de la lumière renvoyée par la Lune. La lumière qui éclaire la Lune provient du Soleil et elle n'est pas polarisée; elle est diffusée par la surface du sol lunaire en proportion inégale selon l'azimuth de la direction de vibration lumineuse. La lumière diffusée est partiellement polarisée. A l'aide d'un polarimètre visuel très sensible mis au point en 1924, B. Lyot a déterminé la courbe de polarisation de la Lune. Celle-ci est obtenue en reportant en abscisse l'angle de phase et en ordonnée la proportion de lumière polarisée[2], Fig. 3. Lyot a montré que la polarisation ne varie pas sensiblement suivant l'inclinaison de la normale au plan de la surface avec la direction d'observation, mais au contraire très fortement en sens inverse de l'albedo de la région examinée.

En comparant ces résultats avec les mesures obtenues sur des échantillons minéraux, il a recherché par analogie la nature de la surface du sol lunaire, à l'échelle de sa structure microscopique. Le sol lunaire est trouvé semblable à un *dépot finement granulaire de cendres volcaniques de petites dimensions.*

[1] Voir note 5, p. 180.
[2] B. Lyot: Thèse 1929. — Ann. Obs. Meudon, Fasc. 1.

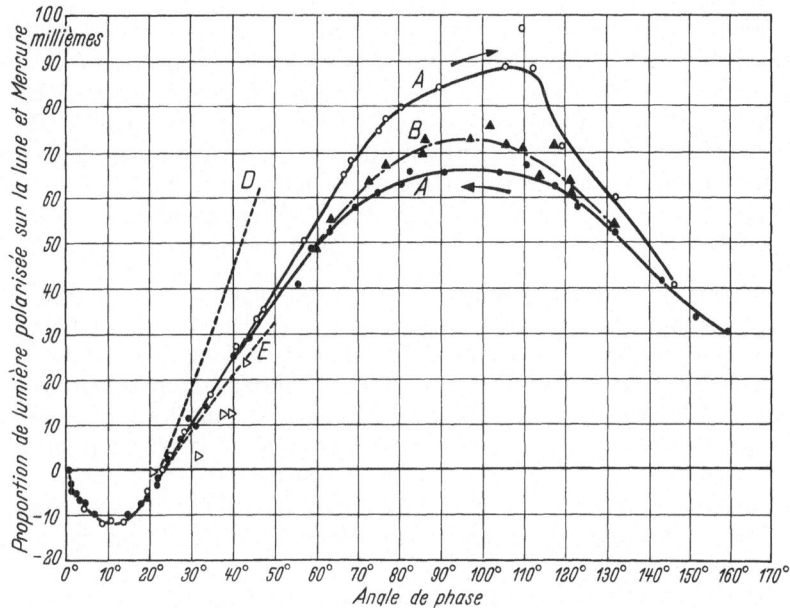

Fig. 3. Courbe de polarisation de la surface lunaire, en fonction de la phase, d'après B. Lyot[1].

3. Dépolarisation de la lumière par la Lune.
Le résultat précédent se trouve confirmé par les observations que l'auteur de l'article a effectuées de la façon suivante (A. Dollfus[2]): lorsque la Lune se présente en quartier, l'hémisphère non éclairé par le Soleil reçoit une partie de la lumière diffusée par la Terre. Cette

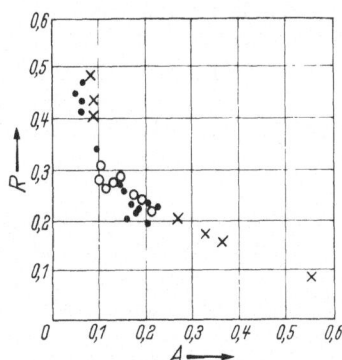

Fig. 4. Polarisation résiduelle du sol lunaire en fonction du pouvoir réflecteur. D'après A. Dollfus[2]. Cercles: mesures sur la Lune. Croix et points: mesures sur des centres volcaniques.

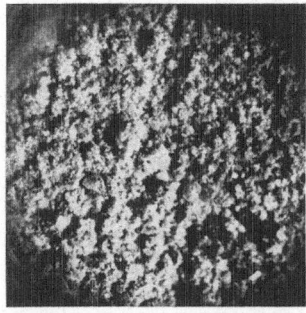

Fig. 5. Aspect microscopique d'un échantillon de cendres volcaniques ayant les mêmes propriétés polarisantes que le sol lunaire. D'après A. Dollfus[2]. champ = (3 mm)[2].

lumière est fortement polarisée, par suite de la diffusion dans l'atmosphère terrestre; en se réfléchissant sur la Lune, elle subit une dépolarisation partielle, dont la proportion dépend de la nature de la surface lunaire.

A l'aide d'un coronographe, l'auteur a mesuré la polarisation résiduelle, sous différents angles de phase. La mesure directe de la polarisation de la lumière de la Terre, effectuée à grande altitude en ballon libre a permis de déduire le facteur de dépolarisation du sol lunaire.

---

[1] Voir note 2, p. 182.
[2] A. Dollfus: Thèse 1952. — Ann. d'Astrophys. Suppl. 1952.

La comparaison des régions lunaires de différents éclats avec des échantillons minéralogiques mesurés au laboratoire montre (Fig. 4) que *le sol lunaire est semblable à une poudre de grains petits, sombres et complètement opaques*. La Fig. 5 donne une reproduction microscopique de l'aspect d'une telle poudre. Les propriétés dépolarisantes de cette structure ne sont pas altérées si, au lieu d'être représentée régulière comme sur la Fig. 5, la surface est modelée d'une façon encore plus tourmentée par des cavités et des sillons; cette dernière structure est en effet nécessaire pour expliquer le résultat des mesures photométriques.

4. Cas particulier de polarisation. Il existe un angle de pente limite au-delà duquel une poudre déposée sur un solide s'éboule en dénudant le support.

L'auteur a examiné la polarisation de la lumière le long de la falaise abrupte du «Mur Droit» et sur les lèvres de la «Vallée de Schörter». La polarisation devient nulle exactement sous le même angle de phase que dans les régions voisines; sous l'angle 94° correspondant à la valeur maximum de la polarisation, les mesures sont identiques à celles des contrées ordinaires de même brillance. Un socle de roches ignées dégagé de poussières aurait donné l'annulation de la polarisation pour un angle beaucoup plus faible, et une polarisation maximum nettement plus accusée.

Il faut en conclure que la poudre adhère au sol en remplissant partiellement les poches constituées par les cavités rugueuses du socle, sous l'action de forces plus importantes que la pesanteur[1].

5. Emission thermique de la surface lunaire à 10 microns. La surface de la Lune émet des radiations, à la façon d'un corps noir; le maximum de l'émission a lieu vers la longueur d'onde 10 microns. En adaptant un thermocouple très sensible au télescope de 2,50 m du Mont Wilson, S.B. NICHOLSON et E. PETTIT sont parvenus à mesurer avec précision cette faible émission[2].

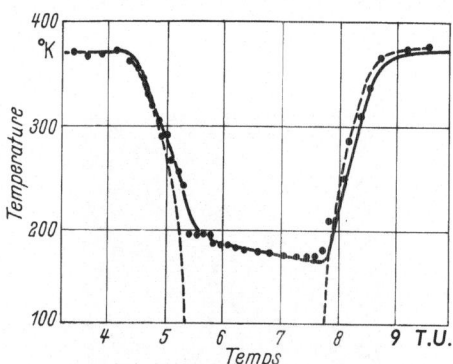

Fig. 6. Variation de la température de la Lune pendant une éclipse, mesurée au centre du disque pour la radiation 10 microns. En pointillé, variation théorique en l'absence de conduction ($k = 0$). En trait continu, variation théorique pour $(kcr)^{-\frac{1}{2}} = 1000$ cal cm^{-2} sec$^{-\frac{1}{2}}$ (°C)$^{-1}$.

Les points de la surface lunaire ayant le Soleil au zénith donnent des températures voisines de 380° K. La partie obscure du disque atteint 120° K. La température suit la variation de l'éclairement solaire selon la phase de la Lune sans aucun retard appréciable.

Des mesures de la radiation thermique de la Lune ont été effectuées pendant l'éclipse du 14 Juin 1927, et du 28 Octobre 1939[3]; elles montrent que la température de la surface de la Lune suit les variations de l'éclairement incident, pourtant très rapides, avec seulement un très faible retard, Fig. 6.

Il faut en déduire que les variations de température pénètrent très peu dans le sol; le matériau qui recouvre la surface lunaire doit être peu conducteur de la chaleur; il doit, de plus, posséder une faible capacité calorifique. La propagation de la chaleur dans la profondeur des couches lunaires dépend principalement du coefficient $(krc)^{-\frac{1}{2}}$ dans lequel k représente la conductivité thermique, r la densité

[1] Voir note 2, p. 183.
[2] S.B. NICHOLSON et E. PETTIT: Astrophys. J. **71**, 102 (1930).
[3] E. PETTIT: Astrophys. J. **91**, 408 (1960).

et c la chaleur spécifique. Selon P. EPSTEIN[1], puis A.J. WESSELINK[2], ce coefficient est voisin de 1000 en unités C.G.S. pour le matériau superficiel lunaire.

Les principales roches terrestres en blocs compacts ont un coefficient $(krc)^{-\frac{1}{2}}$ voisin de 20. Les roches poreuses, comme la lave, donnent des valeurs voisines de 100. Les matières finement divisées en poudre, ont dans les conditions terrestres, un coefficient plus élevé, mais dépassent rarement 200. Pour trouver des valeurs aussi fortes que 1000, il faut faire appel à des poudres fines mesurées dans le vide. Les mesures de M. SOMOLUCHOWSKI[3] rapportées par WESSELINK donnent des résultats très semblables à ceux qui caractérisent le matériau lunaire. Les transports de chaleur par l'air intersticiel deviennent négligeables et la conductivité calorifique s'opère seulement par les très petits points de contact que les grains conservent entre eux. Il faut supposer une couche de plusieurs milli-mètres d'épaisseur, dont les grains ne mesureraient pas plus de 0,1 mm de diamètre moyen. Ce résultat ultérieurement complété et commenté par J.C. JAEGER et par plusieurs auteurs[4 à 9] est en parfait accord avec la structure déduite des mesures de polarisation et représentée Fig. 5.

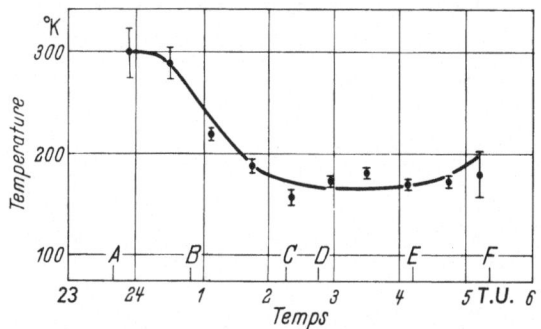

Fig. 7. Variation de la température de la Lune pendant une éclipse, mesurée pour l'ensemble du disque avec la radiation 1,5 mm. D'après W.M. SINTON[10, 11]. A entrée dans la pénombre; B entrée dans l'ombre; C début de la totalité; D fin de la totalité; E sortie de l'ombre; F sortie de la pénombre.

6. Emission thermique de la Lune vers 1,4 mm.

L'atmosphère terrestre opaque pour les longueurs d'onde plus longues que 14 microns, redevient partiellement transparente dans l'intervalle spectral compris entre 1 mm et 1,6 mm. Le rayonnement émis thermiquement par la Lune à la longueur d'onde 1,5 mm a pu être observé par W.M. SINTON[10, 11]. Pendant l'éclipse du 18 Janvier 1954, la température s'est lentement abaissée de 300° K à 270° K seulement avec un retard d'environ 45 mm sur la variation de l'énergie solaire reçue, Fig. 7.

Comparés aux résultats obtenus par PETTIT avec les radiations voisines de 10 microns, cette diminution de l'amplitude et ce retard indiquent que le rayonnement reçu à 1,5 mm provient d'une profondeur plus grande, atteinte plus tard et à un moindre degré par les variations thermiques de la surface. Le sol lunaire est opaque pour les longueurs d'onde visibles et jusque vers 14 microns, mais partiellement transparent pour les radiations voisines de 1,5 mm. Une interprétation théorique simple permit à W. SINTON d'évaluer le rapport du coefficient d'absorption de la lumière par la densité du matériau, dans le domaine 1,5 mm soit 2,9 cm² g⁻¹. Comparé à quelques échantillons poudreux, ce coefficient parait très supérieur à celui des obsidiennes, dont certains échantillons sont

[1] P. EPSTEIN: Phys. Rev. **33**, 269 (1929).
[2] A.J. WESSELINK: Bull. astronom. Inst. Netherl. **10**, 351 (1948).
[3] M. SMOLUCHOWSKI: Bull. Acad. Sci. Cracovie A 1910, 129; A 1911, 548.
[4] J.C. JAEGER et A.F. HARPER: Nature, Lond. **166**, 1026 (1950).
[5] J.C. JAEGER: Austral. J. Phys. **6**, 10 (1953).
[6] J.C. JAEGER: Proc. Cambridge Phil. Soc. A **9**, 355 (1953).
[7] A.J. WESSELINK: Observatory **74**, 215 (1954).
[8] K. BUETTNER: Publ. Astronom. Soc. Pacific **64**, 11 (1952).
[9] H.C. UREY: The Planets. New Haven, Conn.: Yale University Press 1952.
[10] W.M. SINTON: J. Opt. Soc. Amer. **45**, 975 (1955).
[11] W.M. SINTON: Astrophys. J. **123**, 325 (1956).

déjà transparents pour les longueurs d'ondes optiques; il a été trouvé supérieur à celui d'un échantillon de pierre ponce, voisin de celui d'un basalte broyé et un peu plus faible que celui d'un météorite pulvérisé.

7. Emission radio-électrique par la Lune. La surface lunaire émet également des radiations thermiques dans le domaine radio-électrique. Les longueurs d'onde centimétriques proviennent des couches encore plus profondes que celles qui émettent les ondes millimétriques. Les premières mesures radio-électriques de l'émission de la Lune ont été publiées en 1949 par J.H. Piddington et H.C. Minnett[1]. Elles ont été effectuées avec la longueur d'onde 1,25 cm. L'émission thermique recueillie varie sinusoïdalement avec l'angle de phase, ainsi que le montre la Fig. 8; le maximum est décalé par rapport à la pleine Lune et se produit avec 3,7 jours de retard. L'amplitude de la variation de température varie de 30 à 200° K, elle n'est que 0,4 fois seulement celle observée à 10 microns.

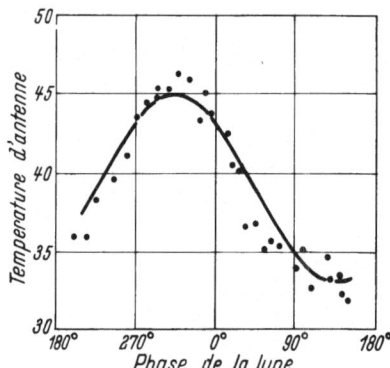

Fig. 8. Variation de la température de la Lune avec la phase, sur onde 1,25 cm. D'après Piddington et Minnett[1].

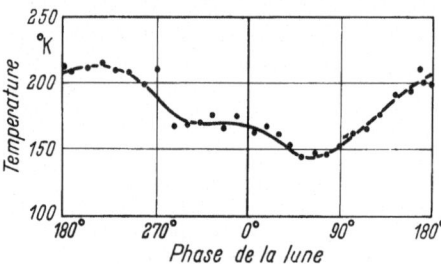

Fig. 9. Variation de la température de la Lune, avec la phase, sur onde 0,8 cm, au centre du disque. D'après J.E. Gilson[6].

Cet amortissement, ce changement de phase et le modelé sinusoïdal de la courbe nous montrent que les variations thermiques ne se propagent que très lentement dans la profondeur de la surface lunaire. La théorie de Fourier de la propagation de la chaleur s'applique à ce cas[2,3]. Les mesures ont été ensuite étendues par différents radio-observateurs à des domaines de longueur d'onde comprise entre 33 cm et 8 mm[4 à 6]. Les plus grandes longueurs d'onde pénètrent le plus pre-fondément dans le sol.

Sur 33 cm, J.F. Denisse a trouvé une température de 220° K ± 33° a peu près indépendante de la phase.

Sur 0,8 cm J.E. Gilson[6] a examiné la Lune avec un faisceau donnant une résolution de 12′ et balayé la surface lunaire diamétralement. La Fig. 9 reproduit les variations de l'émission avec la phase au centre du disque; pendant les éclipses du 29 Janvier 1953 et 18 Janvier 1954, aucune variation de l'émission radio-électrique n'a été trouvée.

Les essais d'interprétation supposant une couche de poussière mince étendue sur un socle de roches ignées ne conduisent pas à des conclusions bien nettes; elles n'améliorent pas sensiblement les données obtenues en supposant simple-ment la surface lunaire recouverte d'une couche poudreuse de très fins granules, sur plusieurs millimètres d'épaisseur. L'épaisseur du dépot de poudre, ainsi que la nature du socle, échappent donc encore à nos moyens d'investigation.

[1] J.H. Piddington et H.C. Minnett: Austral. J. Sci. Res. A **2**, 63 (1949).

[2] Voir notes 2 et 4, p. 185.

[3] J.L. Pawsey et R.N. Bracewell: Radio Astronomy. Oxford: Clarendon Press 1955.

[4] K. Akabane: Proc. Japan Acad. **31**, 160 (1955).

[5] M.R. Selinskaya et V.C. Troitski: URSS. Acad. Sci. Trans. 1955, 99.

[6] J.E. Gilson: Proc. Inst. Radio Engrs. **46**, 280 (1958).

8. Echos de radar sur la Lune. En 1949 pour la première fois le U.S. Army Signal Corps a pu envoyer un signal radar, sur 111,5 Mc/s, dont l'écho a été décelé après réflection sur la Lune. Des expériences plus nombreuses ont été effectuées dans la suite; certains résultats apportent des données sur la nature du sol lunaire[1, 2]. Les échos multiples recueillis se comportent comme si la Lune réfléchissait une grande partie de l'énergie par un petit nombre de points brillants seulement, répartis au voisinage du centre du disque, et correspondant aux petites régions du modelé lunaire orientées perpendiculairement à la direction de la Terre.

Les ondes multiples renvoyées de la sorte interfèrent entre elles; le léger mouvement de rotation du globe lunaire dû à la libration déplace les facettes et combine les interférences en rapides maxima et minima de lumière[2].

L'expérience de l'émission de pulses très brefs, avec l'analyse détaillée de la forme du signal reçu en retour a été effectuée par différents expérimentateurs, entre les longueurs d'onde 10 cm et 2,50 m[3 à 5]. Les impulsions brèves émises sur 198 Mc/s retournent 50% de l'énergie dans les premières 50 microsecondes après le début de l'écho, ce qui correspond à une différence de percours de 8 km, dans le sens du rayon visuel. Le diamètre de la Lune étant de 3470 km, cette énergie est renvoyée par une calotte de surface lunaire de 340 km de diamètre, centrée sur le milieu du disque apparent[3]. La Lune se comporte comme une sphère partiellement polie qui, éclairée par une source lointaine présente une tache lumineuse diffuse sous-tendant un dixième environ de son diamètre apparent. Cette tache est vue de la Terre sous un angle de 2′ et correspond à la diffraction produite par 1000 longueurs d'onde, c'est-à-dire par une étendue parfaitement plane du sol lunaire limitée à 1500 mètres de diamètre. Si, au lieu d'être dû à la diffraction, ce halo est produit seulement par la réflexion sur les pentes inclinées du sol irrégulier de la Lune, les pentes moyennes des talus ne doivent pas dépasser 6°.

Les échos de radar confirment ainsi les observations visuelles en montrant que, à l'échelle du kilomètre, la planéité moyenne du sol lunaire n'entraîne pas fréquemment de pentes nettement supérieures à 10%. Au-dessous de cette limite, une rugosité exagérée de la surface entrainerait une diffraction de l'écho supérieure à l'angle de 2′ observé. Il faut admettre que, à l'échelle inférieure, la surface lunaire demeure peu tourmentée, et privée d'accidents notables de diamètres supérieurs à quelques longueurs d'onde, c'est-à-dire à quelques mètres.

La rugosité du sol décelée par les mesures de photométrie doit donc concerner des cavités et des accidents du sol à une échelle assez petite.

Toutefois cette interprétation est compliquée par la transparence relative du sol pour les radiations concernées; l'écho ne caractérise que les propriétés de diffusion de l'ensemble des couches intégrées sur une certaine profondeur, ce qui atténue l'effet des irrégularités.

La mesure de l'intensité des échos reçus permet de déterminer dans une certaine mesure les constantes électromagnétiques du sol lunaire[1]. Les valeurs, encore peu précises, sont de l'ordre de $9,5 \cdot 10^{-12}$ farads/m pour la constante diélectrique et 20 MΩ/cm pour la résistivité.

---

[1] F. J. Kerr et G. A. Shain: Proc. Inst. Radio Engrs. **39**, 230 (1951). Voir aussi l'article de F. J. Kerr, au vol. LII de cette Encyclopédie, p. 452.

[2] I. C. Browne, J. V. Evans, J. K. Hargreaves et W. A. S. Murray: Proc. Phys. Soc. Lond. B **69**, 901 (1956).

[3] H. Trexler: Proc. Inst. Radio Engrs. **46**, 286 (1958).

[4] B. S. Yaplee, R. H. Bruton, K. J. Craig et N. G. Roman: Proc. Inst. Radio Engrs. **46**, 293 (1958).

[5] T. B. A. Senior et K. M. Siegel: University of Michigan Circ., June 1958.

9. Résumé des propriétés du sol lunaire. L'ensemble des résultats précédents nous apprend que le sol lunaire n'est pas chaotique au-dessous de l'échelle de la vision télescopique, mais assez modérément tourmenté. La texture fine est cependant très rugueuse, perforée de cavités innombrables et enchevêtrées; cette structure, indépendante du caractère tectonique de la région, se retrouve semblable à elle-même en tout point du sol de la Lune.

La surface est complètement couverte d'une poudre formée de grains complexes, de diamètres inégaux généralement inférieurs au millimètre, qui s'étend et adhère en tout point du sol pour former un revêtement dont l'épaisseur inconnue dépasse probablement plusieurs millimètres.

La matière est complètement opaque pour les radiations visibles et infra-rouges; sa teinte rappelle celle des roches ignées, sans lui être identique. Elle est semi-transparente sous l'épaisseur d'un grain pour les radiations de longueur d'onde 1,5 mm, son coefficient d'absorption par unité de masse étant 2,9 cm²/g. Elle devient transparente pour les ondes radio-électriques de 0,8 cm.

La structure granulaire de la surface rend la matière très peu conductrice de la chaleur; sa capacité calorifique est faible; sa constante diélectrique au-dessous des premières couches est de l'ordre de 10^{-11} farads/m, et la résistivité 20 MΩ/cm.

Nous allons indiquer comment on peut expliquer l'origine de cette morphologie.

10. Origine de la texture poudreuse du sol lunaire. Le caractère pulvérulent de la surface d'un sol comme celui de la Lune demande une explication générale. Nous verrons dans la suite de cette étude que les sols de Mercure, de Mars, de Cérès, de Vesta ont également une structure pulvérulente; cette propriété semble caractériser tous les astres du système solaire dépourvus d'une atmosphère épaisse.

K. Buettner[1] et T. Gold[2] principalement ont invoqués l'effet des rayonnements ultra-violets et des rayons X. Ces radiations pénétrantes détruiraient les mailles de certains réseaux cristallins, provoquant de temps à autre des ruptures, et à longue échéance un effritement superficiel. Nous n'avons pas de preuves de l'existence d'un tel mécanisme.

On a souvent invoqué l'éclatement des roches sous l'effet des variations thermiques. Calculons les tensions dans une roche brusquement refroidie: soit ΔT la variation de température, dans le temps t, à la surface du corps. L'intégration de l'équation de la propagation de la chaleur donne, en fonction de la profondeur z à partir de la surface de la roche supposée plane:

$$\frac{\partial T}{\partial z} = \sqrt{\frac{Cr}{2K}} \frac{\Delta T}{\sqrt{t}} .$$

K est la conductivité calorifique pour les roches ignées, soit 0,005 cal cm⁻¹ sec⁻¹, C est la chaleur spécifique 0,30 cal g⁻¹ et r la densité 3,3 g cm⁻³.

La variation de température avec la profondeur z entraîne une dilatation que nous supposerons provisoirement linéaire, $dl/l = D\,dt$; le coefficient de dilatation D est voisin de $7 \cdot 10^{-6}$ C.G.S. pour les roches ignées. La dilatation entraîne une tension F supposée linéaire telle que $dl/l = M\,F$, le module de Young M valant $0,5 \cdot 10^{-12}$ lorsque la tension est exprimée en dynes/cm².

---

[1] Voir note 8, p. 185.
[2] T. Gold: Monthly Notices Roy. Astronom. Soc. London **115**, 585 (1955).

En identifiant:

$$D \, dt = M F, \qquad F = \frac{D}{M} \sqrt{\frac{C \, r}{2 \, K}} \; \frac{\Delta T}{\sqrt{t}} \, dz,$$

on obtient numériquement:

$$F = 1{,}4 \cdot 10^8 \frac{\Delta T}{\sqrt{t}} \, dz \quad \text{(dynes/cm}^2\text{)}.$$

En fait, la tension n'est pas dirigée et cette valeur doit être à peu près doublée:

$$F = 3 \; \frac{\Delta T}{\sqrt{t}} \, dz \; \text{(kg/mm}^2\text{)}.$$

Pendant une éclipse la température mesurée au thermocouple varie de $\Delta T = 100°$ en $t = 30$ minutes. Cette variation correspond à un sol déjà fracturé, dans lequel les petits grains de la surface reçoivent peu de chaleur par conduction de la part des couches inférieures.

Dans une dalle mince posée au sol sur cette poudre isolante, la tension varierait approximativement selon le gradient: $F = 7 \, dz$ (kg/mm^2).

Lors du refroidissement cette dalle se déformera pour devenir concave, et comme la charge de rupture par traction des roches ignées n'est pas très élevée, il n'est pas exclu que le bloc puisse dans certains cas se briser.

Les incertitudes du calcul ne permettent pas de conclure si les refroidissements brusques du sol lunaire doivent entraîner seulement des fracturations occasionnelles, ou bien un morcellement plus général; il ne semble pas probable qu'il conduise à un effritement systématique de la surface.

Pendant les éclipses, les effets du refroidissement ne se manifesteraient que sur la face de la Lune tournée vers la Terre, et plus fortement au centre du disque où l'éclairement est maximum.

Pendant les lunaisons le refroidissement est moins rapide, mais plus fréquent, une région creuse est gagnée par la pénombre, puis par l'ombre portée d'une saillie voisine en 45 minutes; les effets seraient moins accusés près des pôles, malgré la rugosité.

La planète Mercure serait dépourvue de tels effets car elle présente toujours la même face au Soleil.

Nous pensons que la nature pulvérulente du sol lunaire doit au contraire provenir en grande partie du saupoudrage du sol par la poussière projetée lors de la formation des cratères.

Nous verrons que les cratères semblent formés par l'explosion de météores. Certains des plus grands cirques sont entourés d'une vaste auréole de matière poudreuse plus claire que les mers, projetée horizontalement jusqu'à plusieurs diamètres du cratère. Ces auréoles prouvent que l'apparition des cratères s'accompagne de formation de poussière.

En outre, la théorie des impacts développée par ÖPIK[1] prévoit que l'explosion donne naissance à un violent souffle vers le haut.

L'inspection de la surface de la Lune montre qu'environ $5 \cdot 10^{20}$ cm^3 de matière ont été remués au sol lors de la formation des gros cratères, sur l'ensemble de la face visible. Cet hémisphère couvre $2 \cdot 10^{17}$ cm^2. Si un millième seulement du matériau déplacé était projeté au loin pour retomber en un dépôt uniforme sur l'ensemble de la Lune, l'épaisseur de la couche serait déjà de 2,5 cm.

L'effet aurait surtout été appréciable dans la période précédant l'apparition des mers, lors de la formation de la plupart des très gros cratères.

---

[1] E. J. ÖPIK: Contrib. Armagh Obs. **5**, 24 (1957).

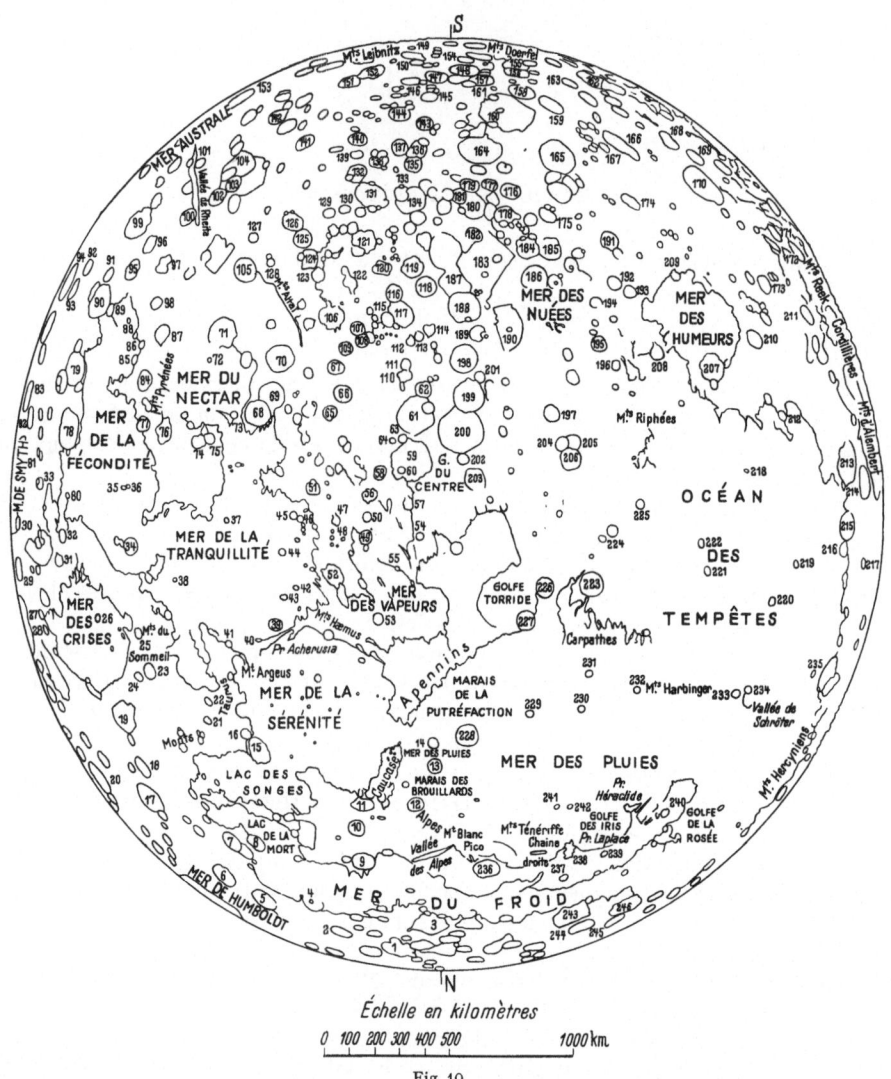

Échelle en kilomètres

0 100 200 300 400 500   1000 km.

Fig. 10.

Ce processus de saupoudrage à distance n'explique pas la couleur différente des mers, dont les contours sont parfaitement tranchés; il n'explique pas non plus la rugosité générale de la surface de toutes les régions.

Dans la période actuelle, l'effet principal serait le criblage par les petits météores.

Selon Wylie, il tombe actuellement sur la Terre environ 6 météores de plus de 5 kg par an et par 10^6 km². En un milliard d'années, il se serait formé au taux actuel environ 1 impact tous les 12 mètres. Si la vitesse de chute est 20 km/sec, la diamètre moyen des cratères formés est au moins de 4 mètres. La plus grande partie du sol serait couverte de tels trous.

Les météores 10 fois plus petits créent des cavités en forme de bols de quelques centimètres de diamètre. Selon les estimations de Wylie, ils sont beaucoup plus de 10 fois plus nombreux; leurs cratères doivent se chevaucher et cribler le sol absolument en tout point.

Fíg. 10. Carte générale de la Lune. D'après l'Astronomie Populaire (Paris 1955).

1 Méton	50 Godin	99 Furnerius	148 Moret	197 Guericke
2 Arnold	51 Delambre	100 Rheita	149 Schlumberger	198 Arzachel
3 W. C. Bond	52 Jules César	101 Mallet	150 Simpelius	199 Alphonse
4 Gartner	53 Manilius	102 Metius	151 Mutus	200 Ptolémée
5 de la Rue	54 Triesnecker	103 Fabricius	152 Manzinus	201 Alpetragius
6 Endymion	55 Hyginus	104 Janssen	153 Pontécoulant	202 Herschel
7 Atlas	56 Lade	105 Piccolomini	154 Short	203 Flammarion
8 Hercule	57 Rheticus	106 Sacrobosco	155 Cassatus	204 Parry
9 Aristote	58 Saunder	107 Azophi	156 Klaproth	205 Bonpland
10 Eudoxe	59 Hipparque	108 Abenezra	157 Gruemberger	206 Fra Mauro
11 Alexandre	60 Horrocks	109 Geber	158 Blancanus	207 Gassendi
12 Cassini	61 Albategnius	110 Argelander	159 Scheiner	208 Agatharchide
13 Aristillus	62 Parrot	111 Airy	160 Clavius	209 Vittelo
14 Autolycus	63 Halley	112 Donati	161 Rutherfurd	210 Mersenne
15 Posidonius	64 Hind	113 Faye	162 Bailly	211 Byrgins
16 Chacornac	65 Descartes	114 Delaunay	163 Kircher	212 Hansteen
17 Messala	66 Abulfeda	115 Playfair	164 Maginus	213 Grimaldi
18 Geminus	67 Almanon	116 Apian	165 Longomontanus	214 Riccioli
19 Cléomède	68 Theophile	117 Krusenstern	166 Schiller	215 Hevelius
20 Gauss	69 Cyrille	118 Werner	167 Bayer	216 Cavalerius
21 Stephanides	70 Catherine	119 Aliacansis	168 Phocyclide	217 Olbers
22 Roemer	71 Fracastor	120 Poisson	169 Wargentin	218 Flamsteed
23 Macrobe	72 Rosse	121 Gemma Frisius	170 Schickard	219 Reiner
24 Tisserand	73 Madler	122 Pontanus	171 Piazzi	220 Marius
25 Proclus	74 Capella	123 Walter	172 Lagrange	221 Képler
26 Picard	75 Isidore	124 Zagut	173 Viete	222 Encke
27 Condorcet	76 Gutemberg	125 Rabbi Lévy	174 Hainzel	223 Copernic
28 Hansen	77 Goclenius	126 Riccius	175 Heinsius	224 Reinhold
29 Neper	78 Langrenus	127 Stiborius	176 Tycho	225 Landsberg
30 Schubert	79 Vendolin	128 Rothmann	177 Pictet	226 Stadius
31 Firmicus	80 Webb	129 Busching	178 Sasseride	227 Eratosthène
32 Apollonius	81 Kastner	130 Buch	179 Saussure	228 Archimède
33 MacLaurin	82 La Pérouse	131 Maurolycus	180 Orontius	229 Timocharis
34 Toruntius	83 Ansgarius	132 Barocius	181 Huggins	230 Lambert
35 Messier	84 Colomb	133 Faraday	182 Lexell	231 Pytheas
36 Pickering	85 Cook	134 Stoefler	183 Deslandres	232 Euler
37 Maskelyne	86 Monge	135 Licetus	184 Gauricus	233 Aristarque
38 Cauchy	87 Santbeck	136 Héraclite	185 Wurzelbauer	234 Hérodote
39 Pline	88 Biot	137 Cuvier	186 Pitatus	235 Otto Struve
40 Dawes	89 Wrottesley	138 Clairaut	187 Regiomontanus	236 Platon
41 Vitruve	90 Petavius	139 Breislak	188 Purbach	237 La Condamine
42 Maclear	91 Hase	140 Bacon	189 Thebit	238 Maupertuis
43 Ross	92 Legendre	141 Pitiscus	190 Birt	239 Bianchini
44 Arago	93 Phillips	142 Rosenberger	191 Capuanus	240 Mairan
45 Sabine	94 Humboldt	143 Lilius	192 Mercator	241 Le Verrier
46 Ritter	95 Snellius	144 Jacobi	193 Campanus	242 Helicon
47 d'Arrest	96 Stevin	145 Zach	194 Kies	243 John Herschel
48 Tempel	97 Reichenbach	146 Pentland	195 Bouillaud	244 Anaximandre
49 Agrippa	98 Borda	147 Curtius	196 Lubiniezki	245 Pythagore
				246 Babbage

Ce criblage sur place explique le caractère poudreux du sol, le maintien de la différence de teinte entre les mers et les régions claires, ainsi que la rugosité complexe de toutes les régions.

Nous verrons que cette action contribue aussi à l'érosion des cratères, et à l'affacement des auréoles.

b) Structure télescopique de la surface de la Lune.

11. Conditions des meilleures observations. L'observation au télescope permet de reconnaître facilement sur la surface de la Lune toutes les formations orographiques de dimensions plus grandes que 1 kilomètre. Avec les instruments les plus puissants que nous possédons et à la faveur des nuits où l'atmosphère terrestre est particulièrement calme, on peut discerner sur la Lune avec un grossissement voisin de 1000, des petites taches individuelles séparées par 400 mètres seulement. Des taches isolées, très contrastées, de 200 mètres de diamètre peuvent être décelées. On aperçoit des trainées rectilignes noires aussi étroites que 100 mètres. L'observateur peut discriminer la formation «cratère» pour des cirques de 600 mètres de diamètre.

En éclairement rasant, de très faibles changements de niveaux de quelques dizaines de mètres seulement, peuvent être perçus grâce aux ombres allongées qu'ils portent sur le sol.

Les meilleures photographies actuellement disponibles ont un pouvoir séparateur inférieur à celui des observations visuelles les plus poussées; on peut cependant
discriminer un cratère d'une colline pour un diamètre de l'ordre de 1 kilomètre.

Toutes les meilleures photographies obtenues dans le monde ont été groupées
et sélectionnées par le Dr. G. P. Kuiper en un «Atlas Photographique de la Lune»
(University of Chicago Press). Ce groupement constitue la documentation topographique la plus complète actuellement disponible.

12. Caractères de l'orographie lunaire. Les éléments dominants de la topographie lunaire sont les mers, les cratères, les chaines de sommets. Leur nomenclature est donnée par la carte et la liste
de la Fig. 10. Leurs aspects s'observent
directement sur les figures qui illustrent
cet article. On découvre de nombreuses
formations topographiques secondaires,
telles que les auréoles, les rainures, les
stries, etc. Nous n'en donnerons pas une
classification morphologique. L'examen
des photographies jointes en donne une
description plus nuancée.

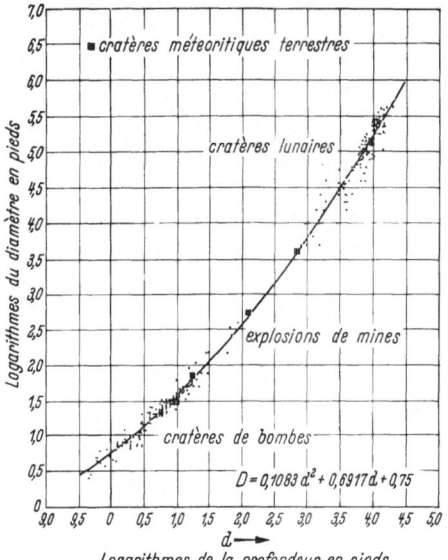

Fig. 11. Relation entre le diamètre et la profondeur pour
les trous de bombe, les cavités d'explosion de mine, les
cratères météoritiques terrestres et les cratères lunaires.
D'après R.B. Baldwin[1].

13. Origine des cratères. On a longtemps cherché des origines volcaniques
aux cratères, sans parvenir à des données numériques et morphologiques rai-
sonnables. Les cratères semblent plutôt
produits par l'explosion des météores qui
tombent sur la surface de la Lune.

Les météores qui tombent sur la
Terre y forment des cratères de forme
identique à ceux de la Lune. La cratère
de l'Arizona mesure 1,2 km de diamètre;
tombé sur la Lune, l'aérolithe qui l'a
produit aurait formé, sous la gravi-
tation plus faible, une cavité de 2 km de diamètre, facilement observable au
télescope comme un objet déjà bien reconnaissable.

Sur Terre, l'érosion résultant de la présence de l'atmosphère efface très rapidement, les cratères à l'échelle de temps des phénomènes géologiques, et presque
au fur et à mesure de leurs formations successives; au contraire, la Lune accumula
les traces des impacts pendant probablement plus d'un milliard d'années[1 à 3].

R.B. Baldwin[1] a montré, Fig. 11, qu'il y a continuité dans la loi de variation
du diamètre avec la profondeur entre les trous de bombes explosives, dont les
plus petits mesurent 10 cm de diamètre, les cratères d'explosion de mines, les
cratères météoritiques terrestres et les cratères lunaires non altérés par l'érosion,
de toutes dimensions jusqu'à 100 km de diamètre.

La répartition des cratères sur la surface de la Lune répond aux lois du hasard:
sur les régions claires du sol qui n'ont pas été recouvertes par les mers sombres
ou par les montagnes de déblais, grands et petits cratères se distribuent au hasard;
des plus petits, beaucoup plus nombreux, criblent la surface, quelles que soient
les formations sous jacentes, en particulier sur les remparts ou les pitons centraux

[1] R.B. Baldwin: The Face of Moon. Chicago, Ill.: Chicago University Press 1949.
[2] A. Dollfus: L'Astronomie Populaire. Paris: C. Flammarion (A. Danjon, etc.) 1955.
[3] R.S. Dietz: Jl. Geology **54**, 359 (1946).

des plus grands (Fig. 12). Les régions appelées mers ont recouvert, lors de leur apparition, tous les cratères préexistants, à l'exception des plus grands d'entre eux dont certains contours apparaissent encore, partiellement ensevelis, refondus ou comblés. Les nouveaux cratères qui recouvrent ces mers sont évidemment plus récents, ils sont moins nombreux que sur les régions claires, et répartis tout à fait au hasard (Fig. 13, 14 et 16).

La vitesse des météores peut atteindre plusieurs dizaines de km/sec. Cette vitesse est très notablement plus grande que celle de la propagation des vibrations mécaniques dans le matériau constituant la surface de la Lune. Lors de l'impact, il se forme donc une onde de choc qui se propage dans le corps même du météore

Fig. 12. Région continentale couverte de cratères anciens et de nombreux cratères plus récents. Région des cratères Maurolycus et Stofler. Cliché B. Lyot (Pic du Midi).

et l'échauffe jusqu'à accroître suffisamment la vitesse de propagation du son pour que la chaleur se dissipe. L'échauffement considérable entraîne la volatilisation instantanée de la matière, sous forme d'une violente explosion.

Ce mécanisme a été étudié par T. Gold[1], par Gillvarry et Hill[2] et par Öpik[3]. La masse de la matière chassée par l'explosion dans le volume du cratère formé dépend très fortement de la vitesse du choc; sur Terre, pour une vitesse de chute de 20 km/sec, elle représenterait environ 50 fois celle du météore. Sur la Lune, où la gravité est 6 fois plus faible, cette masse est peut être plus grande encore. Ces données sont très incertaines.

Lorsque le choc engendre des ondes largement supersoniques, le cratère formé est circulaire, quelle que soit la vitesse; son aspect ne dépend pas de l'angle de chute. Si la vitesse est inférieure à celle de la propagation des vibrations acousti-ques dans le sol, le météore s'enterre plus profondément; il donne naissance à un puits dont les contours très différents dépendent de la vitesse et de l'inclinaison de la trajectoire. On observe de telles cavités autour des grands cratères récents,

[1] Voir note 2, p. 188.
[2] J. J. Gillvarry et J. E. Hill: Astrophys. J. **124**, 610 (1956).
[3] Voir note 1, p. 189.

tels que Copernic, qui a projeté lors de sa formation une pluie de blocs secondaires dont on retrouve les points de chutes des plus lents sous forme de trous profonds et irréguliers.

14. Erosion de la topographie lunaire. Il est dans une large mesure possible d'ordonner dans le temps l'apparition successive des différents cratères:

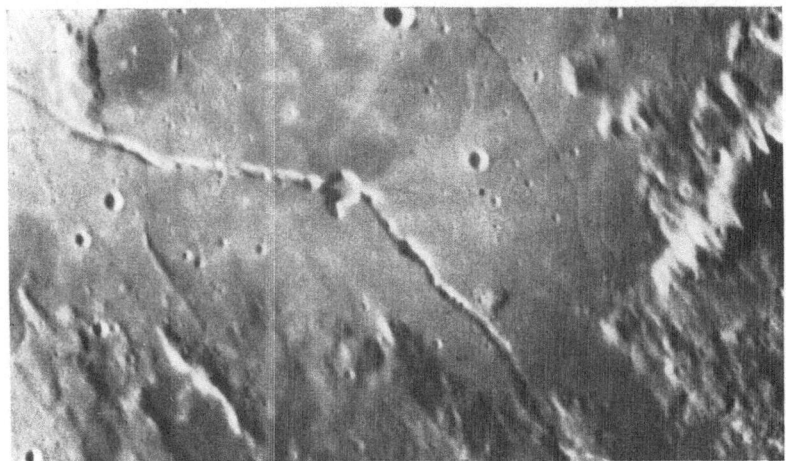

Fig. 13. La grande rainure de Hyginus. Cliché H. CAMICHEL.

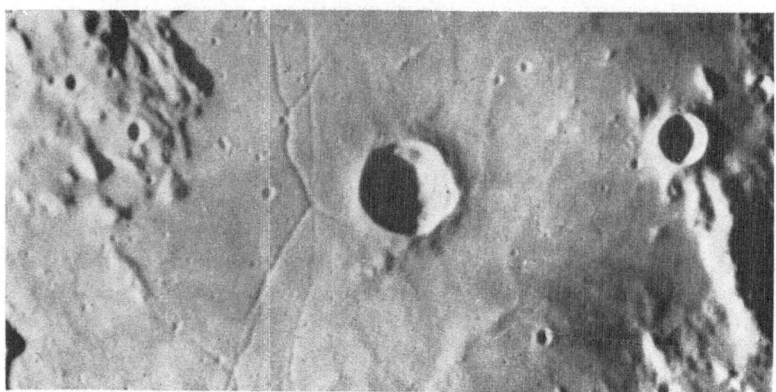

Fig. 14. Rainures et cratères auteur de Triesnecker. Cliché H. CAMICHEL.

Certains grands cirques empiètent sur d'autres plus anciens et détruisent les parties qu'ils recouvrent (Fig. 12).

Des formations de petites dimensions s'impriment sur les remparts ou les talus de cirques formés antérieurement. Les cirques les plus marqués sont évidemment les plus anciens.

Des cratères récents ont projeté des débris dont on retrouve les impacts sur les remparts des cirques, des collines ou des mers dont les existences sont antérieures.

Certains cratères s'entourent de grandes auréoles brillantes qui recouvrent d'une poudre blanche les formations plus vieilles.

Toutes ces propriétés nous permettent de classer les cratères par ancienneté. On peut ensuite comparer le modelé des formations anciennes, au regard de celles apparues plus récemment[1,2].

Les cratères les plus récents, comme Copernic, possèdent des remparts effilés, des sommets francs. Au contraire, les cratères dont l'apparition remonte aux temps les plus reculés, comme Calvius, manifestent toujours un modelé arrondi; ils ne montrent aucune arrête vive.

Cette désagrégation du relief, dont l'action est très importante, se retrouve à la manière d'une règle générale sur toutes les formations anciennes de l'orographie lunaire (Fig. 12 et 16).

On a recherché l'origine de cet adoucissement dans la déformation plastique naturelles des roches. La plasticité croît très vite avec la température: un réchauffement passager de l'ensemble de la croûte lunaire superficielle, sous l'action de la radioactivité intérieure primitive, a été invoqué par G. P. KUIPER[2].

D'autre part le relief semble avoir été émoussé le long des contours de certaines mers, comme si la chaleur communiquée avait refondu la roche.

Il nous semble qu'un effet notable d'érosion doit provenir de l'action indéfiniment répétée des impacts des petits météores. Chaque impact produit dans le sol une cavité; seules les plus grandes d'entre elles sont perceptibles au télescope sous forme de cratères. Les petites météorites sont infiniment plus nombreuses. Leurs explosions accumulent des cavités sur le sol pendant des milliards d'années; celles-ci doivent produire un déplacement de matériau appréciable.

Ce mécanisme d'érosion n'agit pas sur la Terre; l'atmosphère terrestre arrête tous les météores d'un poids inférieur à 5 kg, et les autres causes d'érosion sont largement prépondérantes.

15. Fonds brillants de certains cratères. A la pleine Lune, certains fonds de cratères paraissent éblouissants. Dans certains cas, seuls les contours revêtent cette propriété et on observe un anneau brillant. Le grand éclat s'atténue sous un angle d'éclairement prononcé et il disparaît sous l'incidence oblique le long du terminateur.

Les variations relatives de brillance sont à peu près indépendantes de l'inclinaison de la surface avec la direction d'observation; elles ne dépendent que de l'angle d'incidence du rayon éclairant.

Le phénomène peut s'expliquer en supposant le fond de ces arènes tapissé par une matière dont les propriétés optiques ont été modifiées par la haute température lors de la formation par les météores suffisamment rapides.

G. P. KUIPER a montré[2] que seuls les cratères jeunes manifestent cette propriété.

Il explique le phénomène par un réchauffement général du sol produit par la radioactivité intérieure. L'état des roches a pu être transformé, ce qui aurait permis une modification de la couleur des cratères formés ulterieurement.

Il peut se faire aussi que la poussière projetée lors de la formation postérieure de gros cratères ait uniformisé les teintes. Les impacts directs des petits météores dans la région brillante remuent également le sol et peuvent mélanger les couches.

16. Auréoles entourant les grands cratères. A la pleine Lune, certains grands cratères de formation récente se montrent entourés par de vastes auréoles, très claires formées des jets radiaux ou de filements complexes. Les plus grands jets,

---

[1] Voir note 2, p. 182.
[2] G. P. KUIPER: Proc. Nat. Acad. Sci. **40**, 1096 (1954).

issus du cratère Tycho, s'étendent sur plus de la moitié du disque visible.
Ces formations, comme les fonds de certains cratères, ne sont pas visibles le long

Fig. 15. Rainures parallèles et rides sur e contour de Mare Humorum. (Cliché Observatoire du Pic du Midi.)

Fig. 16. Bordure de Mare Tranquillitatis et rainure de Ariadaeus. (Cliché Observatoire du Pic du Midi.)

du terminateur et leurs éclats ne dépendent que de l'angle entre le rayon éclairant
et la normale à la surface.

Ces auréoles proviennent de l'éjection de poudres au moment de la formation
du cratère central. Projeté dans le vide et sous la faible pesanteur lunaire, le

matériau décrit des trajectoires elliptiques et se dépose dans la direction du lancer initial à des distances plus ou moins grandes.

L'auréole de Tycho montre de grandes trainées radiales rectilignes, fruits de directions d'éjectiées privilégiées; le cratère Copernic s'entoure d'une auréole plus diffuse, composée d'enchevêtrements sinueux, comme si un résidu d'atmosphère avait modifié les trajectoires. Le grand cratère récent Théophile est dépourvu d'auréoles.

Selon N.P. BARABASHEV[1], l'indicatrice de diffusion de ces auréoles est du même type que celle des régions voisines, ce qui révèle une identité de la structure rugueuse du sol. Cependant G.P. KUIPER[2] a décelé au télescope des éraflures et cavités dans le sol assez nombreuses aux emplacements des auréoles, attribuables aux impacts par les blocs projetés les plus gros.

Les mesures de polarisation indiquent une morphologie poudreuse[3]. Au laboratoire, nous avons sélectionné des cendres volcaniques, ayant les mêmes propriétés de polarisation que les mers et que les auréoles, et des éclats comparables. En saupoudrant les dépôts sombres avec les poudres claires, nous sommes parvenus à reproduire les effets de contraste observés sur la Lune, en conservant leurs propriétés photométriques et polarisantes.

Les auréoles qui entourent les cratères consistent donc en un dépôt superficiel généralement mince, recouvrant le sol presque sans changer sa structure.

Ces auréoles ne s'observent qu'autour des cratères très récents. R.B. BALDWIN a suggéré qu'une faible atmosphère, dans le passé, aurait pu empêcher leur formation[4].

Il semble cependant que ces auréoles puissent s'effacer relativement rapidement. Le criblage du sol par les petits météores produit des cavités dans de nombreux cas plus profondes que l'épaisseur du dépôt, et mélange par conséquent le sol rapidement, à l'échelle des temps géologiques.

17. Pitons au centre des cratères. Certains cratères montrent des petites collines au centre de leurs cavités; d'autres en sont dépourvus. Le piton n'est pas toujours placé exactement au centre; il est souvent irrégulier, quelquefois multiple. La pente des contours excède rarement $5°$.

Sur Terre, certains cratères d'explosion de bombe montrent le phénomène. Le Meteor Crater de l'Arizona en est complètement dépourvu.

On a cherché à expliquer la formation de ces pitons centraux comme le résultat direct de l'impact[5,6]. Les météores les plus rapides plongeraient profondément dans le sol, ce qui entraînerait un rebond de la matière. Les météores lents, au contraire, exploseraient à fleur de la surface et la déflagration priverait de toute aspérité la cavité formée.

G.P. KUIPER pense que les pitons des centres des cratères sont des extrusions de lave résultant de la fracture de la croûte superficielle, au point de l'impact. La lave sortirait sous la pression d'un réajustement isostatique. Les pitons ne se montreraient, dans ce cas, que dans les cratères formés à l'époque de l'apparition des mers, lorsque la matière interne liquéfiée est proche de la surface.

---

[1] Voir note 3, p. 180.
[2] G.P. KUIPER: Vistas in Astronautics, Vol. II. London: Pergamon Press 1959.
[3] Voir note 2, p. 183.
[4] Voir note 1, p. 192.
[5] M. HOYAUX: Ciel et Terre **66**, 121 (1950).
[6] Voir note 2, p. 188.

18. Origine des mers. Les grandes étendues sombres, improprement appelées «mers», suggèrent un écoulement de matière, qui se serait solidifié. Cependant, leur origine n'est pas clairement interprétée. Des hypothèses assez diverses ont été avancées.

En 1955, T. Gold[1] a développé l'idée d'un épanchement de poussière. L'action énergique du rayonnement ultra-violet et X dans les mailles cristallines des silicates du sol, ainsi que la chute des micrométéores, produiraient un éclatement et une pulvérisation de la surface. Les petits grains, de temps à autre, sautilleraient sur le sol, sous l'effet de vaporisations ou d'actions électrostatiques; ils se déplaceraient de la sorte le long des pentes. Pendant la longue durée des temps

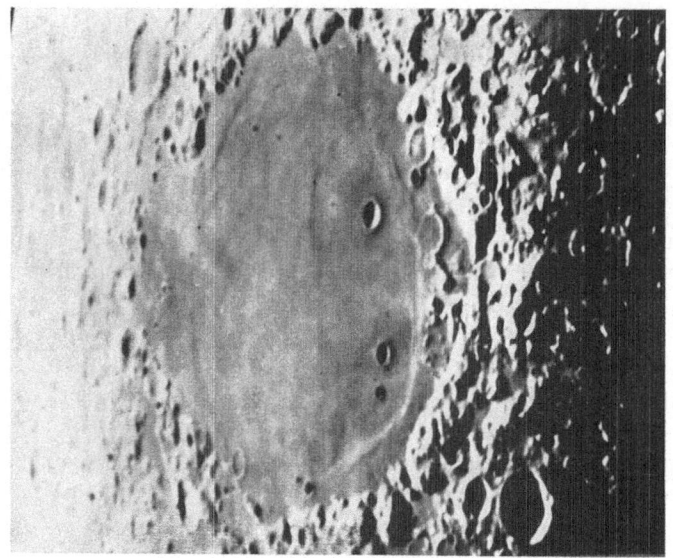

Fig. 17.   Etendue circulaire de laves. Mare Crisium. Cliché B. Lyot (Pic du Midi).

géologiques, la poudre immigrerait progressivement dans les fonds des vallées et s'accumulerait en de grandes étendues plates qui constituent les mers.

Cependant, le caractère rugueux du sol lunaire paraît nécessaire aux interprétations des mesures photométriques et l'effet d'une telle migration des poussières serait de combler les cavités pour constituer des étendues parfaitement lisses.

En 1893, G. K. Gilbert avait déjà indiqué que les mers pouvaient provenir de la chute de très gros météores d'un volume nettement supérieur à ceux qui forment les cratères. La chute entraînerait la formation d'une très grande cuvette, immédiatement comblée par le flot de laves résultant de la fusion du météore et du sol.

Cette idée a été reprise en détail par R. Baldwin, dans un livre publié en 1949[2].

Les mers de la Lune sont volontiers circulaires et entourées d'un rempart de déblais montagneux (Fig. 17), ce qui rapproche leurs structures de celles des cratères. Des stries radiales divergent du centre des mers; nous verrons qu'on peut les attribuer à l'incision du sol par des projectiles métalliques éjectés hori-

---

[1] Voir note 2, p. 188.
[2] Voir note 1, p. 192.

zontalement à grande vitesse. La Fig. 19 due au travail de G. FIELDER[1] reproduit les positions de ces stries. Certaines d'entre elles divergent de Mare Imbrium, d'autres de Mare Nectaris, d'autres de Mare Humorum. Ces stries nous montrent qu'un phénomène de choc très violent a accompagné la formation de ces mers.

Fig. 18. Amas de déblais résultant de la formation de Mare Imbrium. Monts Apennins. Cratère Archimèdes.

BALDWIN décrit ainsi les événements qui ont pu procéder à la formation de Mare Imbrium, la mer la plus étendue de la surface visible de la Lune: Un bloc de fer et nickel de plus de 15 km de diamètre s'est abattu sur la Lune, au centre de l'actuel Mare Imbrium, avec une vitesse voisine de 30 km/sec. Le projectile pénètre sous le sol, un grand dôme s'élève, se pulvérise en projetant une grande quantité de débris en cercle, dont les restes constituent maintenant la chaîne des Apennins (Fig. 18), le Caucase, les Alpes, les Monts Harbinger et les Karpathes lunaires. Des débris de fer et nickel sont projetés à grande vitesse et incisent les remparts ainsi que le sol lunaire sur de grandes longueurs. La vibration générale de l'écorce lunaire produit un rebondissement élastique, amorti par les tensions internes et suivi d'un effondrement général du grand disque constituant le fond du cercle ainsi créé.

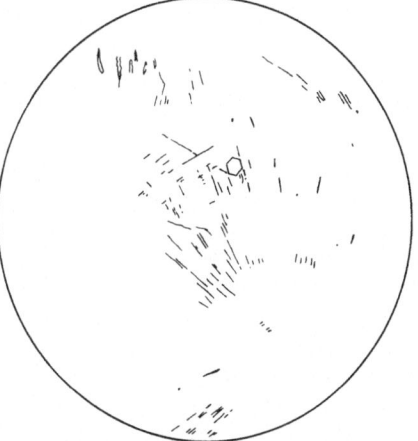

Fig. 19. Répartition des stries sur la surface lunaire. D'après G. FIELDER[2].

Cet effondrement entraîne l'extrusion d'un volume équivalent de la matière liquide à haute température supposée constituer l'intérieur de la Lune. BALDWIN estime à $4 \cdot 10^{15}$ m³ le volume de laves ainsi épanché. Ce matériau parfaitement

[1] Voir note 4, p. 203.
[2] Voir note 3, p. 198.

liquide comble aussitôt l'étendue de Mare Imbrium, se répand à l'Est pour envahir l'actuelle région Oceanus Procellarum et Mare Nubium et s'écoule entre les Monts Apennins et Caucase, à l'Ouest, pour remplir les cavités préexistantes de Mare Serenitatis, Tranquilitatis, Faecunditatis et Nectar.

R. Baldwin calcule que, si la lave sort à 3500° K, l'épanchement ne deviendra visqueux qu'au bout de 4 jours, ce qui laisse un temps suffisant pour cet épanchement.

Selon les valeurs numériques données par l'auteur, le débit de laves chaudes dans le goulet délimité par les Monts Apennins et Caucase aurait entraîné pendant 4 jours un courant de l'ordre de 100 km/heure. Les contours montagneux de ces régions ne montrent pas les traces qu'aurait nécessairement laissées un tel passage. Il semble nécessaire de supposer une formation de lave autonome pour chacune des mers, telles que Mare Serenitatis, Tranquillitatis, Nectaris, Crisium, Humorum, etc.

Ces autres mers bien circulaires auraient résulté de chocs analogues avec des corps de plus petits diamètres selon des directions d'impacts en général moins inclinés (Fig. 17).

D'après l'étude détaillée des aspects lunaires, il semble que la plupart des mers se sont formées vers la même époque, en général plus tardives que celles de beaucoup de grands cratères, excepté Copernic, Tycho et quelques autres dont les auréoles recouvrent les mers.

G. P. Kuiper a exposé en 1956 une interprétation de ses observations télescopiques selon laquelle le matériau liquide semblerait provenir de l'intérieur de la Lune selon un mécanisme analogue à celui invoqué par Baldwin[1,2].

Au début de l'histoire de la Lune, les impacts auraient produit les cratères classiques anciens que nous observons.

Lorsque la radioactivité intérieure de la Lune aurait suffisamment échauffé le noyau pour amener la liquéfaction jusque près de la surface, les météores les plus gros auraient suffi à perforer la croûte et donner naissance à des épanchements de laves provenant de l'intérieur.

Ensuite, le refroidissement résultant de l'épuisement de la radioactivité aurait de nouveau consolidé la croûte et les gros météores produiraient à nouveau de simples cratères.

H. Urey invoque cependant des objections théoriques aux interprétations des observations précédentes. Un réchauffement de la Lune ayant amené la liquéfaction jusque près de la surface aurait diminué exagérément la rigidité de la croûte. Les déformations irrégulières du globe lunaire, de plus de 1 kilomètre, observées sur des vastes régions de la surface auraient difficilement pu subsister[3]. De plus, la croûte siliceuse superficielle, probablement plus dense à l'état solide que sous la forme liquide sous-jacente, ne pourrait pas flotter; elle devrait se disloquer[4].

Selon H. Urey[5 à 8], l'intérieur de la Lune aurait toujours été dans l'ensemble à peu près complètement solide. Le phénomène qui donna naissance au Mare

---

[1] Voir note 2, p. 195.
[2] Voir note 2, p. 197.
[3] H.C. Urey: Endeavour **19**, 87 (1960).
[4] H.C. Urey: J. Geophys. Res. **65**, No. 1, 358 (1960).
[5] Voir note 9, p. 185.
[6] H.C. Urey: Sky and Telescope **15**, 108, 161 (1956).
[7] H.C. Urey: Observatory **76**, 232 (1956).
[8] H.C. Urey: Vistas in Astronomy, Vol. 2. London: Pergamon Press 1956.

Imbrium résulterait du choc d'un objet d'environ 200 km de diamètre principalement formé de silicates, contenant des noyaux métalliques de fer et de nickel. La vitesse relative de ce corps et de la Lune serait faible, comprise entre 2 et 3 km/sec. Le corps aurait heurté la Lune très obliquement, le point d'impact étant le centre de l'arène Sinus Iridum.

Sous le choc, des blocs de silicates auraient été projetés en avant pour constituer l'amas de déblais représenté par les Monts Apennins (Fig. 18) et Caucasus. Les blocs métalliques lancés au même temps, d'environ 1 km de diamètre, auraient incisé le sol et formé les stries observées.

Fig. 20. Débordement de l a matière sombre remplissant le cirque de Wargentin. Cliché H. CAMICHEL.

Un énorme accroissement de température résulterait du choc. Si la densité de l'objet est 3,5 g/cm³ et la vitesse 2,38 km/sec correspondant à la vitesse de libération pour la Lune, l'énergie cinétique dégagée au choc pour 1 gramme de matière vaut $2,8 \cdot 10^3$ cal. L'énergie nécessaire pour échauffer et fondre 1 gramme vaut $2,0 \cdot 10^3$ cal. Il y a donc liquéfaction du météore lui-même et d'une partie du sol lunaire sous-jacent, créant un vaste champ de laves immédiatement répandues dans toute la cavité formée. Cet étalement aurait en grande partie refondu localement les versants des Monts Apennins et de l'ensemble du pourtour de déblais. Cette lave devrait sa teinte sombre au changement de couleur des chondrites sous l'effet de la chaleur.

Selon J. J. GILVARRY et J.E. HILL, le mécanisme invoqué paraît conforme à la théorie des chocs rapides[1].

Z. KOPAL[2] remarque que l'ébranlement du sol lunaire sous l'effet du choc doit être considérable. L'énergie mise en jeu, de l'ordre de 10^{25} ergs, dépasse

---

[1] Voir note 2, p. 193.
[2] Z. KOPAL: Nature, Lond. **183**, 169 (1959).

plusieurs centaines de fois l'énergie séismique de nos plus gros tremblements de terre. La fusion absorbe cependant une grande partie de l'énergie incidente.

Les ondes de surface convergent au point antipode de celui du choc et y accumulent une énergie séismique 0,1 fois celle produite au point d'impact, valeur encore assez considérable.

Z. KOPAL examine le cas des collisions avec de gros noyaux de comètes. Ces noyaux sont aussi fréquents que les gros météores et ils sont constitués par

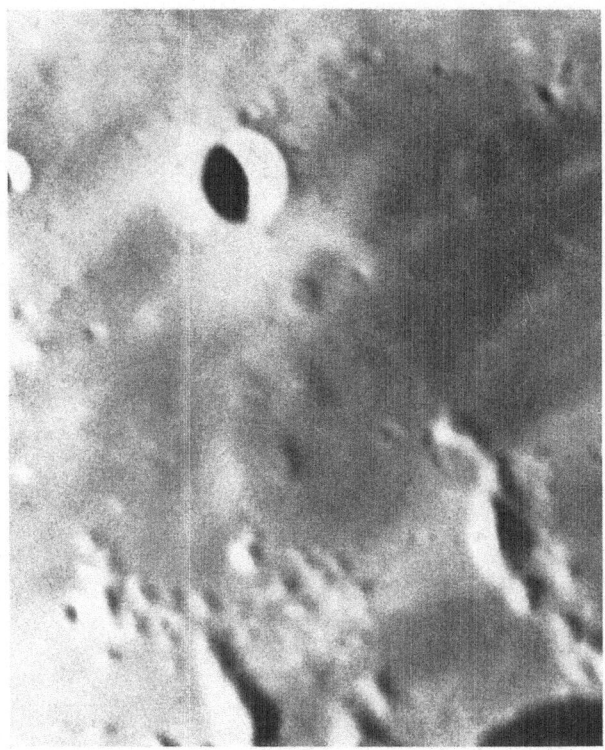

Fig. 21. Intumescences près du cirque Hortensius. Cliché H. CAMICHEL.

l'agglomération non rigide d'hydrocarbones solides. Le choc sous une vitesse de 2 à 3 km/sec produirait également la fusion, avec un ébranlement séismique faible. De telles collisions pourraient expliquer la formation des grands cratères remplis de laves, comme Plato, Archimèdes ou Wargentin (Fig. 20) ou même certaines mers.

19. Petits volcans. W. PICKERING, puis G. P. KUIPER[1] ont signalé sur la surface visible de la Lune une douzaine de petites collines montrant toutes les apparences de volcans éteints.

Ces formations sont des mamelons d'environ 8 km de diamètre. Leur pente est d'environ 5°. Ils possèdent à leur sommet un petit cratère central d'environ 800 mètres de diamètre.

La Fig. 21 montre quatre petits volcans; le cratère visible dans le champ est Hortensius; Copernic est en dehors du champ en bas à gauche.

---

[1] Voir note 2, p. 197.

Ces formations ne sont observées que dans les mers. Elles sont souvent groupées. Leur couleur est identique à celle des mers de sorte qu'elles disparaissent complètement à la pleine Lune.

L'interprétation volcanique de ces monticules paraît montrer qu'il y a eu des réserves de laves liquides sous la surface des mers. La rareté du phénomène montre aussi que les phénomènes volcaniques directs ne paraissent pas très fréquents.

20. Stries et sillons. La formation des mers s'est probablement accompagnée de la projection de matériaux rapides, dont les chutes ont profondément marqué la surface lunaire.

R.B. BALDWIN[1] et H.C. UREY[2] ont étudié les éraflures produites par les projectiles rasant le sol sur un long parcours. Les blocs métalliques ont une masse et une tension de rupture supérieures à celles des roches silicateuses du sol lunaire; des calculs montrent que de tels projectiles peuvent produire des incisions très longues. G.P. KUIPER[3], puis G. FIELDER[4,5] ont signalé des cas où les fragments responsables sont encore visibles à l'extrémité des sillons creusés dans le sol. En général, les blocs semblent allongés par l'usure, en cigare, avec le grand axe dans le sens de la trajectoire.

Comme les impacts sous l'incidence rasante ne représentent qu'une faible proportion des angles d'impacts possibles et que le nombre des cas observés est appréciable, il semble que la quantité de blocs métalliques projetés lors de la formation des mers ait été dans certains cas très grande.

21. Rainures craquelées. La surface du sol lunaire parait craquelée en certains emplacements par des fractures (Fig. 13 et 14).

Ces rainures apparaissent comme de véritables cassures du sol. Elles serpentent de façon complexe et montrent des angles vifs ou de brusques changements de direction. Les failles les plus longues ont leur rebords séparés par plusieurs kilomètres. Leurs lèvres sont généralement assez abruptes. Leur fond est habituellement partiellement comblé.

Souvent ces failles cheminent entre de petits craterlets, ou émanent d'un petit cirque, comme si ces formations affaiblissaient le sol et facilitaient leur propagation (Fig. 13). Ces rainures apparaissent aussi dans les fonds de certains grands cratères comme par exemple dans Gassendi.

Elles ont été produites évidemment par des tensions dans la surface. Celles-ci ont pu résulter soit du refroidissement général de la Lune ou des champs de laves, soit plutôt des vibrations séismiques violentes ayant accompagné la chute de météores.

22. Rainures régulières. D'autres rainures ne présentent ni cassures, ni angles vifs. Elles sont allongées en lignes droites ou en grands arcs de cercle, souvent à peine sinueux (Fig. 15 et 16).

Leur largeur est constante; leur profondeur est régulière et leurs fonds sont plats.

Souvent ces failles semblent s'être formées en écartant ou déprimant les remparts des cratères ou les blocs qu'elles rencontrent sur leur passage. G.P. KUIPER les

---

[1] Voir note 1, p. 192.
[2] Voir note 9, p. 185.
[3] Voir note 2, p. 197.
[4] G. FIELDER: J. Brit. Astronom. Ass. **67**, 60 (1957).
[5] G. FIELDER: J. Brit. Astronom. Ass. **66**, 26 (1955).

compare aux formations géologiques appelées «Graben» et constituées par deux failles parallèles avec affaissement du terrain compris entre elles[1].

Une carte de la répartition de ces failles sur le disque lunaire a été dressée par R. Baldwin[2]. Ces rainures occupent nettement des emplacements priviligiés à la limite de séparation entre les mers et les continents. Elles courent le plus généralement parallèlement aux rivages. Fréquemment des fractures s'alignent parallèles entre elles, par exemple à la bordure Ouest de Mare Humorum (Fig. 15).

Elles semblent dues aux échauffements et aux refroidissements liés à la formation des grands champs de laves constituant les mers.

23. Rides. La surface de toutes les mers est parcourue par de nombreuses rides, sorte de plissements allongés en relief. Les dénivellations sont faibles, les sommets dépassant rarement quelques centaines de mètres. Elles ne sont observables que sous l'incidence rasante, près du terminateur (Fig. 15 à 17). On observe beaucoup plus de rides parallèlement à la direction du terminateur; une ride perpendiculaire à la ligne des cornes ne pourrait être décelée.

Les rides serpentent de façon complexe; elles cheminent quelquefois parallèlement (Fig. 15), mais il arrive qu'elles se croisent. Dans l'ensemble, elles apparaissent plutôt vers les bords des mers qu'au voisinage de leurs centres; elles sont assez souvent parallèles aux rivages (Fig. 15 et 17).

En examinant ces rides avec une grande résolution, G. P. Kuiper a décelé souvent de petites fissures le long du sommet des crêtes, certaines d'entre elles montrant même des sortes d'extrusion[1].

Ces rides suggèrent un phénomène de compression.

G. P. Kuiper explique comme il suit la formation des rides ainsi que celle des rainures: Le refroidissement de la Lune a commencé par la surface. La contraction qui a résulté a ouvert de nombreuses failles. Un certain nombre d'entre elles se sont partiellement comblées, par l'extrusion de laves sousjacentes qui se sont ensuite solidifiées. En se poursuivant, le refroidissement a intéressé des couches plus profondes qui se sont contractées à leur tour. La tension de la surface s'est annulée, puis a fait place à une compression. Celle-ci a refermé certaines failles, mais a laissé subsister celles qui étaient partiellement comblées; cette compression aurait donné naissance aux stries observées.

Toutefois les dégagements de chaleur produits localement au moment de la formation cataclysmique des mers et de l'épanchement des laves pourraient aussi être invoqués, ainsi que les secousses séismiques correspondantes.

24. Falaises. On peut expliquer aussi les falaises que l'on observe quelquefois.

Le «Mur droit» et la dénivellation au Nord d'Aristarchus paraissent être d'anciennes failles dont les lèvres ont été déportées et qu'une compression aurait refermées.

Le long du rivage de la mer des Humeurs, au Sud de Gassendi, une longue saignée apparaît en certains points sous forme d'une rainure étroite, en d'autres points selon l'aspect de falaises; une faille a dû s'ouvrir, sous l'effet de tensions, puis se refermer par compression, après que des changements de niveaux aient apparu en certains endroits de part et d'autre.

c) Conclusion.

25. Privée d'atmosphère et dépourvue des causes d'érosion correspondantes, la Lune conserve sur son sol les marques de son long passé. Les phénomènes

---

[1] Voir note 2, p. 197.
[2] Voir note 1, p. 192.

de cataclysmes s'y sont accumulés; leurs vestiges superposés rendent leur histoire difficile à dégager.

Il n'y a pas nettement les traces d'une atmosphère primitive. L'histoire de la Lune est dépourvue du rôle des grands plissements qui ont travaillé l'écorce terrestre. On n'y observe que peu d'effets nettement volcaniques.

La morphologie de la surface lunaire semble dominée par l'effet des collisions dont la Lune a été l'objet. La Lune a accumulé sur son sol les impacts de tous les corps célestes rencontrés.

Les très petits météorites, de beaucoup les plus nombreux, ont ravagé le sol par leurs explosions, n'épargnant aucun endroit. Ils ont pulvérisé la surface et creusé en toutes régions d'innombrables cavités superposées. Les accidents de terrain se sont trouvés de la sorte dégradés: les plus vieilles formations ont été les plus ravagées.

Les explosions des plus gros météores ont creusé les cratères que nous observons. De grandes auréoles de poussières ont été projetées ainsi que des fragments.

Les collisions exceptionnelles ont provoqué les champs de laves solidifiées que nous appelons les mers. Le mécanisme détaillé de ces épanchements ne semble pas recueillir encore l'unanimité des opinions. Des amas de déblais ont été projetés; ils constituent maintenant des chaînes de sommets. Des fragments métalliques, violemment éjectés, ont ricoché et incisé le sol sur de longs parcours. Les vibrations du sol out ouvert des cassures à la surface.

Des effets de chaleur ont dans certains cas pu se manifester. Le refroidissement du sol a entraîné des tensions qui ont fracturé la surface selon des failles. Les compressions ont refermé certaines fentes et ridé les contours des mers selon de longs cordons.

La Lune nous révèle la manifestation des phénomènes les plus simples que peut subir un corps céleste, privé d'atmosphère et dépourvu d'effets intérieurs trop intenses mais ayant gravité longuement dans le système solaire.

II. La Planète Mercure.

La Planète Mercure est un astre petit, très proche du Soleil, et difficile à bien observer. Par suite nos connaissances sur la surface de son globe restent encore fragmentaires.

26. Diamètre du globe de Mercure. Le diamètre de Mercure est particulièrement mal connu. La liste de ses principales déterminations est donnée page 206.

Les mesures qui semblent les plus exactes donnent dans leur ensemble une moyenne de 6″45, mais la précision ne dépasse pas 5%. En admettant que les deux dernières déterminations reportées ont bénéficié du procédé d'observation le plus approprié, le diamètre apparent à la distance de 1 U.A. serait environ 6″45, soit à peu près 4700 km. Le diamètre de la Lune est 3473 km.

27. Carte topographique de la surface de Mercure. Dans les conditions d'observation assez rarement réalisées où la turbulence atmosphérique est suffisemment faible pour laisser aux lunettes astronomiques une certaine résolution, la planète Mercure apparaît sous forme d'un croissant marqué de quelques taches légères. La persistance de ces taches permit de conclure que la durée de rotation de la planète autour de son axe polaire est de 88 jours; par suite, la planète montre toujours la même face au Soleil. L'autre hémisphère n'est pas éclairé. Toutefois la libration en longitude atteint la forte valeur de 23°7, ce qui permet d'observer beaucoup plus de la moitié de la surface du globe.

Procédé d'observation	Auteur	Référence	Diamètre apparent à 1 U.A.
Micromètre à fil	Le Verrier	Liouville's J. **8**, 279 (1843)	6″65
	Ball	Astronom. Nachr. No. 2458	7″34
	See	Astronom. Nachr. No. 3732	5″90
	Barnard	Astronom. Nachr. No. 3760	6″59
	Wirtz	Ann. Obs. Strasbourg **4**	6″43
Micromètre à fil lors d'un passage devant le disque solaire	Mädler	Astronom. Nachr. No. 538	6″54
	Campbell	Astronom. J. No. 331	5″73
	Barnard	Astronom. J. No. 335	6″04
	Rabe	Astronom. Nachr. No. 5600	6″32
	Jonckheere	Monthly Notices Roy. Astronom. Soc. London **75**, 31 (1914)	6″43
Héliomètre	Kaiser	Seiden Ann. **3**	6″61
	Ambronn	Astronom. Nachr. No. 3034	6″58
	Hartwig	Naturforsch. Ges. Bamberg **21**	6″78
Héliomètre lors d'un passage devant le disque solaire	Bessel	Astronom. Untersuch. **2**	6″68
	Todd	Astronom. Nachr. No. 2208	6″60
	Schur	Astronom. Nachr. No. 3038	6″38
	Hartwig	Naturforsch. Ges. Bamberg **21**	6″72
Instants des contacts lors d'un passage devant le disque solaire	Discussion des mesures de Innes et de Stroobant par Danjon		6″16
Micromètre à double image lors d'un passage devant le Soleil	Dollfus	L'Astronomie 1954, 338	6″45
Micromètre à double image	Müller	L'Astronomie 1955, 61	6″42

Les observateurs les plus assidus sont parvenus à dresser des cartes planisphères de ces taches. Voici la liste de celles publiées à ce jour:

Schiaparelli: Astronom. Nach. 246 (1889).
Lowell: Mem. Amer. Acad. of Arts and Sci. **12** (1897).
Jarry Desloges: Observations des Surfaces Planétaires VII.
Antoniadi: «La Planète Mercure», Gauthier Villars 1934.
Rudaux: «L'Astronomie» 1928. 301.
Bidault de l'Isle: «L'Astronomie» 1928. 301.
McEwen: J. Brit. Astronom. Ass. **45**, 582 (1936).
Dollfus: «L'Astronomie» 1953. 62.

La Fig. 22 reproduit les plus intéressants de ces planisphères. On remarque qu'ils se divisent en deux groupes, les trois cartes de la colonne de gauche sont semblables. Les deux cartes supérieures de la colonne de droite sont voisines et assez différentes des autres. Le planisphère du bas, à droite présente un intérêt tout particulier. Il a été dressé d'après des clichés photographiques, obtenus à l'Observatoire du Pic du Midi par B. Lyot et H. Camichel. Ces très beaux clichés montrent les taches avec suffisamment d'évidences pour être mesurables[1].

Ce dernier planisphère montre un accord raisonnable avec ceux de la colonne de gauche après rotation de ceux-ci d'environ 15°, dans le sens trigonométrique, autour de leurs centres. Un tel changement d'orientation s'explique de la manière suivante: presque toutes les observations de Antoniadi, et un grand nombre de celles de Schiaparelli, ont été effectuées aux élongations favorables des mois

---

[1] A. Dollfus: L'Astronomie 1953, 61.

d'été, de Juillet à Octobre; cette sélection naturelle des périodes d'observation favorables fait que la planète n'a été observée que sur une partie de son orbite seulement, voisine de l'aphélie, comprise entre les longitudes 200 et 250°. Au contraire, les observations photographiques du Pic du Midi ont été affectuées aux élongations du matin, en 1942 et 1944, entre les mois de Juillet et d'Août,

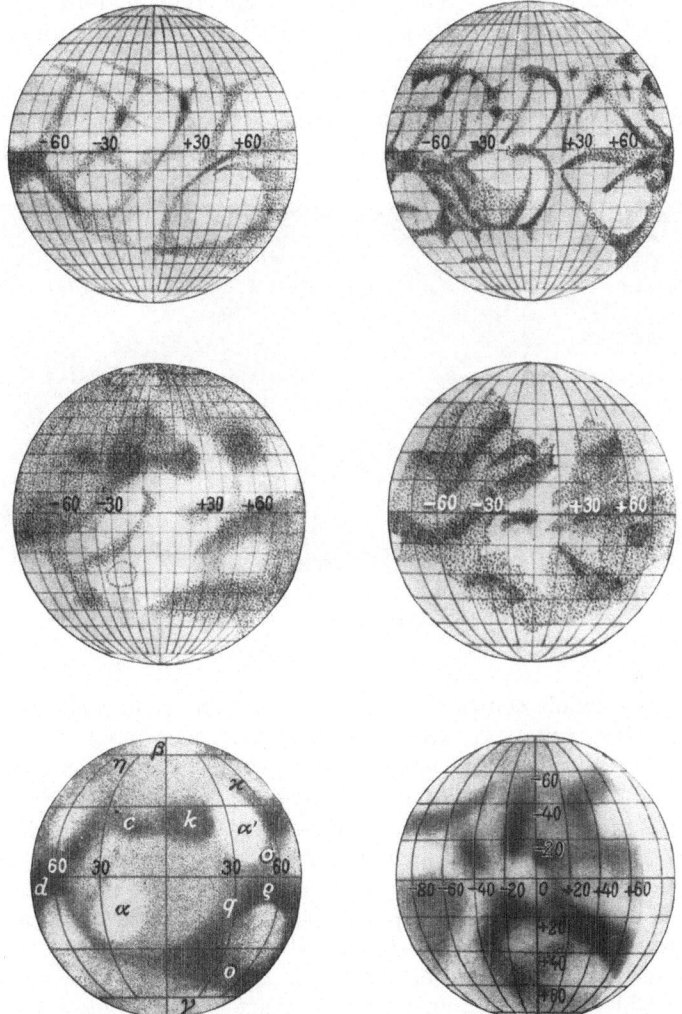

Fig. 22. Cartes topographiques de la surface de Mercure. a) SCHIAPARELLI 1889. b) ANTONIADI 1934. c) RUDEAUX 1928. d) JARRY DESLOGES 1920. e) McEWEN 1936. f) DOLLFUS 1953.

alors que la planète occupait sur son orbite une position presque diamètralement opposée. Une inclinaison de l'axe de rotation de 7° par rapport à la normale au plan de l'orbite suffit à empliquer les écarts observés.

Sur les deux planisphères du haut de la colonne de droite, les parties gauches rejoignent les aspects des cartes de gauche, et le côté droit coïncide avec le planisphère photographique. Les périodes d'observation favorables, différentes pour les élongations du soir et du matin, se sont réparties de sorte que les deux parties de ces planisphères correspondent à des positions moyennes opposées sur l'orbite.

Le planisphère basé sur des documents photographiques, groupés en un court laps de temps, est probablement particulièrement fidèle, l'orientation du réseau de coordonnées superposé n'étant cependant pas exactement précisé.

28. Observation de Mercure avec un grossissement élevé. Les remous de l'atmosphère font qu'il est très difficile d'observer la planète Mercure avec un pouvoir séparateur élevé[1]. Cependant, grâce à une étude détaillée des conditions d'observation à l'Observatoire de montagne du Pic du Midi, A. Dollfus a bénéficié plusieurs fois de conditions exceptionnelles, avec une lunette de 60 cm de diamètre[2]. Il fut alors loisible d'examiner Mercure parfaitement immobile, pendant de longs instants, avec des grossissements tels que 750 et 900, qui amènent la planète à 150 000 km du regard, c'est-à-dire à environ la moitié de la distance

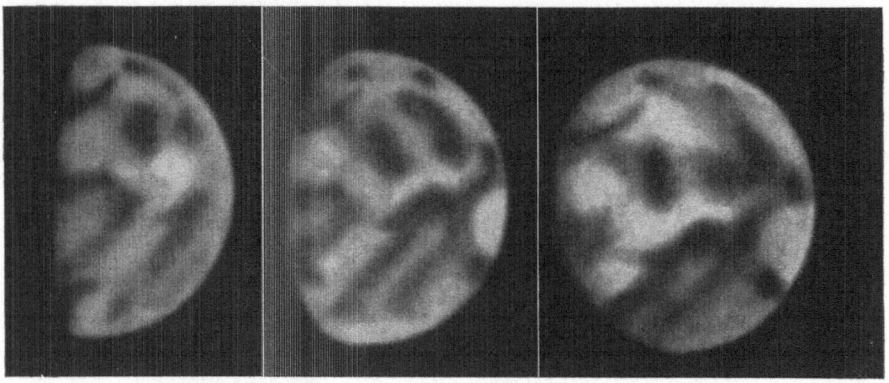

Fig. 23. Dessins de la planète Mercure obtenus par A. Dollfus à l'Observatoire du Pic du Midi les 6, 12 et 19 Octobre 1950.

de la Lune. Des détails contrastés distants de 300 km sur le sol de Mercure sont alors vus séparés. Les trois dessins de la Fig. 23 nous montrent une surface recouverte par endroits de taches sombres, permanentes, dont l'aspect et le contraste sont très semblables à ceux que montre la Lune observée à l'œil nu.

Le contour du terminateur s'est montré privé d'accidents ou de sinuosités attribuables à un relief du sol. Sur la Lune, les montagnes et les principaux grands cirques se devinent avec une lunette de 3 mm d'ouverture, donnant une résolution double de celle offerte par la lunette du Pic du Midi sur Mercure. Un relief comparable à celui de la Lune doit donc échapper sur Mercure à nos moyens d'observation. Des cratères dûs à la chute de très gros météorites ne seraient décelables qu'avec un grossissement deux fois plus puissant encore que de nos meilleurs instruments actuels.

29. Photométrie de la planète Mercure. La magnitude globale de Mercure et sa variation avec la phase ont été déterminées par Müller (1893), J. Hopmann (1922), C. Hoffmeister (1940) et surtout A. Danjon (1949); ce dernier travail résume et dépasse les recherches antérieures[3].

La courbe de lumière complète de la planète Mercure est reproduite Fig. 24. La magnitude visuelle aux distances unités s'exprime, selon A. Danjon, par la

---

[1] E.M. Antoniadi: La Planète Mercure. Paris: Gauthier Villars 1934.
[2] Voir note 1, p. 206.
[3] A. Danjon: Bull. Astron. **14**, 315 (1949).

relation:　　$M = -\,0{,}21 + 3{,}82\,(V/100) - 3{,}37\,(V/100)^2 + 2{,}00\,(V/100)^3.$

Cette variation de lumière est très voisine de celle donnée par la Lune.

En admettant pour diamètre du globe de Mercure la valeur 6″16 à 1 U.A., A. Danjon donne pour l'albedo intégral de Mercure les valeurs suivantes:

Albedo visuel　　　　　$A_v = 0{,}063$.
Albedo photographique $A_p = 0{,}052$.

Ces albedos sont assez voisins de celui de la Lune, déterminé par G. Rougier (Lune $A_v = 0{,}073$), quoiqu'un peu plus faible.

L'intégrale de phase donnée par A. Danjon pour Mercure est $I_p = 0{,}563$. La valeur correspondante pour la Lune vaut 0,584 et elle est relativement très semblable. Le facteur de diffusion moyen de Mercure, rapporté à celui d'un écran blanc parfaitement diffusant, est 0,112, contre 0,125 pour la Lune.

Mercure reste légèrement moins brillant que ne le serait une planète de même diamètre ayant les propriétés superficielles du sol lunaire; la faible différence, d'abord croissante, tend à s'atténuer pour les grandes valeurs de l'angle de phase.

30. Polarimétrie de la planète Mercure.
Les premières mesures de polarisation sur la planète Mercure sont dues à B. Lyot[2],[3]. La courbe reliant la proportion polarisée de la lumière de l'ensemble de l'astre avec l'angle de vision est identique à celle de la Lune (Fig. 3).

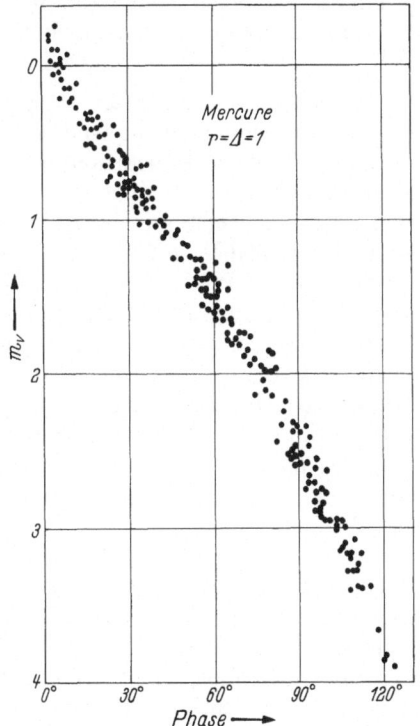

Fig. 24.　Courbe photométrique de la planète Mercure. D'après A. Danjon[1].

En 1950, A. Dollfus a mesuré la polarisation, en lumière verte et rouge, sur les différentes régions de la surface de la planète[4]. Pour les petits angles de phase, on ne peut déceler de différence sensible entre les régions claires et les taches sombres; mais à la quadrature la polarisation est nettement plus forte sur les taches sombres; elle varie grossièrement en sens inverse de l'albedo. Cette propriété s'observe de la même façon sur la Lune; elle confirme la très grande analogie qui doit exister entre les structures des sols de ces deux astres.

Les faibles variations de polarisation décelées près des cornes du disque de Mercure peuvent s'expliquer en supposant la planète enveloppée d'une très faible atmosphère, dont l'épaisseur dans les conditions normales serait environ trois millièmes seulement de celle de l'atmosphère terrestre[4].

31. Emission thermique par la surface de Mercure.
Le sol de la planète émet un rayonnement infra-rouge thermique, à la façon d'un corps noir. Ce rayonnement a été mesuré à l'aide d'un thermocouple très sensible, avec le télescope de 2,50 m du Mont Wilson, par E. Pettit et S. Nicholson[5]. L'énergie

---

[1] Voir note 3, p. 208.
[2] B. Lyot: Thèse de Doctorat, Paris 1929. — Ann. Obs. Meudon **8** (1929).
[3] B. Lyot: C. R. Acad. Sci., Paris **191**, 703 (1930).
[4] Voir note 2, p. 183.
[5] E. Pettit et S. Nicholson: Astrophys. J. **83**, 84 (1936).

recueillie, réduite aux distances unités pour la Terre et le Soleil, est identique à celle que l'on mesure dans les mêmes conditions d'éclairement sur la Lune. Reportée en fonction de l'angle de phase, cette radiation varie de la même façon que dans le cas de la Lune. La courbe rappelle celle que donnerait une sphère sombre, dépolie, non conductrice et ne tournant que lentement. Cependant la plus grande proximité du Soleil donne sur Mercure une amplitude beaucoup plus grande. La température du point ayant le Soleil au zénith est trouvée voisine de 610° K. Cette valeur est 380° K pour la Lune.

32. Conclusions sur l'état de la surface de Mercure. La surface de Mercure nous paraît singulièrement semblable à celle de la Lune.

L'identité de la polarisation indique une contexture microscopique semblable, formée de petits grains opaques.

L'origine de cette pulvérisation superficielle ne peut être une variation thermique car la planète présente toujours la même face au Soleil et la température y est constante. Près du terminateur la libration en longitude donne des variations d'éclairement très fortes, mais très lentes; ces variations thermiques sont atténuées par le plus grand diamètre apparent du Soleil, qui atteint 1°40′, et jette entre les aspérités du sol une pénombre progressive. Le polarimètre ne décèle d'ailleurs pas de variations systématiques entre les régions voisines du terminateur et celles qui ne sont pas atteintes par la libration.

Ces remarques confirment le calcul développé à la section 10 et doivent nous faire exclure également les variations thermiques pour expliquer la pulvérisation du sol sur la Lune.

La structure poudreuse du sol de Mercure nous semble provenir, comme dans le cas de la Lune, de la projection de débris lors des impacts par les petits météores. La similitude des courbes photométriques de Mercure et de la Lune indique un façonnement morphologique superficiel comparable. On peut en conclure que l'atmosphère très faible qui semble envelopper la planète est suffisamment ténue pour ne pas reboucher les cavités du sol, au moins les plus grosses, par le transport de poussière volante. Il semble aussi que les proportions des météores qui creusèrent les très nombreuses cavités de ces sols furent à peu près les mêmes dans les deux régions du système solaire voisines des orbites de Mercure et de la Terre.

La planète Mercure tourne toujours la même face au Soleil et offre par conséquent toujours le même hémisphère dans la direction du déplacement sur l'orbite. Cet hémisphère doit collecter plus de météores et les impacts doivent y être plus rapides. Mais on n'a pas décelé de différences de propriété attribuables à ce comportement.

Les étendues sombres appelées «mers» ont même contraste et même physionomie sur Mercure et sur la Lune. S'ils sont dûs, sur la Lune, aux impacts par de très gros corps célestes, il semble raisonnable de les attribuer, sur Mercure, à la même cause, et de déduire que de très gros météores existaient aussi dans cette partie du système solaire.

Les vitesses orbitales autour du Soleil étant 1,6 fois plus rapides à la distance de Mercure, les chocs y développent en moyenne, pour une même masse, 2,6 fois plus d'énergie. Selon les interprétations précédentes, des cratères analogues à ceux de la Lune devraient exister sur Mercure et pouvoir, dans l'avenir, être décelés.

Mercure est plus gros que la Lune; son échauffement intérieur par radioactivité a dû être plus fort et les effets sur la surface résultant de cet échauffement ont dû créer des actions plus prononcées. Mais ces phénomènes échappent à nos moyens d'observation.

III. Les astéroïdes.

Les petits corps célestes qui gravitent entre les orbites de Mars et de Jupiter ont une trop faible masse pour retenir une atmosphère. Ils montrent donc la surface de leur sol; bien que leurs diamètres ne soient généralement pas décelables, des données peuvent être recueillies sur l'état de leur surface par l'intermédiaire de la photométrie et de la polarisation.

33. Dimension des astéroïdes. Avec les meilleures lunettes, dans les conditions d'observation exceptionnelles, les petits astres les plus brillants ont quelquefois montré un aspect légèrement différent de celui des étoiles voisines. Au siècle dernier, BARNARD put estimer les diamètres apparents de quatre d'entre eux, avec la grande lunette de LICK[1]. Les ordres de grandeur trouvés sont les suivants:

Cérès 770 km, Pallas 490 km, Junon 190 km, Vesta 380 km.

Des mesures interférométriques pratiquées par HAMY peu avant 1900 confirmèrent le diamètre de 380 km attribué à VESTA[2]. Les méthodes interférométriques appliquées à nos grands réflecteurs modernes permettraient de mesurer encore d'autres astres; ce travail n'a pas suffisamment attiré l'attention des observateurs.

Les nombreuses mesures photométriques effectuées au siècle dernier à Potsdam par MÜLLER[3], puis ensuite à Harvard par PARKHURST[4] fixèrent la magnitude d'un grand nombre d'astéroïdes. En admettant que ces astres aient un pouvoir réflecteur du même ordre que celui des quatre astéroïdes de diamètre connu, la mesure de leur éclat donne leur diamètre approximatif. On dénombre actuellement environ 200 astéroïdes de diamètre compris entre 100 km et 150 km, et 670 entre 50 km et 20 km.

Selon G. P. KUIPER et ses collaborateurs[5-7] le nombre d'astéroïdes connus double à peu près pour chaque accroissement de magnitude, entre la 10ème et la 16ème magnitude. Les astres de très petites dimensions sont donc extrêmement nombreux.

34. Forme des astéroïdes. L'étude photométrique des astéroïdes révèle des changements d'éclats, assez rapides et périodiques. Ceux-ci doivent nécessairement résulter d'une rotation de l'astre sur lui-même. Ils proviennent soit de taches sur la planète, soit d'une forme notablement différent de la sphère.

Dans le premier cas les variations d'éclat sont faibles et se reproduisent avec la même périodicité que celle de la rotation. Ce doit être le cas de Vesta qui tourne en 5^h20^m5 et dont les variations d'éclat sont seulement de 0,11 mag, tandis que les taches de la surface de Mars donnent des variations de 0,4 mag. De faibles variations de l'indice de couleur, voisines de 0,01 mag, confirment cette interprétation[5]. Cérès, dont les variations d'éclat ne sont que de 0,03 mag, et probablement Pallas, rentrent aussi dans cette catégorie.

Les astres qui ne sont pas sphériques montrent deux maxima et deux minima de lumière par période de rotation, et ceux-ci sont variables selon la présentation de l'axe de rotation par rapport à la Terre. L'amplitude de la variation lumineuse dépasse souvent 0,5 mag et elle peut devenir très grande pour les corps en forme

---

[1] E.E. BARNARD: Astronom. Nachr. **159**, 261.
[2] M. HAMY: C. R. Acad. Sci., Paris **128**, 583 (1899).
[3] G. MÜLLER: Publ. Astr. Obs. Potsdam **8** (1893).
[4] M.H. PARKHURST: Ann. Harvard College Obs. **18**, 29; **29**, 65. — Astronom. J. **9**, 127.
[5] G.P. KUIPER: Astrophys. J. **120**, 200, 529, 551 (1954).
[6] T. GEHRELS: Astrophys. J. **123**, 331; **125**, 550.
[7] I. VAN HOUTEN et J. VAN HOUTEN: Astrophys. J. **127**, 253 (1958).

de cigare tournant autour d'un axe perpendiculaire aux directions de l'allonge-
ment et de l'observateur. Les variations d'éclat de Eros sont considérables; lors
du grand rapprochement de cet astre et de la Terre, en 1931, Van den Bos et
Finsen ont observé la planète sous forme d'un petit batonnet tournant avec la
même période que les variations d'éclat[1].

Une étude photoélectrique systématique des astéroïdes a été entreprise
récemment à l'Observatoire de Yerkes et l'Université de Indiana, par G.P.
Kuiper, T. Gehrels, I. et C. Van Houten, etc.[2-4]. 90% des astres observés
ont montré une variation de lumière périodique décelable. Les périodes de rotation sont comprises le plus généralement entre 4^h et 20^h, sauf la petite planète
((32)) qui tourne en 2^h52^m. Les axes de rotation semblent orientés à peu près
au hasard sur la sphère céleste. Dans certains cas le sens de rotation a pu être
déterminé; l'astéroïde ((15)) Eunomia tourne dans le sens rétrograde; ((39)) Lacti-
tia tourne dans le sens direct.

Les axes de rotation semblent occuper indistinctement toutes les positions
par rapport aux axes des ellipsoïdes d'inertie de ces corps.

Les astéroïdes semblent avoir les formes les plus complexes, comme le voudrait
la fragmentation brutale d'un corps solide beaucoup plus gros. Aux températures ordinaires la force de cohésion des corps à l'état solide est très supérieure
aux forces de gravité pour ces petits astres; ceux-ci doivent conserver leurs formes
primitives. L'astre dont ils sont probablement les fragments devait donc être
en grande partie à l'état solide lors de sa fracturation. Junon mesure 190 km de
diamètre et semble déjà rentrer dans la catégorie des astres à forme complexe
non sphérique.

35. Pouvoir réflecteur moyen de la surface des astéroïdes. Le facteur de
diffusion moyen de la surface est la fraction de la lumière renvoyée dans la direction du Soleil, rapportée à celle que donnerait un écran blanc parfaitement diffusant et normal à la direction de l'éclairement. Pour les quatre astéroïdes de
diamètre apparent connus, H.N. Russell déduisit les facteurs de diffusion
suivants:

Cérès 0,10, Pallas 0,13, Junon 0,22, Vesta 0,48.

L'albedo est la fraction de l'énergie lumineuse renvoyée dans tout l'espace,
rapportée à celle reçue par le corps. Les valeurs sont approximativement les suivantes:

Cérès 0,06, Pallas 0,07, Junon 0,12, Vesta 0,26.

On pouvait s'attendre à trouver des valeurs comparables pour ces différents
astres, qui ont probablement une même origine. Cérès, Pallas et même Junon
ont des facteurs de diffusion et albedo voisins de ceux de la Lune qui valent 0,12
et 0,073. Les estimations de diamètre sont très approximatives et l'imprécision
sur le calcul de l'éclat est considérable. Toutefois la valeur 0,48 attribuée à
Vesta, beaucoup plus élevée que les autres, ne semble pas pouvoir s'expliquer
entièrement par de telles erreurs. Cet astre est probablement sphérique et présente sur sa surface des taches d'éclat différent. L'ensemble de son sol doit
être beaucoup plus clair que celui de la Lune. Peu de substances possèdent le
facteur de diffusion 0,48 qui lui est attribué. La craie, certaines argiles très
claires, des cendres exceptionnellement brillantes ou un dépôt de givre peuvent
cependant être invoquées.

---

[1] van den Bos et Finsen: Astronom. Nachr. **241**, 329 (1931).
[2] Voir note 5, p. 211.
[3] Voir note 6, p. 211.
[4] Voir note 7, p. 211.

A. S. STARKANOV et E. A. LUBLINOVA d'une part[1], G. P. KUIPER d'autre part, ont calculé que, selon certaines hypothèses, la température interne d'un corps de diamètre comparable à Veste, Cérès et Pallas a pu s'élever au centre sous l'effet de la radioactivité, à plus de 1000°. Une modification de l'état de surface aurait pu éventuellement en résulter.

36. Couleur de la surface. De nombreuses mesures d'indice de couleur ont été réalisées, tant par la méthode photographique que par photométrie photoélectrique.

G. P. KUIPER et ses collaborateurs[2] ont pratiqué des mesures photoélectriques dans les trois intervalles spectraux, U, B et V définis par JOHNSON et MORGAN[3]. La relation entre les indices de couleur $B - V$ et $U - V$ est à peu près linéaire et voisine de celle des étoiles de la série principale; la plupart des astéroïdes se groupent autour des indices du couleur des étoiles $K0$ à $K2$.

Leurs surfaces sont donc dans l'ensemble de couleur comparable et légèrement jaunes. Aucune relation n'apparaît nettement entre la couleur et la magnitude. Ces résultats traduisent une certaine uniformité du matériau constitutif. A l'exception de la légère variation de couleur avec la phase reconnue sur Vesta, et précédemment mentionnée, aucun astéroïde n'a montré de relation nette entre la variation périodique de brillance et l'indice de couleur; les différentes faces des blocs doivent donc avoir des propriétés semblables.

37. Rugosité de la surface. Les irrégularités et cavités qui marquent la surface d'un astre éclairé obliquement se portent ombre mutuellement; elles se traduisent par une décroissance prononcée de la magnitude avec l'angle de phase.

Au siècle dernier, G. MÜLLER a déterminé les coefficients de phase de nombreux astéroïdes[4]. Il trouva une croissance de magnitude proportionnelle à la phase, le coefficient moyen pour l'ensemble de ses déterminations étant 0,030 mag/degré. Les mesures plus récentes de T. GEHRELS[5] donnent une moyenne de 0,029 mag/degré, mais l'auteur trouve une chute d'éclat plus rapide au voisinage immédiat de la phase zéro. Mercure nous montre une propriété analogue. Correction faite de ce petit effet, le coefficient de phase entre les angles 10 et 20° serait en moyenne 0,023 mag/degré seulement. Aucune corrélation n'est trouvée entre cette valeur et la magnitude absolue, ou les éléments de l'orbite.

Le coefficient de phase moyen de la Lune est 0,028 mag/degré. La rugosité de la surface des astéroides doit donc être du même ordre que celle de la surface lunaire, c'est-à-dire très prononcée.

Sur 50 cas, nous avons trouvé la répartition suivante pour les valeurs du coefficient de phase:

de 0,015 à 0,018 mag/degré	1 cas
0,019 à 0,022 mag/degré	6 cas
0,023 à 0,026 mag/degré	11 cas
0,027 à 0,030 mag/degré	5 cas
0,031 à 0,034 mag/degré	6 cas
0,035 à 0,038 mag/degré	3 cas
0,039 à 0,042 mag/degré	5 cas
0,043 à 0,046 mag/degré	7 cas
0,047 à 0,050 mag/degré	5 cas
0,051 à 0,054 mag/degré	1 cas

La statistique laisse deviner deux faibles maxima autour des valeurs 0,025 et 0,045. Aucune corrélation n'apparaît entre ces deux maxima et la magnitude.

[1] A. S. STARKANOV et E. A. LUBINIOVA: Astronom. J. URSS. **31**, No. 5 (1954).

[2] Voir notes 5, 6 et 7, p. 211.

[3] Voir C. FEHRENBACH, vol. L de cette Encyclopédie, p. 84.

[4] Voir note 3, p. 211.

[5] Voir note 6, p. 211.

Quoiqu'il en soit, les astéroïdes doivent avoir des surfaces extrêmement rugueuses. Ces blocs doivent être criblés d'aspérités et de cavités. Celles-ci doivent être attribuables d'une part à l'irrégularité des cassures, d'autre part, comme dans le cas de la Lune et de Mercure, aux explosions des petits météores qui ont bombardé la surface.

38. Contexture de la surface. La structure fine de la surface nous est révélée par l'analyse de la lumière polarisée.

En 1934, Lyot détermina la courbe de polarisation de la lumière de Vesta, en fonction de l'angle de phase (Fig. 25)[1]. La courbe est assez semblable à celle de la Lune; le minimum est un peu plus faible, l'angle d'inversion légèrement plus grand. La structure correspondante est celle d'une poudre de granules opaques déposée en couche irrégulière. Des cendres volcaniques éclaircies par de la poudre de craie donnent ce comportement.

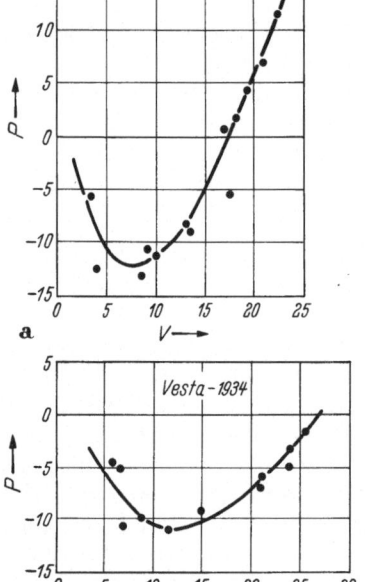

Fig. 25. Courbes de polarisation des astéroïdes (a) Cérès et (b) Vesta. D'après les observation de B. Lyot[1].

Une courbe de polarisation non publiée fut obtenue par Lyot en 1935 sur Cérès. La Fig. 25 a la reconstitue. En 1936, Lyot a trouvé sur Cérès une polarisation un peu plus forte et variable. Cette différence pourrait indiquer un changement dans l'orientation de l'axe de rotation relativement à la Terre, dévoilant des régions différentes. L'angle d'inversion est de 17°5. La polarisation n'est pas celle de granules en poudre; elle se rencontre sur les cassures de laves et sur des grés.

En 1955, S. Provin réalisa quelques nouvelles mesures, photoélectriquement, sur Cérès, Pallas et Iris. La courbe de Iris rappelle celle de la Lune et correspond à des granules. Iris montre une variation périodique d'éclat de 0,2 mag qui ne se traduit par aucun changement de polarisation; la structure superficielle sur ce corps est donc de même nature sur les différentes faces et cassures qui constituent probablement la surface.

Iris et Vesta semblent avoir une structure superficielle pulvérisée semblable à celle rencontrée sur les autres astres dépourvus d'atmosphère importante, tels que la Lune, Mercure et Mars. Le cas de Cérès paraît assez différent.

39. Conclusions sur l'état de surface des astéroïdes. Les résultats précédents semblent indiquer que tous les astéroïdes plus petits que 200 km sont des fragments de forme complexe.

Ces petits astéroïdes renvoient la lumière en produisant à peu près la même couleur jaune, ce qui laisse supposer une certaine uniformité du matériau constitutif. Les fragments de moins de 200 km sont trop petits pour que la surface ait pu subir les effets d'un échauffement interne. Par contre, des collisions entre astéroïdes ont probablement dû altérer considérablement la surface. Le sol est extrêmement rugueux comme sur la Lune et Mercure; les explosions de petits météores ont dû cribler ces corps de cavités et de petits cratères. La petite

---

[1] B. Lyot: C. R. Acad. Sci., Paris **199**, 774 (1930).

planète Iris rentre dans la catégorie des fragments; sa courbe de polarisation semblable à celle de la Lune et de Mercure indique probablement une poussière formée de grains opaques; une pulvérisation superficielle a peut-être pu résulter de la projection de débris par le bombardement météorique.

Les astres plus gros que 300 km sont Cérès, Pallas et Vesta. Vesta possède une forme voisine de la sphère; sa surface possède la même rugosité générale que les petits fragments et même structure poudreuse; elle est en partie marquée par des taches de couleur légèrement différentes. Il en est probablement le même de Cérès et Pallas. Ces astres ont peut-être pu subir un échauffement radio-actif interne suffisant pour permettre la mise en forme sphérique et même modifier l'état de surface. Les taches pourraient être dues, comme dans le cas de Mercure et de la Lune, à la collision avec de gros météores, entraînant l'échauffement ou la fusion de la matière au lieu de l'impact.

Le pouvoir réflecteur de la surface de Vesta est nettement plus élevé que celui de tous les autres corps du système solaire dépourvus d'atmosphère. La courbe de polarisation de Cérès diffère de celle de tous les autres corps analogues et caractérise une matière rugueuse ou fraîchement cassée. Ces deux anomalies pourraient résulter peut-être de collisions exceptionnelles; la poussière déposée à la surface de corps si petits n'est retenue que par une très faible force de gravité.

IV. Les satellites de Jupiter.

Le satellite principal de Saturne, Titan, est entouré d'une atmosphère chargée de nuages opaques. Les quatre gros satellites de Jupiter montrent au contraire distinctement la surface de leurs sols et rentrent dans la catégorie des objets célestes examinés dans cette étude. Io et Europe n'ont probablement pas d'atmosphère très dense. Ganymède révèle des voiles très légers qui ne gênent pas sensiblement l'observation de la surface.

40. Diamètre des satellites de Jupiter. Ces petits corps célestes sous-tendent un diamètre apparent voisin de 1″ et la détermination précise de leurs dimensions est une tâche délicate. Les premières mesures ont été relevées à l'aide du micro-mètre à fil; en voici la liste réduite à la distance 5 U.A.:

Observateur	Lunette	Io	Europe	Ganymède	Callisto	Références
Burnham . . .	91 cm	1″11	1″00	1″78	1″61	1
Bigourdan . .	32 cm	0″80	0″75	1″38	1″24	2
Barnard . . .	91 cm	1″09	0″88	1″57	1″49	3
T. J. J. See . .	65 cm	0″95	0″86	1″48	1″30	4
Moyenne		0″99	0″87	1″55	1″41	

L'agitation et le bouillonnement des images affectent les mesures données par le micromètre à fil et doivent donner dans l'ensemble des valeurs trop élevées.

Les mesures réalisées à l'aide de micromètres interférentiels sont affectées surtout par l'assombrissement des bords de l'astre et les taches de la surface; elles donnent probablement des valeurs un peu faibles, dont voici la liste:

---

[1] Burnham: Nature, Lond. **45**, 160 (1891).
[2] Bigourdan: Ann. Obs. Paris **21**.
[3] E. E. Barnard: Monthly Notices Roy. Astronom. Soc. London **55**; **56**.
[4] T. J. J. See: Astronom. Nachr. No. 3745—3868.

Observateur	Io	Europe	Ganymède	Callisto	Références
MICHELSON 1891 .	0″95	0″88	1″25	1″21	1, 2
HAMY 1898	0″86	0″78	1″14	1″15	3, 2
DANJON 1933 . . .	0″90	0″78	1″20	1″08	2
Moyenne	0″90	0″81	1″20	1″14	

Des mesures nombreuses ont été réalisées vers 1944 par H. CAMICHEL à l'Observatoire du Pic du Midi, en projetant dans le champ de la lunette un disque artificiel de comparaison dont le diamètre, la couleur, l'éclat et le flou du bord sont ajustables[4]. Les valeurs obtenues sont:

$$0″90, \quad 0″78, \quad 1″35 \quad \text{et} \quad 1″26.$$

Un assombrissement au bord du disque conduirait à des valeurs légèrement trop faibles.

Le micromètre à double image fut utilisé pour la première fois vers 1906 par SALET et BOSLER[5], qui obtinrent pour les quatre satellites:

$$0″98, \quad 0″91, \quad 1″43, \quad 1″33.$$

Des mesures beaucoup plus nombreuses furent obtenues de 1948 à 1953 avec la lunette de 60 cm du Pic du Midi par A. DOLLFUS[6]. Les déterminations furent corrigées dans une certaine mesure de l'assombrissement qui semble se manifester au bord du disque lorsque l'on pratique les mesures à l'aide d'un choix de diaphragmes; l'effet du passage au bord des taches de la surface mis en valeur sur Ganymède a été corrigé. Les valeurs les plus probables déterminées de la sorte sont:

$$0″97, \quad 0″85, \quad 1″53, \quad 1″38.$$

Io mesurerait 3550 km, soit presque exactement la dimension de la Lune. Europe donne 3100 km. Ganymède, un peu plus petit que Mars, donne 5600 km. Callisto, plus difficile à bien mesurer, ferait 5050 km.

41. Observation télescopique de la surface. Les premières observations des satellites de Jupiter avec les grandes lunettes remontent au siècle dernier. HOLDEN (1888), puis BARNARD (1890—1891) observèrent à Lick des taches sur Io et Ganymède lors de leurs passages devant le disque de Jupiter. En dehors du disque, les observations sur le fond sombre du ciel sont plus difficiles. Vers 1894 cependant, PICKERING réalisa avec une lunette de 45 cm de curieuses observations[7], examinées plus tard par B. LYOT[8].

Au début de notre siècle, E.M. ANTONIADI obtint à Meudon de nouvelles observations lors des passages devant Jupiter, et découvrit des taches sur Europe[9]. En dehors du disque, G. FOURNIER fit plusieurs dessins entre 1917 et 1920[10], puis A. DANJON obtint en 1934 à Strasbourg 7 observations détaillées de Ganymède et 3 dessins de Callisto, avec un objectif de 48,6 cm[11]; un premier planisphère de Ganymède fut ébauché.

---

[1] MICHELSON: Nature, Lond. **45**, 160 (1891).
[2] A. DANJON: Ann. d'Astrophys. **7**, 125 (1944).
[3] M. HAMY: Bull. Astr. **16**, 257 (1899).
[4] H. CAMICHEL: Ann. d'Astrophys. **16**, 42 (1953).
[5] SALET et BOSLER: Bull. Astr. **23**, 325 (1906).
[6] A. DOLLFUS: C. R. Acad. Sci., Paris **238**, 1475 (1954).
[7] W.H. PICKERING: Ann. Lowell Obs. **2**, 1 (1900).
[8] B. LYOT: L'Astronomie, Avril 1943, 49.
[9] E.M. ANTONIADI: J. Roy. Astronom. Soc. Canada **33**, 273 (1939).
[10] G. FOURNIER: Observations des Surfaces Planétaires. Paillart Abbeville, Fasc. **6** et **7**.
[11] A. DANJON: L'Astronomie, Mars 1944, 33.

Les premières observations systématiques furent réussies au Pic du Midi par B. Lyot. Celui-ci obtint, en 1941, avec une lunette de 38 cm, 25 dessins de Io,

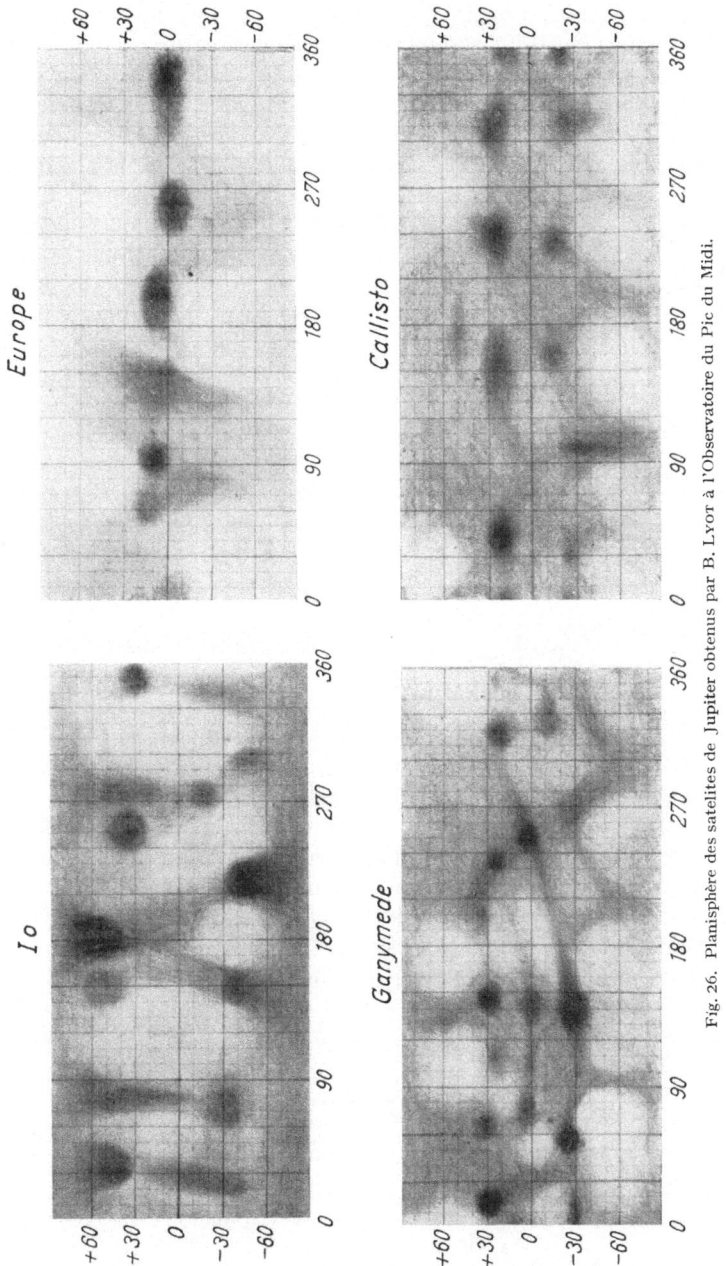

Fig. 26. Planisphère des satelites de Jupiter obtenus par B. Lyot à l'Observatoire du Pic du Midi.

19 dessins d'Europe, 26 de Ganymède et 26 de Callisto, d'où furent déduits des planisphères complets pour chacun de ces satellites[1]. Dans la suite, l'installation

---

[1] Voir note 8, p. 216.

d'une lunette de 60 cm permit d'accroître considérablement la finesse des détails sur les planisphères constitués[1] (Fig. 26).

Les quatre satellites tournent toujours la même face vers Jupiter, comme il en est de la Lune pour la Terre.

Le satellite le plus proche de Jupiter, Io, donne une couleur légèrement jaune. Les deux pôles sont assombris; la région équatoriale plus claire montre des taches et des trainées méridiennes, reproduites sur le planisphère de la Fig. 26.

Le deuxième satellite, Europe, est beaucoup plus brillant, et très blanc. Son aspect semble complémentaire de celui de Io, avec des pôles très clairs analogues à de véritables calottes polaires, et des taches moins brillantes le long de l'équateur.

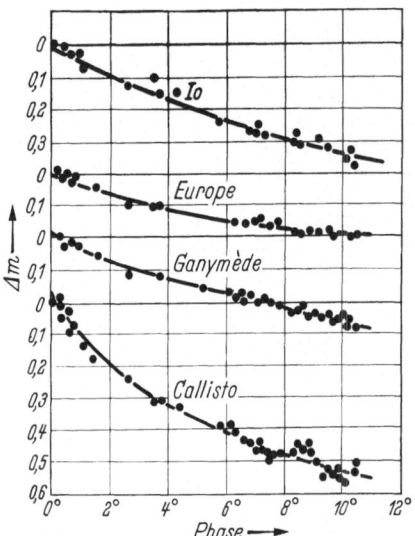

Fig. 27. Courbes photométriques des satellites de Jupiter. D'après les observations de J. Stebbins[3].

Ganymède, le plus gros des satellites, montre de nombreux détails permanents. Il est légèrement jaune, comme Io, et son aspect rappelle, avec un grossissement de 1000, le disque de Mars, vu à une opposition périhélique avec une petite lunette de 5 cm. Les taches sont aussi contrastées que sur Mars, complexes et irrégulières. Des blancheurs persistantes ont été décelées sur le bord levant; elles semblent masquer quelquefois les taches sombres qui devraient occuper ces régions; elles pourraient être attribuées à des nuées légères, ou à un dépôt volatil, et semblent indiquer la présence d'une atmosphère.

Callisto, presque aussi grand que Ganymède, est beaucoup plus sombre; ses taches sont également réparties au hasard, et nettement moins accusées.

42. Rugosité de la surface des satellites de Jupiter. De nombreuses mesures photométriques ont été effectuées sur les satellites de Jupiter vers 1914, par P. Guthnick[2]. En 1927, J. Stebbins réalisa à l'Observatoire Lick des comparaisons photoélectriques précises, qui donnèrent l'accroissement de magnitude des quatre satellites en fonction des angles de phase, entre 0 et 10°[3].

La Fig. 27 reproduit certaines mesures.

La baisse d'éclat est rapide immédiatement de part et d'autre de la phase nulle; puis l'assombrissement devient plus lent. Voici les valeurs déduites des courbes, pour ces satellites, Mercure, la Lune et l'astéroïde ((20)) Massalia particulièrement bien étudié:

Objet	Phase nulle mag/degré	Phase 10° mag/degré
Io	0,042	0,020
Europe	0,028	0,013
Ganymède . .	0,042	0,022
Callisto	0,100	0,028
Mercure. . . .	0,038	0,029
La Lune . . .	non mesurable	0,028
Astéroïde ((20))	0,100	0,030

[1] B. Lyot: L'Astronomie, Janvier 1953, 1.
[2] P. Guthnick: Astronom. Nachr. **206**, 157.
[3] Stebbins et Jacobsen: Lick Obs. Bull. No. 401.

Callisto se comporte comme Mercure, la Lune et les astéroïdes; sa perte initiale d'éclat autour de la phase nulle est particulièrement accentuée. La surface doit être très rugueuse et probablement criblée des mêmes cavités très nombreuses et profondes. Io et Ganymède ont une surface moins tourmentée. Les valeurs attribuables à Europe sont très voisines de celles relevées sur la planète Mars et supposent une érosion ou un comblement notable des cavités.

43. Pouvoir réflecteur de la surface. Les mesures de diamètre ci-dessus rapportées, nettement plus fortes que celles admises dans le travail photométrique classique de G. ROUGIER[1], conduisent aux déterminations suivantes pour les facteurs de diffusion moyen des satellites, rapportés à celui d'un écran parfaitement diffusant:

Io: 0,40,   Europe: 0,50,   Ganymède: 0,26,   Callisto: 0,11.

Les facteurs de diffusion pour la Lune et Mercure sont respectivement 0,125 et 0,112. Callisto est donc très semblable à ces deux astres.

Les sols de Io et Europe paraissent extraordinairement réfléchissants. Les valeurs données correspondent à l'ensemble du disque et les taches les plus claires ont un facteur de diffusion encore plus élevé.

Très peu de roches renvoient la lumière dans une proportion voisine de 0,5. Aucune matière ignée et aucune lave n'atteignent cette valeur. L'auteur a sélectionné un échantillon exceptionnel de cendres volcaniques en sphérules donnant 0,55. Parmi les roches sédimentaires, l'argile très blanche donne 0,56 et la craie fraîchement cassée 0,78.

Par contre, les poudres de substances claires ou transparentes broyées donnent des valeurs élevées. La neige fraîche et le givre conviendraient parfaitement. Ces substances donnent sous les petits angles de phase une polarisation très faible, tandis que les cendres et l'argile avantagent fortement la vibration contenue dans le plan de vision. Un brouillard de gouttelettes donnerait une polarisation très différente. Nous ne possédons malheureusement encore aucune mesure de polarisation sur les satellites de Jupiter.

44. Conclusions sur la surface des satellites de Jupiter. Callisto a même éclat que la Lune, même couleur, une rugosité comparable et des taches semblables. Ce fait est très remarquable car la densité moyenne, déduite de la valeur de la masse donnée par les perturbations, est voisine de 1,3, contre 3,3 pour la Lune, ce qui traduirait, malgré la grande imprécision de cette donnée, une structure interne, probablement très différente.

Callisto ne montre aucune trace d'érosion ou de comblement de cavités du sol par le transport de poussières. L'atmosphère ne peut donc être que très faible, ou bien stratifiée en couches parfaitement stables par suite de la basse température, du faible rayonnement et de la rotation très lente.

Io et Europe ont pour densité 2,7 et 2,9 et se rapprochent de la Lune pour leur constitution interne. Mais leurs surfaces éblouissantes n'ont d'équivalent dans le système solaire que les couches nuageuses parfaitement blanches recouvrant Vénus, Jupiter et certaines régions de la Terre. Bien que l'hypothèse de légères brumes persistantes, locales, ne soient pas exclue, il semble plutôt que ces globes montrent leurs sols sans l'interposition de nuages; il faut alors supposer la surface couverte d'une poudre éclatante, ou de cristaux.

On pourrait songer à un dépôt volatil congelé; le produit cristallisé ne pourrait être l'ammoniac non décalé spectroscopiquement par G. P. KUIPER sous une

---

[1] G. ROUGIER: L'Astronomie, Avril 1937, 165.

teneur limite de 40 cm (cntp)[1]. La neige carbonique et la glace d'eau ne sont pas décelables dans les domaines spectraux explorés. La vitesse d'évaporation du givre serait extrêmement lente à cette distance du Soleil. Mais il faudrait que le dépôt fût très épais, ou renouvelé fréquemment à la surface afin de ne pas être dégradé rapidement par les chutes de météores; sa surface devrait conserver une certaine rugosité pour rendre compte des courbes photométriques observées.

On peut aussi imaginer un matériau blanchi par un très fort échauffement, comme dans le cas des auréoles claires qui entourent certains cratères sur la Lune. Des mesures de polarisation permettraient de discriminer entre ces hypothèses.

Ganymède possède un éclat intermédiaire, et probablement des blancheurs au bord levant qui indiqueraient un dépôt de givre extrêmement mince, ou de légères nuées.

Les actions extérieures sur ces astres ont été probablement les mêmes que sur tous les autres corps célestes dépourvus d'atmosphère épaisse: pulvérisation et perforation de la surface par les petits météores, formations de cratères par les gros météores et création de bassins de lave sous le choc des corps célestes importants.

Les actions internes ont dû être importantes sur Ganymède plus volumineux que la Lune et Mercure. Elles ont peut-être été très différentes sur Callisto, probablement constitué d'un matériau d'une autre nature.

Les faibles atmosphères qui entourent probablement ces astres ne semblent pas avoir donné une érosion importante, sauf sur Europe. Elles semblent avoir pu former, sur le sol, des dépôts volatils congelés.

V. La Planète Mars.

La Planète Mars est le seul astre du système solaire dont nous pouvons examiner en détail la surface du sol et qui possède une atmosphère relativement importante.

L'enveloppe gazeuse est transparente; la pression au sol est environ le dixième de celle de l'atmosphère terrestre[2]; l'atmosphère recèle souvent des nuages de types variés, mais ceux-ci ne troublent pas sensiblement la vision de la topographie de la surface.

45. Diamètre du globe. Les déterminations des diamètres équatorial et polaire du globe de Mars avec le micromètre à fil ont été très nombreuses; elles ont été résumées par W. Rabe et leur discussion donne environ 9″50 pour le diamètre équatorial à la distance 1 U.A.[3]. Les mesures plus récentes ont fait appel à deux procédés nouveaux. Le premier, entièrement photographique, consiste à mesurer sur de nombreux clichés la trajectoire d'une petite tache bien caractérisée. Par suite de la rotation de la planète, cette tache décrit autour de la ligne des pôles un cercle qui est vu comme une ellipse dont le grand axe mesure le diamètre du parallèle à la latitude concernée.

Cette méthode, d'abord utilisée par Trumpler, fut mise à profit par H. Camichel, grâce aux nombreux clichés photographiques qu'il recueillit à l'Observatoire du Pic du Midi depuis 1941[4,5].

[1] G.P. Kuiper: The Atmosphere of the Earth and Planets. Chicago, Ill.: Chicago University Press 1957.

[2] A. Dollfus: C. R. Acad. Sci., Paris **232**, 1066 (1951).

[3] W. Rabe: Astronom. Nachr. **234**, No. 5600—5601 (1928).

[4] H. Camichel: Thèse 1954. Paris: Gauthier Villars.

[5] H. Camichel: Bull. Astron. **20**, 131 (1956).

La dispersion des mesures est de l'ordre de 1 % ; les oppositions les plus proches de la Terre donnent des valeurs systématiquement plus élevées. La valeur moyenne du diamètre équatorial est voisine de 9″40.

La seconde méthode repose sur l'emploi du micromètre à double image. Un nouvel instrument donne par biréfringence un dédoublement suffisant pour être adapté à une grande lunette[1]. Des mesures visuelles ont été conduites par A. Dollfus avec le réfracteur de 60 cm du Pic du Midi, en plusieurs couleurs. Le diamètre polaire n'est pas altéré par la phase et donne les mesures les plus exactes. Les valeurs trouvées en lumière jaune, lors des oppositions les plus favorables, sont les suivantes:

$$1954 \ldots \ldots 9″29$$
$$1956 \ldots \ldots 9″36$$
$$1958 \ldots \ldots 9″27.$$

La moyenne est 9″31, la précision atteint quelques millièmes. Le diamètre équatorial, $d(\text{éq})$, mesuré au moment de l'opposition exacte donne l'applatissement polaire: $\dfrac{d(\text{éq}) - d(\text{pol})}{d(\text{éq})}$, de l'ordre de 1/95.

Cette valeur semble beaucoup plus forte que celle calculée par le déplacement des orbites des satellites Phobos et Deimos, mais ces orbites sont très mal connues; on a pu penser à l'existence de voiles atmosphériques très élevés et entourant l'équateur de Mars, qui fausseraient les mesures. Cependant le diamètre équatorial déduit des mesures précédentes est 9″41 ; il reste voisin des mesures données par le premier procédé qui concernent directement le sol. Le diamètre équatorial du globe résultant de cette détermination serait de 6810 km.

46. Topographie permanente de la surface. La surface du globe de Mars montre des étendues claires de couleur jaune ocrée. Ces territoires clairs, à peine marbrés par des demi-tons et des trainées, font place par endroits à des étendues plus sombres, grises ou faiblement colorées. Les taches sombres dessinent une topographie caractéristique. Nous verrons plus loin que celle-ci n'est pas constante, cependant les principales configurations sont suffisamment persistantes pour permettre une cartographie de la planète. De nombreux observateurs expérimentés se sont donnés à cette tâche, parmi lesquels Schiaparelli, Lowell, Fournier, Antoniadi. L'achèvement le plus récent dans ce domaine résulte du travail d'un groupe d'observateurs, organisé par une commission de l'Union Astronomique Internationale. Les coordonnées des détails les plus caractéristiques et persistants de la surface ont été mesurées sur de nombreux clichés photographiques couvrant 10 années martiennes, et les territoires permanents ont été redessinés à leurs emplacements corrects sur un planisphère dans une projection voisine de celle de Mercator[2 à 4]. Les Fig. 28 et 29 reproduisent cette carte.

Pour reconnaître les régions de la surface, une désignation est nécessaire. Plusieurs auteurs ont proposé et ajouté des noms successifs à la nomenclature mythologique proposée à la fin du siècle dernier par Schiaparelli. La Commission de l'Union Astronomique Internationale a retenu les désignations les plus nécessaires, groupées en une liste reconnue. Les grandes régions sont désignées selon la liste et les figures des planches 28 et 29. Les petits détails et les formations éphémères sont rapportées à leurs coordonnées, lues sur les cartes des Fig. 28 et 29.

---

[1] A. Dollfus: C. R. Acad. Sci., Paris **235**, 1477 (1952).
[2] Trans. Union. Astron. Internat. **10** (1958).
[3] Sky and Telescope, Nov. 1958, 23.
[4] L'Astronomie, Novembre 1956.

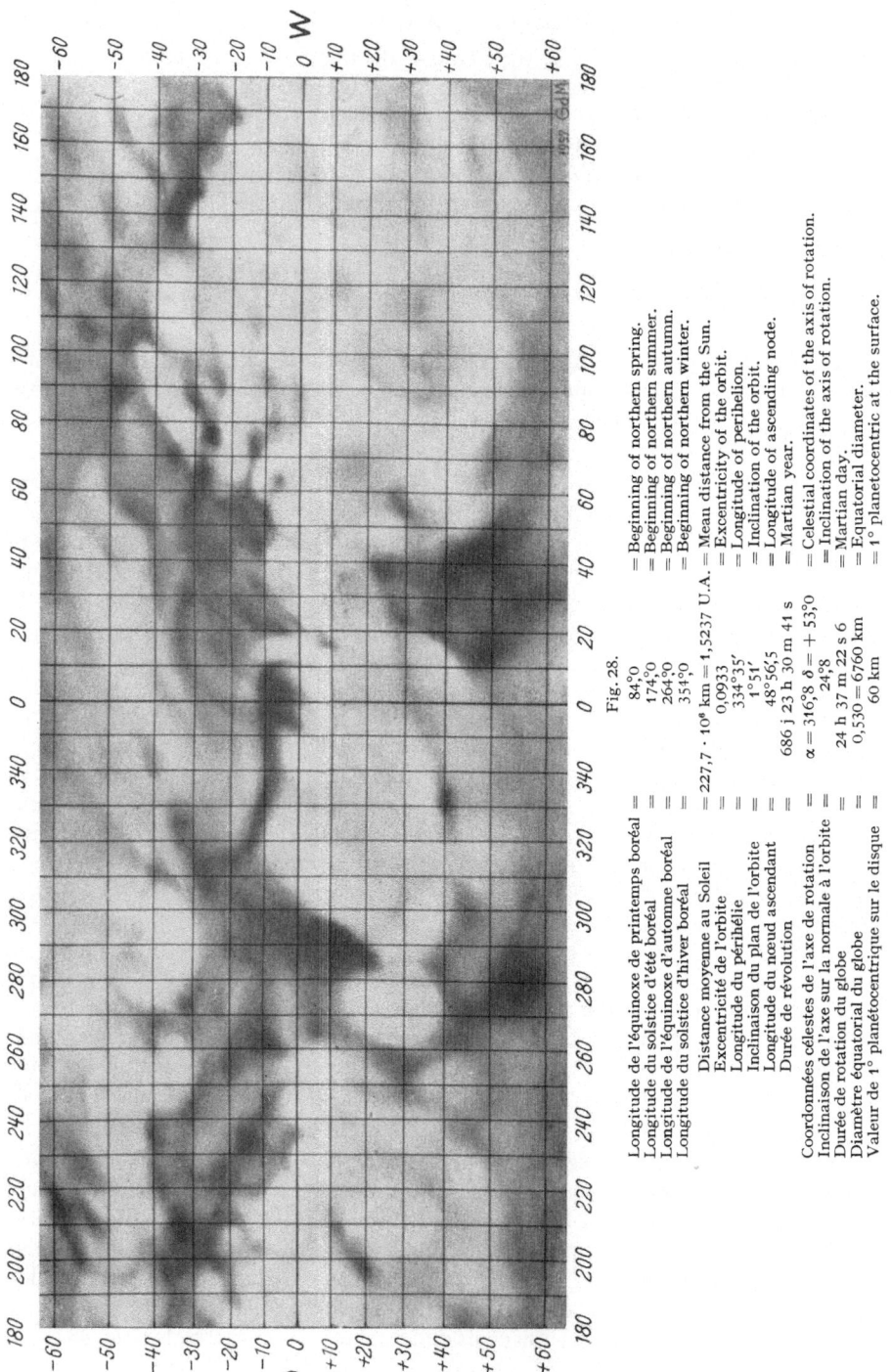

Fig. 28.

Longitude de l'équinoxe de printemps boréal	=	84,°0 = Beginning of northern spring.
Longitude du solstice d'été boréal	=	174,°0 = Beginning of northern summer.
Longitude de l'équinoxe d'automne boréal	=	264,°0 = Beginning of northern autumn.
Longitude du solstice d'hiver boréal	=	354,°0 = Beginning of northern winter.
Distance moyenne au Soleil	= 227,7 · 10⁶ km = 1,5237 U.A.	= Mean distance from the Sun.
Excentricité de l'orbite	=	0,0933 = Excentricity of the orbit.
Longitude du périhélie	=	334°35' = Longitude of perihelion.
Inclinaison du plan de l'orbite	=	1°51' = Inclination of the orbit.
Longitude du nœud ascendant	=	48°56,5 = Longitude of ascending node.
Durée de révolution	= 686 j 23 h 30 m 41 s	= Martian year.
Coordonnées célestes de l'axe de rotation	= α = 316,°8 δ = + 53,°0	= Celestial coordinates of the axis of rotation.
Inclinaison de l'axe sur la normale à l'orbite	=	24,°8 = Inclination of the axis of rotation.
Durée de rotation du globe	= 24 h 37 m 22 s 6	= Martian day.
Diamètre équatorial du globe	= 0,530 = 6760 km	= Equatorial diameter.
Valeur de 1° planétocentrique sur le disque	=	60 km = 1° planetocentric at the surface.

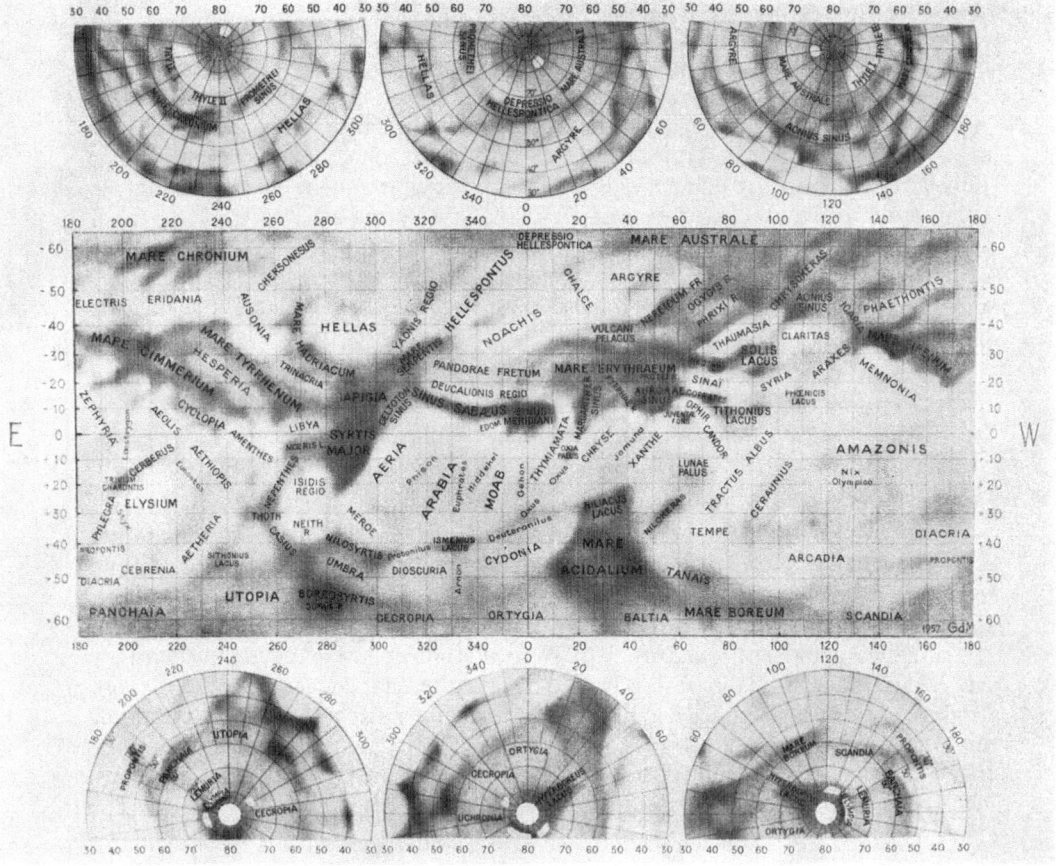

Fig. 29.

Les petits détails sont désignés par leurs coordonnées planétographiques.
Small features are designated by their planetographic coordinates.
Les grandes régions sont désignées par un nom dont voici la liste et les coordonnées.
Main markings are designated by names according to the following record and coordinates.

Acidalium M. (30°, +45°)
Aeolis (215°, −5°)
Aeria (310°, +10°)
Aetheria (230°, +40°)
Aethiopis (230°, +10°)
Amazonis (140°, 0°)
Amenthes (250°, +5°)
Aonius S. (105°, −45°)
Arabia (330°, +20°)
Araxes (115°, −25°)
Arcadia (100°, +45°)
Argyre (25°, −45°)
Arnon (335°, +48°)
Aurorae S. (50°, −15°)
Ausonia (250°, −40°)
Australe M. (40°, −60°)
Baltia (50°, +60°)
Boreum M. (90°, +50°)
Boreosyrtis (290°, +55°)
Candor (75°, +3°)
Casius (260°, +40°)
Cebrenia (210°, +50°)
Cecropia (320°, +60°)
Ceraunius (95°, +20°)
Cerberus (205°, +15°)
Chalce (0°, −50°)
Chersonesus (260°, −50°)
Chronium M. (210°, −58°)
Chryse (30°, +10°)
Chrysokeras (110°, −50°)
Cimmerium M. (220°, −20°)
Claritas (110°, −35°)
Copais Palus (280°, +55°)

Coprates (65°, −15°)
Cyclopia (230°, −5°)
Cydonia (0°, +40°)
Deltoton S. (305°, −4°)
Deucalionis R. (340°, −15°)
Deuteronilus (0°, +35°)
Diacria (180°, +50°)
Dioscuria (320°, +50°)
Edom (345°, 0°)
Electris (190°, −45°)
Elysium (210°, +25°)
Eridania (220°, −45°)
Erythraeum M. (40°, −25°)
Eunostos (220°, +22°)
Euphrates (335°, +20°)
Gehon (0°, +15°)
Hadriacum M. (270°, −40°)
Hellas (290°, −40°)
Hellespontica Depressio
    (340°, −6°)
Hellespontus (325°, −50°)
Hesperia (240°, −20°)
Hiddekel (345°, +15°)
Hyperboreus L. (60°, +75°)
Iapigia (295°, −20°)
Icaria (130°, −40°)
Isidis R. (275°, +20°)
Ismenius L. (330°, +40°)
Jamuna (40°, +10°)
Juventae Fons (63°, −5°)
Laestrigon (200°, 0°)
Lemuria (200°, +70°)
Libya (270°, 0°)

Lunae Palus (65°, +15°)
Margaritifer S. (25°, −10°)
Memnonia (150°, −20°)
Meroe (285°, +35°)
Meridiani S. (0°, −5°)
Moab (350°, +20°)
Moeris L. (270°, +8°)
Nectar (72°, −28°)
Neith R. (270°, +35°)
Nepenthes (260°, +20°)
Nereidum Fr. (55°, −45°)
Niliacus L. (30°, +30°)
Nilokeras (55°, +30°)
Nilosyrtis (290°, +42°)
Nix Olympica (130°, +20°)
Noachis (330°, −45°)
Ogygis R. (65°, −45°)
Olympia (200°, +80°)
Ophir (65°, 10°)
Ortygia (0°, +60°)
Oxia Palus (18°, +8°)
Oxus (10°, +20°)
Panchaia (200°, +60°)
Pandorae Fretum
    (340°, −25°)
Phaethontis (155°, −50°)
Phison (320°, +20°)
Phlegra (190°, +30°)
Phoenicis L. (110°, −12°)
Phrixi R. (70°, −40°)
Prometheī S. (280°, −65°)
Propontis (185°, +45°)
Protei R. (50°, −23°)

Protonilus (315°, +42°)
Pyrrhae R. (38°, −15°)
Sabaeus S. (340°, −8°)
Scandia (150°, +60°)
Serpentis M. (320°, −30°)
Sinaï (70°, −20°)
Sirenum M. (155°, −30°)
Sithonius L. (245°, +45°)
Solis L. (90°, −28°)
Styx (200°, +30°)
Syria (100°, −20°)
Syrtis Major (290°, +10°)
Tanaïs (70°, +50°)
Tempe (70°, +40°)
Thaumasia (85°, −35°)
Thoth (255°, +30°)
Thyle I (180°, −70°)
Thyle II (230°, −70°)
Thymiamata (10°, +10°)
Tithonius L. (85°, −5°)
Tractus Albus (80°, +30°)
Trinacria (268°, −25°)
Trivium Charontis
    (198°, +20°)
Tyrrhenum M. (255°, −20°)
Uchronia (260°, +70°)
Umbra (290°, +50°)
Utopia (250°, +50°)
Vulcani Pelagus (15°, −35°)
Xanthe (50°, +10°)
Yaonis R. (320°, −40°)
Zephyria (195°, 0°)

47. Les calottes polaires. Les régions polaires de Mars sont recouvertes en hiver d'un dépôt blanc éblouissant. Ce dépôt atteint au moins le 60ème parallèle. Au début du printemps le dépôt commence à se résorber; il diminue d'étendue. A la fin de l'été, il est réduit à une minuscule tache brillante, au voisinage du pôle et son éclat paraît très terni. Il semble disparaître quelquefois complètement. Puis la région se couvre de nuages opaques et ceux-ci masquent ensuite les régions polaires durant tout l'hiver. Le dépôt blanc se reforme sous ce couvert, car à la fin de la saison froide, les voiles vont progressivement se déchirer et la calotte blanche apparaîtra complètement réconstituée. Le cycle recommence chaque année. Au pôle de l'hémisphère opposé le phénomène est semblable, mais décalé d'une demi-année.

Une courbe moyenne, établie par Antoniadi[1], donne le diamètre de la calotte blanche en fonction de la longitude héliocentrique sur l'orbite. Les observations individuelles correspondant à chaque année martienne ont été collectées principalement par G. Fournier[2]. Des variations avec l'activité solaire n'ont pas été nettement prouvées.

Il est bien établi que ce dépôt est constitué par de l'eau, déposée sous forme d'une couche de givre peu épaisse. D'une part l'étude spectrophotométrique infra-rouge des taches polaires permit à G. P. Kuiper de retrouver une répartition spectrale semblable à celle de la glace d'eau[3]. D'autre part, les mesures polarimétriques caractérisent la nature de la surface étudiée; A. Dollfus put identifier très exactement au laboratoire la calotte polaire martienne avec des dépôts de givre[4]. Formés dans des conditions de pression atmosphérique et de radiation analogues à celles existant sur Mars, de tels dépôts présentent des structures très particulières. Ils sont formés de cristaux très fins et disparaissent sous l'effet d'un rayonnement, par sublimation, sans apparition d'eau liquide.

L'épaisseur de la couche ne doit pas être grande. Les calculs de la chaleur reçue et absorbée par la couche de givre sur la planète conduisent G. de Vaucouleurs à estimer l'épaisseur d'eau liquide contenue dans la couche de givre à quelques centimètres[5].

48. Les régions claires. Les grandes étendues claires jaune ocré couvrent à peu près les deux tiers de la surface du globe. On y décèle de faibles marbrures, des changements locaux de teinte ou d'éclat, et de fines trainées. Ces trainées se montrent quelquefois dans les conditions d'observation médiocre sous forme de lignes très noires, qui furent appelées «canaux». Avec un instrument puissant sous un ciel homogène, ces formations se résolvent en un grand nombre de petites taches, trainées, accidents naturels, assez volontiers alignés[6,7].

α) *Spectrophotométrie des régions claires.* Les régions claires ont une coloration orangée très marquée. Les mesures photométriques de l'éclat global de l'astre donnent les propriétés des regions claires, car celles-ci envoient le plus de lumière et sont très étendues. Les déterminations photoélectriques à travers différents

---

[1] E.M. Antoniadi: La Planète Mars. Paris: Hermann 1930.

[2] G. Fournier: Observation des surfaces planétaires, 10 tomes de 1908 à 1946. Paris: F. Paillard.

[3] G.P. Kuiper: The Atmosphere of the Earth and Planets. Chicago, Ill.: Chicago University Press 1950.

[4] A. Dollfus: Thèse Paris 1955. — Suppl. Ann. d'Astrophys. 1956.

[5] G. de Vaucouleurs: C. R. Acad. Sci., Paris **216**, 720 (1943).

[6] A. Dollfus: C. R. Acad. Sci., Paris **226**, 996 (1948).

[7] A. Dollfus: L'Astronomie, Mars 1953, 85.

filtres colorés effectués par WOOLLEY[1], JOHNSON et GARDINER[2], DE VAUCOU-LEURS[3], DOLLFUS[4] s'accordent entre elles. Elles sont cependant améliorées par l'étude locale de régions claires isolées. De nombreuses déterminations photo-métriques ont été pratiquées en Union Soviétique, d'après la mesure de clichés photographiques, par N. P. BARABASHEV, par V. V. CHARONOV et par leurs col-laborateurs. Des déterminations les plus directes ont été effectuées à l'aide d'un photomètre visuel sur les différentes régions de Mars, en valeur absolue, par A. DOLLFUS[4].

Le facteur de diffusion des régions claires de Mars, au centre du disque et sous l'éclairement normal à la surface est donné en fonction de la longueur d'onde par la courbe de la Fig. 30. Le pouvoir reflecteur est 4 fois plus élevé en rouge qu'en bleu, circonstance rencontrée seulement sur des corps très colorés. Cette absorption intense dans le proche ultraviolet caractérise la substance recouvrant ces régions.

Les roches très teintées par de l'oxyde de fer donnent une bande de faible diffusion dans ce domaine et semblent convenir par-faitement dans tout l'intervalle spectral con-sidéré.

Cette étude a été complétée dans l'infra-rouge entre les bandes d'absorption telluri-ques. G. P. KUIPER a comparé les spectres de Mars et de la Lune, entre 0,8 et 2,7 mi-crons[5]. Contrairement au comportement dans le domaine visuel, les régions claires de Mars ne présentent pas dans l'infra-rouge de grosses variations d'intensité; l'éclat décroît légère-ment par rapport à celui de la Lune lorsque la longueur d'onde augmente à partir de 0,8 microns. Selon KUIPER, les roches sédimentaires teintées ne donnent pas cet effet; certaines roches ignées, telle que la felsite très colorée (orthoclase avec occlusions de quartz), ou pétrosilex, repro-duisent l'effet observé sur Mars. Il serait intéressant de connaître les propriétés infra-rouges des oxydes de fer du type limonite pulvérisé dont il sera question plus loin. Les mélanges de substances en poudre peuvent produire une grande variéte d'éclats et de variations spectrales progressives; le critère de la spectro-photométrie ne permet une détermination précise de la nature du sol que s'il met en valeur des bandes très marquées.

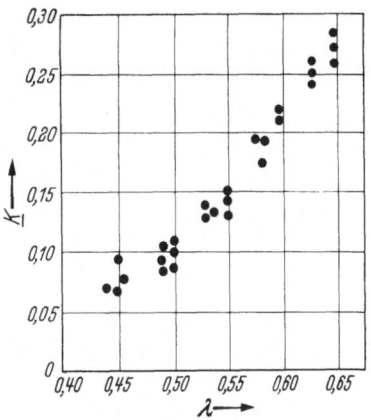

Fig. 30. Facteur de diffusion des régions claires de Mars en fonction de la longueur d'onde. D'après A. DOLLFUS[4].

β) *Structure microscopique et nature des régions claires.* La texture fine du sol est révélée par la manière dont la lumière solaire est polarisée par la diffusion sur la surface.

B. LYOT a déterminé en 1922, 1924 et 1926 la courbe de proportion de lumière polarisée en fonction de l'angle de phase, pour l'ensemble du globe[6]. LYOT trouva une grande similitude avec la polarisation donnée par la Lune. Il conclut

[1] R. V. D. R. WOOLLEY: Monthly Notices Roy. Astronom. Soc. London **113**, 521 (1953); **115**, 87 (1955).

[2] JOHNSON, et GARDINER: Publ. Astronom. Soc. Pacific **67**, 74 (1955).

[3] G. DE VAUCOULEURS: Sky and Telescope **18**, 484 (1959).

[4] A. DOLLFUS: C. R. Acad. Sci., Paris **244**, 162 (1957).

[5] Voir note 1, p. 220.

[6] B. LYOT: Thèse Paris 1929. — Ann. Obs. Meudon 1929.

que Mars, comme la Lune, est recouvert d'une matière très absorbente et pulvérisée.

Des mesures beaucoup plus nombreuses ont été reprises par A. Dollfus depuis 1948[1]. Ces déterminations portent sur de petites régions bien définies de la surface, et elles donnent les propriétés polarisantes des étendues claires sous les divers angles d'éclairement et d'observation.

La polarisation dépend presque uniquement de l'angle de phase; le trait ponctué de la Fig. 31, donne la courbe de polarisation en fonction de cet angle, en lumière orangée, pour les plages claires au centre du disque. En observant près

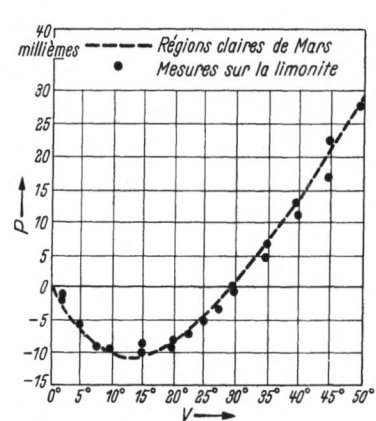
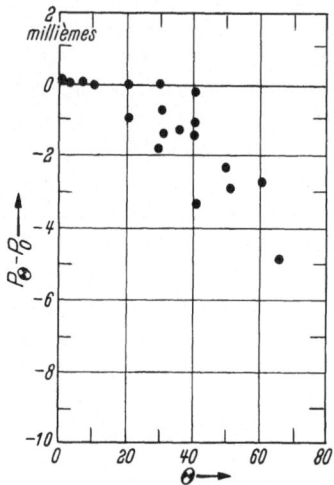

Fig. 31. Polarisation des régions claires de Mars en fonction de l'angle de phase. D'après A. Dollfus[1].

Fig. 32. Polarisation des régions claires de Mars en fonction de l'inclinaison, à la phase nulle. D'après A. Dollfus[1].

du bord du disque la surface du sol inclinée, une polarisation légère apparaît, selon la courbe de la Fig. 32. Cette propriété ne se trouve pas sur la Lune, ce qui indique une différence de structure assez notable.

Très peu de substances possèdent simultanément ces deux propriétés et presque aucune ne les reproduit exactement. Cette heureuse propriété permet une discrimination précise. Il est nécessaire que la substance soit formée de grains parfaitement opaques[2], finement broyée en une poudre très irrégulière. Les cendres volcaniques ne donnent pas la diminution de polarisation observée sous une forte inclinaison. Les oxydes de fer très absorbants semblent convenir beaucoup mieux. La goethite concassée et bien broyée se rapproche des propriétés considérées. La limonite (Fe_2O_3, $3H_2O$) pulvérisée reproduit exactement ces propriétés.

Les points de la Fig. 31 sont relatifs à cette substance. L'aspect jaune orangé rappelle la coloration martienne; la variation spectrale du facteur de diffusion est compatible avec les données de la Fig. 30. La photographie de la Fig. 33 montre ce dépôt à la loupe. La contexture du sol martien doit rappeler cette image[1,3].

---

[1] Voir note 4, p. 224.
[2] A. Dollfus: Ann. d'Astrophys. **19**, 82 (1956).
[3] A. Dollfus: Publ. Astronom. Soc. Pacific **76**, 57 (1958).

γ) Rugosité de la surface. Les mesures de l'éclat global de l'astre donnent la variation de magnitude avec l'angle de phase. Au siècle dernier, G. MÜLLER obtint à Potsdam de nombreuses mesures photométriques visuelles de Mars[1]. Il donna pour coefficient de phase dans le jaune 0,0149 mag/degré. Les mesures photographiques de KING, réalisées à Harvard entre 1916 et 1922 donnent 0,0152 mag/degré, pour la radiation 0,56 microns. En 1952, A. DOLLFUS a trouvé photoélectriquement 0,0145 mag/degré, vers 0,56 microns[2]. R. WOOLLEY donne 0,0115 mag/degré pour 0,598 microns et 0,213 pour 0,543 microns[3].

Ces valeurs sont très différentes de la valeur 0,0282 mag/degré donnée

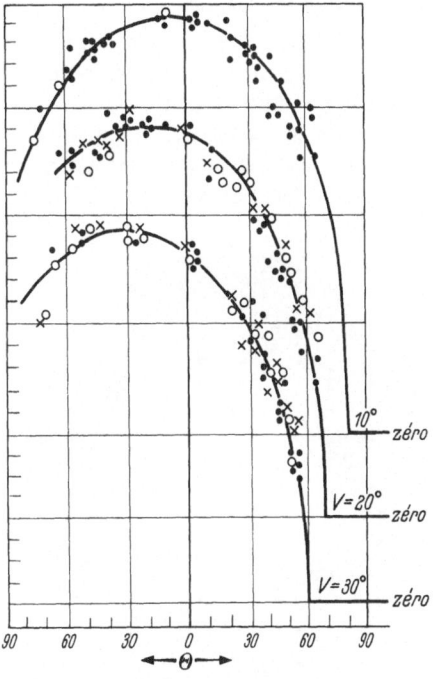

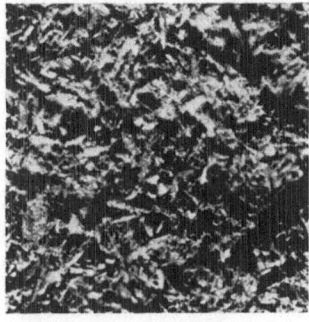

Fig. 33. Aspect microscopique d'un dépôt de Limonite broyée reproduisant la polarisation, l'éclat et la couleur des régions claires de Mars.

Fig. 34. Facteur de diffusion des régions claires de Mars en fonction de l'inclinaison de la surface, pour trois valeurs de l'angle de phase. D'après A. DOLLFUS[2].

par ROUGIER pour la Lune. On peut en conclure que la surface de Mars est beaucoup moins tourmentée que celle de la Lune.

La faible rugosité du sol est confirmée par les mesures photométriques détaillées des régions claires. A l'aide d'un photomètre à franges, A. DOLLFUS a mesuré la brillance des étendues claires de Mars sous différents angles de vision et en divers points de la planète correspondant à différentes inclinaisons[2]. La Fig. 34 résume les mesures, rapportées aux trois angles de vision 10°, 20°, 30°, en fonction de l'inclinaison le long de l'équateur.

Les courbes peuvent être attribuées à plusieurs substances; elles sont compatibles avec les structures de limonite pulvérisées identifiées par la polarisation. Un dépôt de cette substance avec surface tourmentée déforme les courbes correspondant aux plus grands angles de vision et rompt la similitude des trois courbes de la Fig. 34. Il est nécessaire de supposer un sol assez lisse et modérément accidenté.

Cette propriété diffère notablement de celle que l'on observe sur la Lune. Sur celle-ci les chutes de météores de toutes dimensions ont dû cribler la surface de cavités jointives ou juxtaposées. Sur Mars, nous connaissons l'existence d'une légère atmosphère qui a pu arrêter les météores les plus petits. Cette atmosphère

---

[1] G. MÜLLER: Publ. Astr. Obs. Potsdam **8** (1893).

[2] Voir note 4, p. 225.

[3] Voir note 1, p. 225.

est animée de vents dont les vitesses mesurées par les déplacements des nuages sont de plusieurs dizaines de km/heure. Nous savons que des mouvements convectifs soulèvent de temps en temps les grains les plus fins de la surface du sol pour former un nuage de sable qui ne se depose à nouveau que lentement. Ce mécanisme doit combler les cavités.

δ) *Le relief sur la planète Mars.* Nos meilleurs télescopes sont encore insuffisants pour montrer sur la planète Mars les montagnes et le relief. Ceux-ci se manifestent indirectement.

L'étude de la calotte polaire avec un grossissement élevé permet de suivre l'évolution complexe du contour du dépôt de givre, au cours de la régression. Ce contour n'est pas rectiligne. Il accuse des failles et des enclaves qui se dégagent

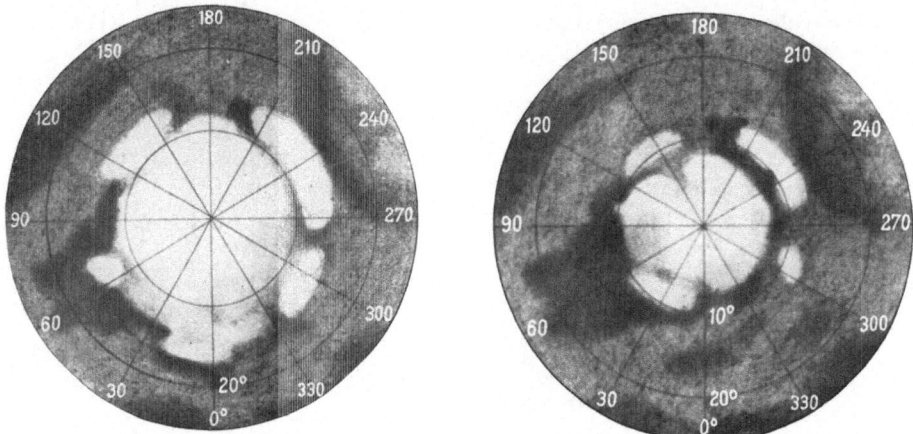

Fig. 35. Région polaire Nord de la planète Mars et dépôt de givre au cours du printemps et de l'été martien. D'après
A. Dollfus[2].

plus vite, ainsi que des promontoires sur lesquels le givre persiste au contraire plus longtemps. Des névés peuvent même s'isoler complètement et persister comme des taches brillantes détachées du dépôt de givre qui s'est rétracté. L'évolution de la calotte Sud a été décrite par Antoniadi[1]. Celle de la calotte Nord, plus difficile à bien observer, a été étudiée par A. Dollfus[2]. La Fig. 35 montre deux aspects des brèches et ilots laissés par la calotte Nord. Il paraît évident que les brèches représentent des vallées, tandis que les taches isolées sont des plateaux plus élevés qui retiennent le givre plus longtemps.

Cette interprétation est confirmée par l'aspect de la calotte au début de la régression du givre, alors que le dépôt est encore très étendu. On observe alors des taches encore plus blanches que le givre, et qui persistent assez longtemps aux mêmes emplacements. La polarisation de la lumière indique qu'elles sont dues à des nuages de cristaux de glace. Ces régions couvertes de nuées s'identifient avec celles où des névées persistent un certain temps. Il apparaît donc que les voiles de cirrus enveloppent les sommets de collines, qui retiennent le givre plus longtemps[3].

On observe également les conséquences du relief dans d'autres régions de la planète. Au milieu du grand territoire clair, Amazonis, fig. 28 (p. 222), l'emplace-

---

[1] Voir note 1, p. 224.
[2] Voir note 7, p. 224.
[3] Voir note 4, p. 224.

ment (135°, +30°) montre quelquefois une petite tache circulaire, blanche, éblouissante, qui est soit du givre, soit un nuage toujours localisé au même emplacement.
Cette tache fut baptisée Nix Olympica[1]. Il s'agit là probablement d'un sommet
élevé.

Aux coordonnées (345°, 0°), la tache Edom montre par moments de très
petites blancheurs dont les formes évoluent d'un jour sur l'autre en restant sur
place, bien que les nuées voisines soient déplacées par le vent. La Fig. 36 a—d

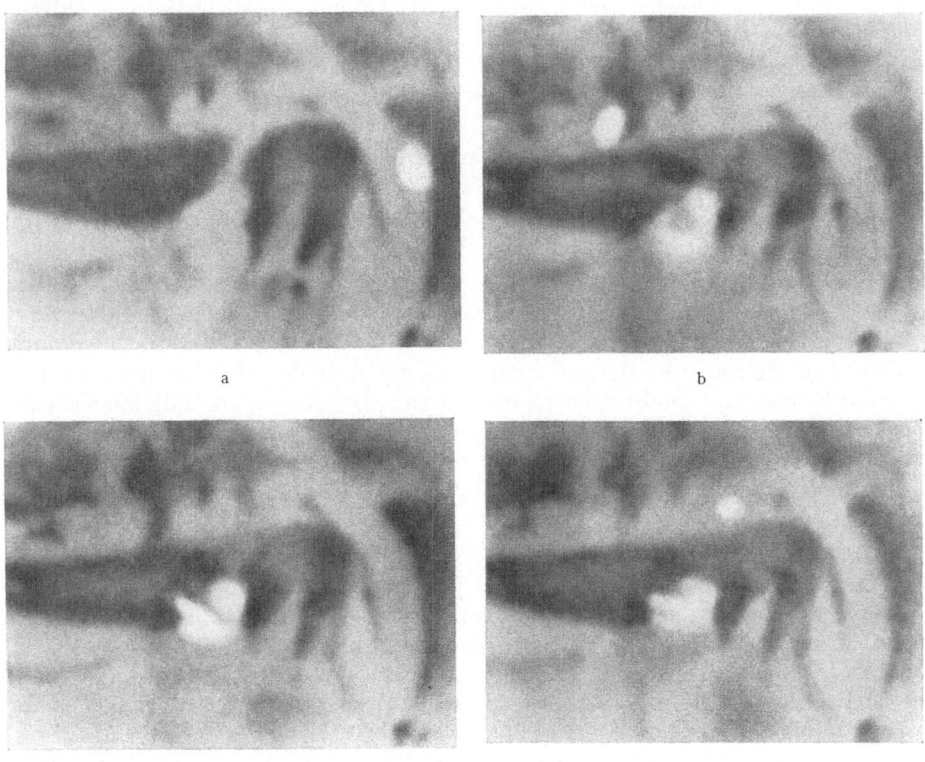

　　　　　　　　　　a　　　　　　　　　　　　　　　　　　　　　　　b

　　　　　　　　　　c　　　　　　　　　　　　　　　　　　　　　　　d

Fig. 36 a—d. Petit nuage persistant au sommet d'une colline sur la planète Mars. Observations par J. H. Focas et
A. Dollfus au Pic du Midi. 7 Septembre 1956; 8 Octobre 1956; 12 Octobre 1956; 13 Octobre 1956.

montre l'évolution d'une telle formation. Cette nuée persistente est probablement
localisée au sommet d'un massif élevé. D'autres petits nuages analogues s'observent
en d'autres endroits, par exemple à (35°, −15°), à (45°, −20°), dans la région
Ophir (70°, −10°), etc.

Tous ces phénomènes traduisent une orographie légèrement accidentée,
comportant quelques montagnes caractérisées. Il semble que les régions élevées
se trouvent de préférence dans les régions claires, ou bien entre les taches sombres.

49. Les régions sombres. α) *Aspect et colorations.* Les configurations sombres
de la surface martienne montrent un contraste accusé Examinées avec un pouvoir
séparateur élevé, elles révèlent une structure incroyablement compliquée. On
distingue un enchevêtrement de petites taches et trainées, diversement colorées,
ainsi que des grisailles aux contours déchiquetées. Une vision moins acérée
montre des structures plus émoussées, résultat de la dégradation de ces aspects.

---

[1] Voir note 1, p. 224.

Les plus petites taches que l'on peut isoler distinctement les unes des autres sont distantes d'environ 60 km.

Les régions sombres apparaissent diversement colorées; on note souvent les teintes vertes, bleues et brunes, ainsi que des tonalités incertaines. En comparant ces couleurs à des échelles de teinte, G. P. Kuiper a montré qu'il s'agissait de très faibles tonalités, fortement renforcées par le contraste avec les régions ocrées très colorées voisines. En particulier, les impressions de vert vif que l'on recueille par instant sont accentuées par l'effet de la turbulence atmosphérique qui, lorsqu'elle brouille les images, laisse subsister sur la rétine la couleur complémentaire de celle des étendues jaunes qui l'entourent[1].

Mais certaines couleurs sont plus prononcées et les différentes teintes correspondent à des états physiques du sol différents.

β) *Photométrie des régions sombres.* De nombreux travaux photométriques ont été réalisés par la mesure de clichés photographiques, en Union Soviétique. Ils sont dûs principalement à N. P. Barabashev et à V. V. Charonov (cf. [2]). Les mesures visuelles sont effectuées directement à l'oculaire de la lunette[3,4]. Un photomètre à franges, réalisé par A. Dollfus, permet l'examen individuel des différentes régions de la surface.

Les contrastes relevés entre les taches sombres et les régions claires donnent toutes les valeurs à partir du temi-ton le plus léger. Les valeurs maximum relevées atteignent 0,50, dans le rouge. Les myriades de petites taches qui composent les étendues sombres, et qui ne sont pas juxtaposées, ont individuellement un contraste plus élevé que la moyenne de la région examinée. La variation de brillance du centre au bord du disque, correction faite de l'effet de l'atmosphère, suit la même loi que pour les étendues claires. Il faut en conclure une analogie de structure entre ces taches sombres et les régions claires.

Les mesures effectuées à travers divers filtres colorés montrent que la variation spectrale de brillance est seulement un peu moins accusée que celle des régions claires donnée fig. 30. Dans tous les cas l'éclat croît avec la longueur d'onde. Les taches sombres, si elles pouvaient être vues isolées, apparaîtraient de couleur orangée. Leurs aspects colorés résultent bien du fond très coloré sur lequel elles se détachent. L'absorption dans le bleu due à l'oxyde de fer y est seulement moins prononcée que sur les régions claires. On obtient des surfaces ayant les caractères relevés en saupoudrant une étendue de roches pulvérisées analogues aux grands territoires orangés avec des poudres plus ou moins teintées et clairsemées[4].

Nous verrons plus loin que les mesures de polarisation ne sont pas opposées à cette interprétation; elles apporteront de curieuses précisions.

γ) *Etude spectrophotométrique infrarouge des régions sombres.* L'étude de la réflectivité des étendues sombres de Mars a pu être étendue loin dans l'infrarouge, jusque vers 4,2 microns, grâce aux cellules photoconductrices. Les premiers résultats furent obtenus par G. P. Kuiper en 1948, en mesurant entre les bandes d'absorption telluriques le contraste des taches sombres par rapport aux régions claires.

Le facteur de diffusion de ces étendues sombres semble à peu près constant dans l'infra-rouge. Il diffère notablement, sous ce rapport, de celui des étendues

---

[1] G. P. Kuiper: Astrophys. J. **125**, 307 (1957).
[2] G. de Vaucouleurs: Physique de la Planète Mars. Paris: Albin Michel 1951. — Physics of the Planet Mars. London: Faber & Faber 1954.
[3] Voir note 4, p. 225.
[4] A. Dollfus: C. R. Acad. Sci., Paris **244**, 1458 (1957).

végétales de phanérogames. De nombreuses substances donnent un comportement analogue, parmi lesquelles entre autre des cryptogames, des petites algues
et des lichens.

Grâce aux ressources offertes par des cellules encore plus sensibles, W. Sinton
obtint en 1958 avec le grand réflecteur de 5 m de Palomar de très beaux spectres
plus détaillés[1]. Ceux-ci révèlent trois faibles minima de lumière, l'un à 3,67 microns, l'autre à 3,56 microns, le dernier assez faible vers 3,43 microns (Fig. 37).
La mise en valeur de telles bandes réduit beaucoup les variétés possibles d'interprétation. Les carbonates et plusieurs composés minéraux ont des bandes d'absorption dans ce domaine. Cependant les composés avec groupements CH_2 sont
caractérisés par de fortes absorptions
vers 3,4 et 3,55 microns. Sur Terre,
ces molécules se trouvent surtout
chez les êtres animés. La bande
3,67 microns pourrait être explicable
par l'adjonction de molécules d'oxy-
gène, en carbohydrates, telles que le
Dr. Sinton les a observés en mesurant
la lumière renvoyée par certains
lichens.

Le remarquable critère spectro-
scopique précédent indique l'existence
de molécules voisines de celles trouvées
chez les organismes animés. Il fa-
vorise beaucoup, comme nous le verrons plus loin, l'hypothèse d'une vie
sur la planète.

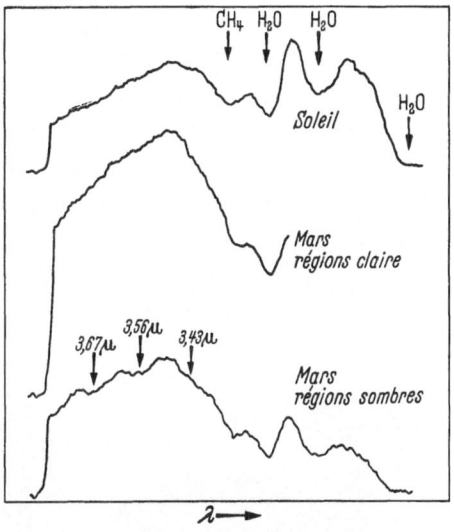

Fig. 37. Courbes spectrales infrarouges de Soleil (en haut),
des régions claires de Mars (au centre) et des régions sombres
de Mars (en bas). D'après W. Sinton[1].

δ) *Variations des régions sombres.*
Les taches de la surface de Mars
présentent d'étranges phénomènes de
variations. Les étendues sombres ne
conservent pas les mêmes configurations. D'une année martienne à l'autre,
quelquefois d'un mois sur l'autre, elles montrent souvent de complètes transformations. Ces modifications concernent aussi les colorations.

Une lunette à très grande résolution montre des variations encore plus prononcées dans les petites structures très déliées qui composent les contrées sombres.
Les petites taches s'évanouissent ou bien apparaissent sous forme d'un bourgeonne-
ment général; elles acquièrent de nouvelles teintes; il en résulte des structures
différentes, même quelquefois dans les territoires qui, globalement, ne semblent
pas s'être modifiés. La fig. 38 donne quelques exemples.

Les observateurs de la planète Mars se sont attachés à la laborieuse tâche
de cataloguer ces variations. Schiaparelli, Lowell[2], Flammarion[3], Fournier[4], Antoniadi[5] en particulier ont dressé des monographies qui permettent
d'en reconstituer les étapes.

Depuis 1907, une importante collection de photographies a été constituée
par E.C. Slipher, à l'Observatoire Lowell.

---

[1] W. Sinton: Lowell Obs. Bull. **103** (1959).

[2] P. Lowell: Mars and its Canals. New York 1909.

[3] C. Flammarion: La Planète Mars. Paris: Gauthier Villars 1892 et 1909.

[4] Voir note 2, p. 224.

[5] Voir note 1, p. 224.

Depuis 1941, l'évolution de la planète Mars est suivie entièrement par la photographie. L'Observatoire de Meudon a procédé au rassemblement d'environ

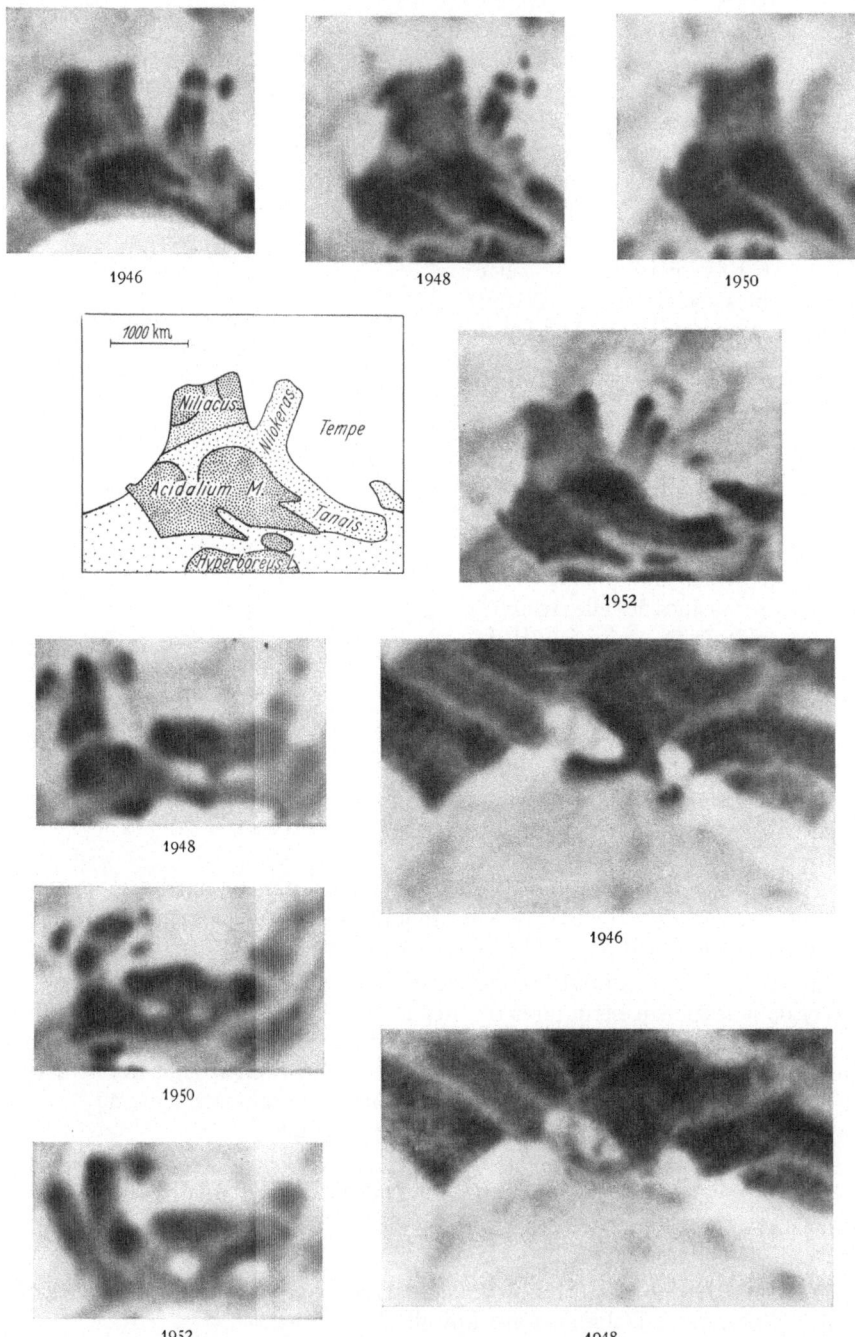

Fig. 38. Variation des petites taches sur la surface de Mars. D'après les observations de A. DOLLFUS.

700 des meilleures images obtenues par H. CAMICHEL au Pic du Midi au cours de chacune des dernières oppositions. En outre, environ 300 images diapositives

viennent de divers autres Observa-
toires. Cette très importante collection est classée dans un ordre per-
mettant la consultation immédiate,
et la comparaison de clichés. G. DE
MOTTONI a constitué des planisphères
récapitulatifs très détaillés pour
chacune des oppositions. La comparaison des 10 planisphères déjà réalisés
donne la description historique de ces
modifications[1].

ε) *Description des variations.* On
peut distinguer dans l'apparente complexité des variations observées quelques lois. La phénomène est saisonnier
et il progresse en latitude. Un assombrissement général apparaît dans les
régions polaires, au début du printemps. Puis cet assombrissement se
propage; en règle générale, il descend
vers l'équateur, et les régions sombres
de latitude de plus en plus faibles sont
successivement intéressées. La vague
dépasse l'équateur au milieu de l'été,
et elle s'affaiblit. Les taches s'éclair-
cissent enfin pendant l'hiver. Simultanément le même phénomène se
produit, à partir de l'hémisphère opposé, en sens inverse et décalé d'une
demi-année.

Les différentes régions martiennes
sont inégalement intéressées par cette
propagation générale et l'on peut
distinguer plusieurs sortes de comportements.

Certaines régions, très sensibles au
phénomène saisonnier, manifestent
chaque année, au moment du passage
de la vague concernée, une intensifica-
tion prononcée. Ainsi, vers la longitude
340° le groupe des formations Depressio Hellespontica, Hellespontis,
Mare Serpentis et Pandorae Fretum
évolue-t-il selon un même processus,
reproductible, et prévisible; la modification est illustrée par les images de
la fig. 39.

D'autres régions présentent aussi
une certaine prédilection au regard
de la variation saisonnière. Mais leurs

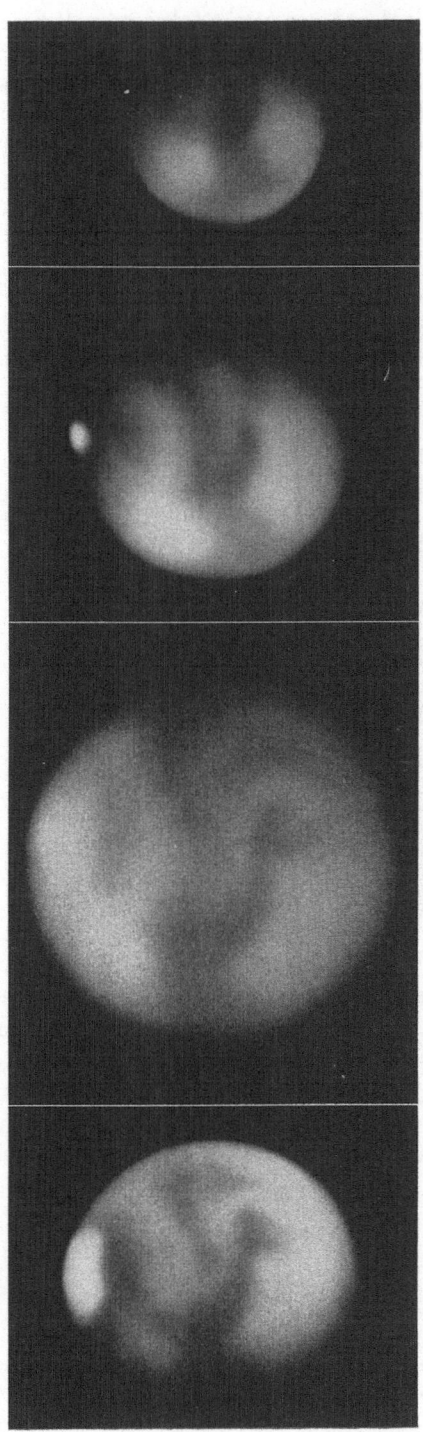

Fig. 39. Variation saisonnière des régions Hellespontus et Pandorae Fretum. D'après les photographies obtenues au Pic du Midi par H. CAMICHEL 31 Juillet 1956; 7 Septembre 1956; 11 Octobre 1956; 17 Novembre 1956.

---

[1] G. P. KUIPER: Planets and Satellites.
Chap. 15. Chicago: Univ. Press 1961.

comportements imprévisibles se traduisent par l'apparition de formations chaque fois nouvelles; elles inaugurent presque chaque été martien des configurations originales. Ainsi, vers 90°, la région de Solis Lacus n'a cessé de se modifier (Fig. 40). Vers 260° les formations Nepenthes et Casius présentèrent à chacune de ces dernières années une topographie renouvelée.

Enfin certaines contrées, apparemment permanentes et peu sensibles au phénomène saisonnier, peuvent tout à coup subir une modification profonde,

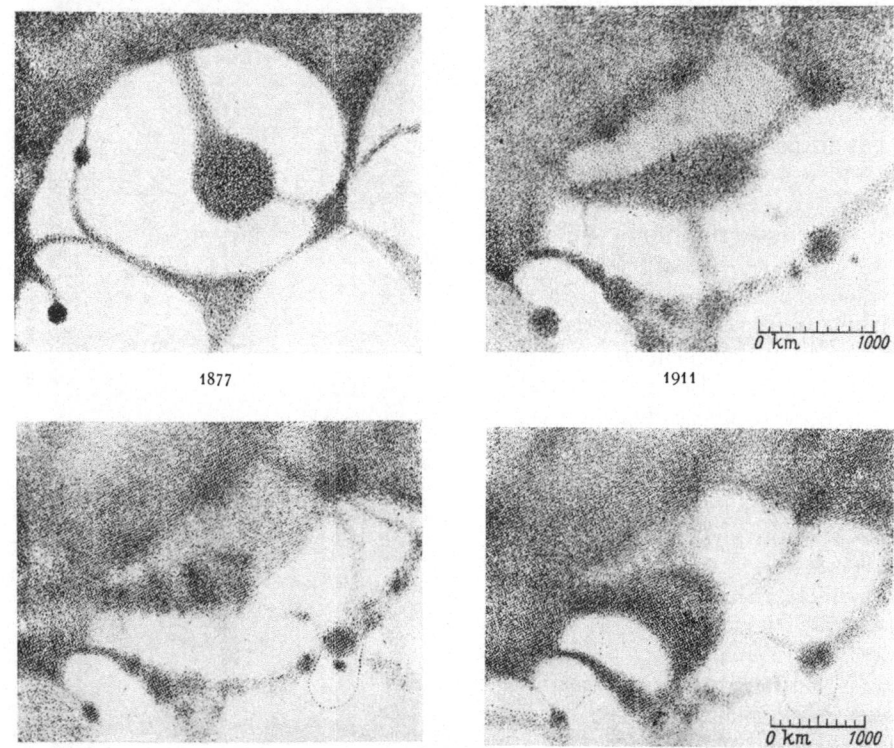

1877

1911

1924

Fig. 40. Variations dans la région de Solis Lacus. D'après E. M. Antoniadi.

1926

laquelle demeure ensuite stable et permanente; toute la région conserve ensuite un aspect nouveau et durable. Tel fut le cas, par exemple, de Boreosyrtis en 1948.

ζ) *Etude photométrique des variations saisonnières.* Malgré cette diversité, le phénomène saisonnier reste bien caractérisé. Il s'offre à des mesures qui permettent de le préciser. Dès la fin du siècle dernier, P. Lowell avait estimé l'intensité des trainées sombres et dressé des graphiques montrant la propagation en latitude de l'assombrissement qui les concernent[1]. G. de Vaucouleurs obtint des résultats analogues sur les taches sombres en 1941[2]. Une étude photométrique plus complète du phénomène fut réalisée par J. H. Focas, avec A. Dollfus[3]. La Fig. 41 montre, sur différentes taches réparties en latitude, la propagation de l'assombrissement depuis les pôles vers l'équateur, en fonction de la saison exprimée par la longitude sur l'orbite. La Fig. 42 montre l'affaiblissement de l'effet en fonction de la latitude, pour les deux vagues issues des deux pôles respectifs. La vitesse de propagation est constante et vaut environ 35 km par jour.

---

[1] Voir note 2, p. 231.
[2] Voir note 2, p. 230.
[3] J. H. Focas: C. R. Acad. Sci., Paris **248**, 626 (1959).

η) *Liaisons entre la variation saisonnière et les phénomènes atmosphériques.* Nous savons que la vapeur d'eau dans l'atmosphère de la planète est le siège d'une circulation annuelle parfaitement rythmée[1]. En hiver une certaine quantité d'eau est déposée sur le sol, autour de l'un des pôles, sous forme de givre. Au printemps le givre se sublime et charge l'air au voisinage du pôle d'une certaine quantité de vapeur. En raison de la faible pression atmosphérique, et du gradient thermique vertical presque adiabatique, la vitesse de diffusion est beaucoup plus grande que sur la Terre. Par suite la vapeur d'eau va se propager très rapidement vers l'équateur; elle va gagner progressivement le pôle de l'hémisphère opposé pour se condenser à nouveau en un dépôt de givre. Le mécanisme va recommencer ensuite, en sens inverse. La plus grande partie de l'eau oscille ainsi constamment, sous forme de vapeur, d'un pôle vers l'autre.

Les deux vagues d'assombrissement des taches sombres du sol alternativement issues des deux pôles semblent étroitement liées à l'accroissement de la teneur de l'air en vapeur d'eau[1,2].

ϑ) *Régénération des taches sombres.* Nous savons que l'atmosphère de la planète est

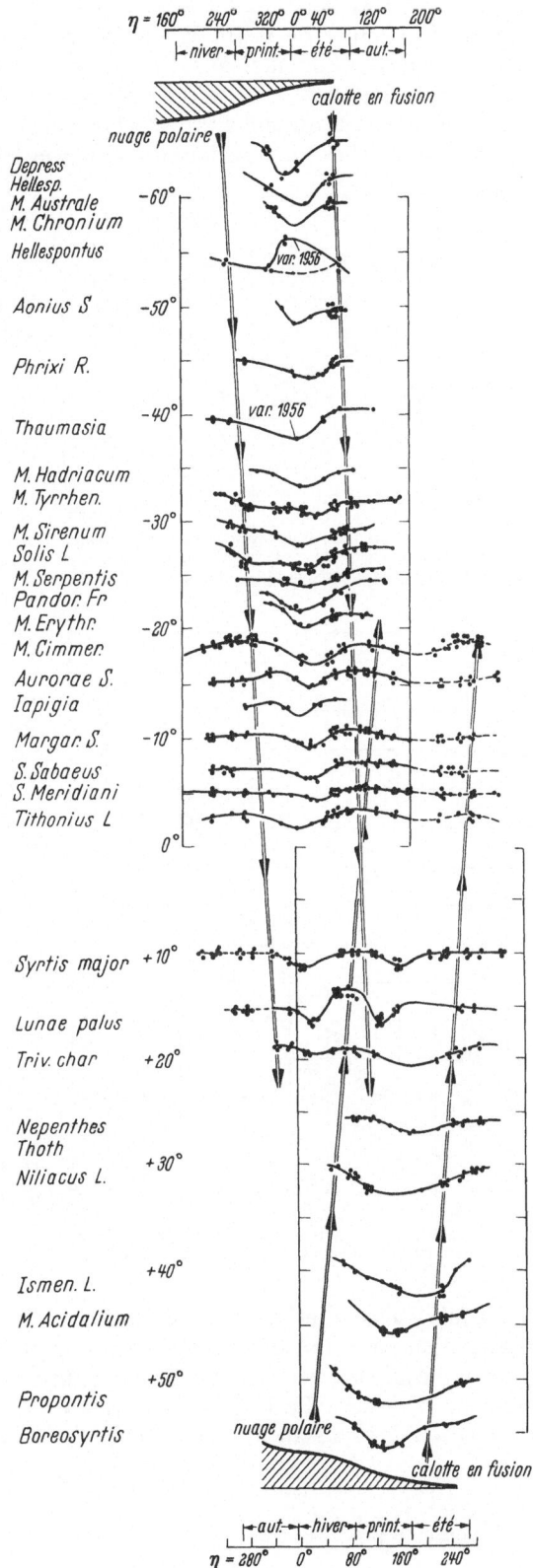

Fig. 41. Mesures photométriques de la brillance des taches sombres de Mars, de leurs variations, de la propagation de l'assombrissement. Horizontalement la longitude héliocentrique sur l'orbite. Verticalement les taches sont ordonnées par latitude décroissante. D'après J.H. Focas[3].

[1] A. DOLLFUS: La Météorologie **42**, 82 (1956).

[2] Voir note 4, p. 224.

[3] Voir note 3, p. 234.

quelquefois l'objet de vents et d'ascendances suffisamment fortes pour soulever les poussières du sol en créant d'importants «vents de sable». Les granules ainsi enlevées se déplacent puis se déposent à nouveau sur le sol. E. Öpik a fait l'ingénieuse remarque qu'un tel transport de poussière devrait rapidement uniformiser la teinte des taches sur la surface martienne et faire disparaître les régions sombres.

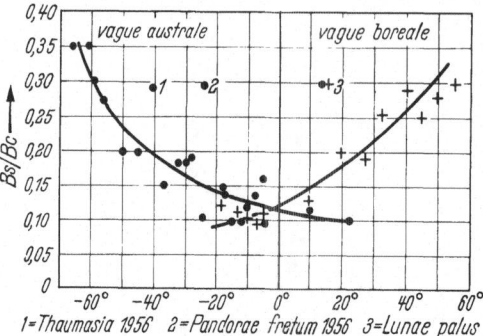

1=Thaumasia 1956   2=Pandorae fretum 1956   3=Lunae palus

Fig. 42. Variation de l'intensité de l'assombrissement saisonnier des taches de Mars en fonction de la latitude. Horizontalement, la latitude areocentrique. Verticalement, la variation de contraste. D'après J.H. Focas.

Or celles-ci persistent, au contraire, même après le passage d'un voile de poussière important. Öpik conclut que les taches sombres ont un certain pouvoir régénérateur, analogue à celui que possèderaient, par exemple, une végétation[1]. On pourrait imaginer encore des structures très lisses sur lesquelles les vents chasseraient les poussières[2]. Ou bien des formations verticales entre lesquelles les grains resteraient accumulés. Nous allons voir que la polarisation de la lumière doit faire exclure ces deux éventualités.

ι) Étude polarimétrique des régions sombres. Le test de la polarisation de la lumière donne la structure microscopique des régions sombres étudiées. Les premières mesures des taches individuelles de la surface martienne ont été effectuées au Pic du Midi par l'auteur, depuis 1948[3]. La polarisation est trouvée assez semblable à celle des régions claires ocrées. Elle doit provenir d'une structure irrégulière de petits grains opaques et très sombres. Toute surface lisse doit être éliminée ainsi que des terrains imprégnés d'eau liquide. Des granules noires et mats saupoudrés sur une étendue de limonite pulvérisée analogue aux régions ocrées reproduisent bien les courbes observées.

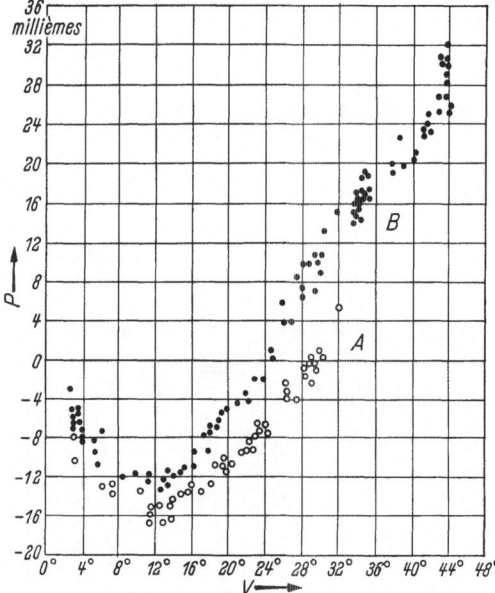

Fig. 43. Courbes de polarisation des taches sombres de Mars. A au printemps; B en hiver. D'après A. Dollfus.

Le phénomène le plus surprenant est la variation de la polarisation. Au moment du passage de la vague d'assombrissement et de transformation de l'aspect des taches, la courbe de polarisation passe brusquement de la forme B à la forme A (Fig. 43). Cet extraordinaire comportement nécessite une modification très importante de l'état physique de la surface. La Fig. 44 donne la variation de la polarisation en fonction de la saison; le changement est exactement lié à la variation saisonnière d'intensité.

[1] E. J. Öpik: Irish Astron. J. 1, 45 (1950).
[2] Voir note 1, p. 230.
[3] Voir note 4, p. 224.

L'action de l'eau sur le sol produirait une variation de polarisation inverse de celle observée. Une altération de la pureté de l'atmosphère serait variable et manifesterait un effet plus accusé près des bords du disque. Il est nécessaire d'invoquer une transformation même de la matière constituant la surface du sol, au moment de son assombrissement saisonnier lié au passage de la vague d'humidité.

Ce très curieux phénomène semble trop complexe pour pouvoir être attribué à des modifications structurales simples; la présence de petits organismes animés est fortement suggérée[1,2]. Les mesures de polarisation de la lumière réalisées également au laboratoire par l'auteur montrent en effet que de petits microorganismes, algues ou champignons en petits sphérules, donnent des polarisations voisines de celles observées sur Mars; une modification de leur croissance sous l'action de la vapeur d'eau expliquerait les changements de polarisation et d'intensité plus facilement que des modifications physiques inanimées que l'on pourrait chercher à imaginer.

x) Interprétation des taches sombres. La présence d'absorption spectrale vers 3,5 microns suggère des molécules CH_2, très fréquentes chez les êtres animés; les variations de forme, d'intensité et de couleur, leurs synchronismes avec la teneur en vapeur d'eau de l'air, la persistance des taches sombres

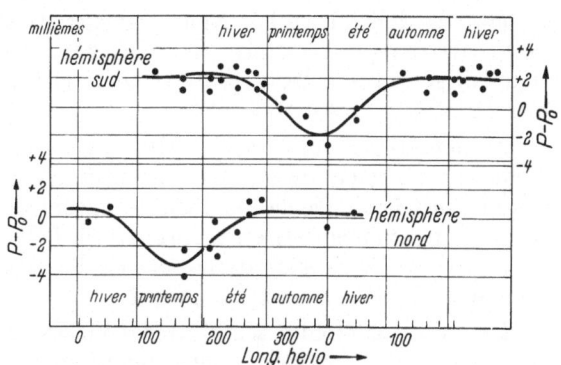

Fig. 44. Variation saisonnière de la polarisation des taches sombres de la surface de Mars. En ordonnée, la différence de polarisation entre les régions sombres et les taches claires, sous la phase 25°. En abscisse, la longitude héliocentrique et les saisons correspondant à chaque hémisphère. D'après A. Dollfus.

malgré les voiles de poussière, les variations saisonnières de l'état physique révélées par la polarisation, sont autant de phénomènes qui doivent être simultanément expliqués.

L'hypothèse de sels hygroscopiques colorés par l'ultra-violet et décolorés par la vapeur d'eau suggérés par Arrhenius[3], puis par Dauvillier[4] ne produirait pas les polarisations observées.

L'hypothèse d'éruptions volcaniques accompagnées de projections de cendres étalées par les vents saisonniers, étudiée par McLaughlin[5], ne rend pas compte de la structure décelée par les télescopes à grande résolution. Des dépôts de poussière alternativement classés par le vent sur des étendues sombres suffisamment lisses s'opposent aussi aux résultats de la polarisation.

On remarque que l'hypothèse d'être animés rend compte de façon très remarquable de la complexité très diverse des faits invoqués; devant l'absence même de toute autre proposition valable, nous pensons que la présence sur le sol martien, en certaines de ses régions, de petits microorganismes vivants, en granules très colorés comme certaines algues, ou en bacilles chromogènes plus ou moins dispersées doit être considérée.

---

[1] Voir note 4, p. 224.
[2] A. Dollfus: C. R. Acad. Sci. Paris **250**, 463 (1960).
[3] S. Arrhenius: Les Athmosphères des Planètes. Paris: Herrmann 1911.
[4] A. Dauvillier: Genèse, Nature et Evolution des Planètes. Paris: Hermann 1947.
[5] D. B. McLaughlin: Astronom. J. **60**, 261 (1955).

VI. Conclusion.

50. L'étude précédente permet de reconstituer les différents comportements qui ont modelé la surface des astres du système solaire.

Les astres privés d'atmosphère subissent en premier lieu les actions des collisions avec les météores et les autres corps célestes. Le bombardement prolongé par les tout petits météores marque la surface par des cavités de toute dimension et le relief se trouve à la longue dégradé par ce martellement. Les collisions avec les corps plus importants créent des cratères ainsi qu'une pulvérisation du sol se traduisant par la projection de poudres, d'auréoles et de grands rayonnement autour des cratères. Les collisions avec de très gros corps célestes produisent des accidents importants avec épanchement de laves. La chaleur dégagée change la couleur de la roche, produit des mers sombres, des auréoles claires ou des fonds de cratères très brillants. Nous connaissons ces propriétés par l'observation de la surface de la Lune privé d'atmosphère et proche de la Terre. Cette action se manifeste sur tous les corps du système solaire, mais elle ne se conserve intacte que sur ceux privés d'atmosphère.

Les actions provenant de l'intérieur des astres, en particulier de leur refroidissement, se manifestent au contraire de façon très inégale suivant le volume du corps céleste considéré. Quelle que soit la nature de l'échauffement intérieur (compression, recombinaisons ou radioactivité), l'effet dépend beaucoup du diamètre de l'astre, car le rapport de la surface de refroidissement au volume intéressé diminue lorsque le diamètre augmente.

Les petits astéroïdes semblent dépourvus de tout effet de cette sorte. Les astres plus gros, comme la Lune, manifestent déjà de tels effets à la surface: la rétractation de la croûte superficielle ouvre des fractures et des failles. Puis la propagation du refroidissement vers l'intérieur donne naissance à une compression à la surface qui referme certaines failles, crée des rides et dans certains cas des plissements.

Sur les corps plus grands, tels que la Terre et probablement déjà la planète Mars, des mouvements latéraux plus importants produisent des plissements avec compression, des effondrements, des réajustements d'une certaine ampleur.

Enfin, des réserves de laves peuvent manifester du volcanisme et des épanchements dont nous avons déjà quelques exemples sur la Lune et dont la Terre est abondamment pourvue.

Lorsque les astres présentent une atmosphère sensible les phénomènes superficiels précédents se compliquent. Les gaz constituant l'atmosphère peuvent provoquer une modification chimique de la surface; la présence d'oxyde de fer sur Mars pourrait s'expliquer de la sorte. Lorsqu'elle n'est pas stratifiée en couches très stables, une atmosphère, même légère, produit une érosion éolienne. La poussière superficielle est soulevée par les vents et elle se dépose en d'autres endroits. Il en résulte un nivellement et un comblement de cavités du sol dont on observe les effets sur la planète Mars et sur Ganymède.

Certains constituants peuvent se déposer sur la surface du sol à l'état de cristaux solides, tels que le givre d'eau dans les régions polaires de la surface de Mars et éventuellement les dépôts invoqués pour expliquer le grand pouvoir réflecteur des satellites Io et Europe de Jupiter. Lorsque ce givre s'évapore par sublimation, il ne produit pas d'érosion sensible.

Les dépôts sur le sol à l'état liquide produisent au contraire des phénomènes très complexes. Nous ne connaissons un tel comportement que sur la Terre: la mer, les fleuves, la pluie créent une érosion très importante et un remaniement superficiel considérable accompagné de sédimentation et de dissolution de roches. Sur la Terre, de plus, l'eau peut prendre l'état solide: le regel crée des éclatements de roches; la glace donne par écoulement des érosions par frottements considérables, accompagnées de mouvements isostatiques verticaux de la croûte superficielle résultant des changements du poids d'eau glacée accumulée.

Le cas de la Terre est donc très particulièrement complexe, mais certains phénomènes de cette nature se produisent peut-être aussi sur Vénus, dont nous ignorons l'état de la surface.

Enfin, la surface du sol des planètes peut être profondément marquée si les circonstances sont favorables à l'éclosion de la vie.

Tel est le cas sur la Terre, peut-être sur Mars, et éventuellement sur Vénus. Les phénomènes superficiels qui en résultent ont été décrits pour le cas de Mars. Ils sont dans tous les cas d'une extrême complexité.

Tous les caractères précédents dépendent de la distance au Soleil qui règle la température et le rayonnement. Ils dépendent aussi de la dimension du corps qui détermine l'importance des effets intérieurs ainsi que le maintien et la composition d'une atmosphère. Le système solaire nous donne des échantillons variés de corps célestes dans différents stades et dans différentes conditions. La Terre, possédant l'eau à l'état liquide et solide ainsi que la vie, représente un cas particulièrement complexe.

Photoelectric Techniques.

By

A. E. WHITFORD.

With 25 Figures.

I. Introduction.

1. General remarks. It is the purpose of this article to examine the role of the photoelectric process in measurements of astronomical radiation, and to describe the procedures that have been successfully used in making these measurements. The rise of astrophysics has, because of its great dependence on the analysis of the radiation from astronomical bodies, turned a great part of observational astronomy into an evaluation of intensities. Determinations of position or angle remain important in astrometry and radial velocity work, but in measurement of magnitudes, of gross spectral distribution as defined by colors or spectrophotometry, or of finer spectrum detail in line profiles or equivalent widths, or of the polarization of starlight, the attainable results have depended on the precision of the photometry.

The principal reason for assigning an increasing fraction of these tasks to the photoelectric method has been the greater precision obtained. The linear response and the spectral range covered by a single cathode are additional reasons. The applications of the method have, however, been limited by the fact that at one instant only a single chosen element of the field under study can receive attention, as opposed to thousands or millions of elements simultaneously recorded on a photographic plate. The development of image tubes, considered in the final sections of this article, gives promise of greatly reducing this limitation.

The treatment in the following sections will emphasize the measurement of radiation, with little attention to the fundamental physics of the detecting devices employed, and only sufficient reference to the astronomical results achieved by photoelectric methods to illustrate the techniques described. Since the range of possible astronomical measurements is often limited by the faintness of the sources it will be important to compare the performance of various radiation detectors with an ideal quantum detector that would faithfully record the arrival of each photon. Extraneous influences introduced by any part of the optics or electronics will also be considered in relation to the error level.

Finally, specific examples of practical instruments that have given satisfactory performance on the telescope will be described, and some of the techniques of observation outlined.

2. Physical photometers. The photoelectric effect has been by no means the only physical process that has been introduced into astronomical photometry as a substitute for the human eye. The limitations of the eye as a light-sensitive organ include its restricted spectral response, and the variation of that response with intensity (Purkinje phenomenon), its decreasing contrast discrimination at low intensities, and its inability to integrate for times longer than a fraction of a second. The subjective and personal character of the judgments made by various observers also constitutes a limitation on the attainable accuracy.

The term *objective photometry* is applied to all systems of measuring intensity in which the human eye is not called upon either for a judgment of equality between adjacent point sources or luminous fields, or for interpolation between standards of known intensity ratio. Photographic photometry qualifies as being objective if the relative blackening of various areas on the plate is evaluated by a microphotometer or other impersonal device which gives a scale reading.

The term *physical photometer* is applied to the family of radiation detectors in which the incident energy produces a thermal or electrical effect that can be read directly on a scale or dial. Unlike the indirect indication of the photographic plate, the response is usually directly proportional to the intensity. Photoemissive and photoconductive cells, the principal concern of this article, have been called *selective* detectors because their response varies considerably with wavelength, and is zero for some parts of the spectrum. This property is shared with other quantum detectors such as the eye and the photographic plate.

Another group of detectors, all of which depend on the transformation of incident energy into heat, are non-selective and respond equally to radiant energy over a wide range of wavelengths. The ambiguous term energy detectors has been applied to these devices; the designation *thermal detectors* seems preferable. Examples of such detectors would include the pyrheliometer, the bolometer, the radiometer, the radiation thermocouple and the pneumatic cell. The range of wavelengths over which each of these is non-selective is determined by the transparency of the entrance window and the spectral absorptivity of the receiving surface.

All of these thermal detectors have been used in astronomical photometry. A discussion of their operating principles, their limitations, and the technique of their use is beyond the scope of this chapter[1]. In general their ultimate sensitivity is considerably poorer than that of photoelectric detectors, and they have been used only on the planets and brighter stars. For laboratory measurement of the spectral response of selective detectors, or for absolute calibration of the spectral emission of standard sources of radiation, thermal detectors are regularly used. In these cases the intensity level can be much higher than that received from astronomical sources.

Until the recent development of photoconductive materials sensitive to wavelengths in the 10 micron region, thermal detectors were the only ones available in this part of the infrared where planetary heat must be measured. A fundamental limitation on the sensitivity of all infrared-sensitive devices is discussed in Sect. 7. Practical thermal detectors have approached but not reached this limit[2]. The detection limit for the best radiation thermocouples and the pneumatic cell is of the order of 1×10^{-11} watts for 1 cycle sec^{-1} bandwidth, using the smallest practical receiver area in each case. This figure, which is defined on the basis of a signal-to-noise ratio of unity, is convenient to keep in mind for comparison with photoelectric detectors.

3. Historical review[3]. The first use of the photoelectric process for the measurement of astronomical radiation came in 1895, when G. M. MINCHIN observed the light of several first-magnitude stars with an electrolytic cell containing a selenium layer as one of the electrodes. The photoconductive effect in selenium had been discovered by WILLOUGHBY SMITH in 1873. Serious astronomical photometry

---

[1] For a more extended treatment see B. STRÖMGREN: Handbuch der Experimentalphysik, Bd. 26, S. 797. Leipzig: Akademische Verlagsgesellschaft 1937; also Ref. [1] at the end of this article.

[2] [2], p. 38. — See also footnote 1, p. 254.

[3] H. F. WEAVER: Popular Astron. **54**, 504 (1946). Complete references to the original papers. Part of a historical review of the development of all types of astronomical photometry.

by photoelectric methods began in 1907, when Joel Stebbins and F.C. Brown undertook observations with a selenium photoconductive cell on a 12-inch refractor at the University of Illinois. They developed the technique sufficiently so that accurate light curves of eclipsing binaries could be obtained. Stebbins' classical observations of the light curve of Algol and the discovery of the secondary minimum were made with this photometer.

The first useful photoelectric cells of the emission type were made by Elster and Geitel in 1890, seven years after Hallwachs discovered the external photoelectric effect. They had found that the alkali metals sodium and potassium, amalgamated with mercury for ease in handling, are sensitive to visible light, and took the important step of forming the sensitive surface of these chemically unstable materials in an evacuated glass bulb. The first astronomical application of this forerunner of the modern photoelectric cell was an attempt to measure the brightness of the solar corona on Mount Aetna during the eclipse of 1899.

Further experimentation by Elster and Geitel led in 1910 to the discovery that the hydrides of the alkali metals are more photosensitive than the pure metals. A hydrogen glow discharge in the bulb containing the freshly deposited alkali cathode produced a colored colloidal layer which was far more sensitive than anything previously tested. The latter development brought the potassium hydride photocell to the form where it could measure the light of a star at the focus of a telescope of moderate size. In 1913 workers at three different observatories succeeded almost simultaneously in obtaining useful measurements. In Urbana, Kunz and Schulz used such a cell with a quadrant electrometer to measure the photocurrent from bright stars. In the same year Guthnick in Berlin and Rosenberg and Meyer at Tübingen achieved similar results with photoelectric photometers of nearly the same construction.

Following these initial successes, there was a fairly rapid advance to a standard technique which was used with little change for about 15 years. Nearly all of the photoelectric cells had an alkali hydride cathode, usually potassium hydride, and were sensitized by the Elster and Geitel glow discharge. There was little improvement in the cathode response, but a gaseous amplication factor of 50 or more was achieved through collisional ionization in an inert gas in the cell envelope. Dark current, i.e. leakage over the surfaces of the cell, was minimized by electrical guard rings, by careful cleaning, and by dessication of the surrounding air. J. Kunz introduced the use of fused quartz as a cell envelope material and produced cells with extremely low dark current.

The photoelectric current was measured by an electrometer carried on the telescope. The Wulf taut-fiber type, which required a gimbal suspension to keep it vertical, was supplanted by the Lindemann quadrant electrometer in 1924. The balanced construction of the latter made it insensitive to the direction of gravity; the useful sensitivity was about the same as for the Wulf string electrometer. The smallest current that could be measured with an accuracy of 1 to 2% was between 10^{-13} and 10^{-14} ampere. This corresponded to an $A0$ star of about magnitude 7.5 for a telescope of 12 inches aperture. Performance of this order was reported by Meyer and Rosenberg as early as 1913.

The standard combination of an alkali hydride photoelectric cell and an electrometer quickly demonstrated the possibilities of photoelectric photometry in precise measurements of starlight. Stebbins achieved high precision in studies of variable stars, particularly eclipsing stars. As early as 1916 Guthnick and Prager showed the possibility of color measurement by comparison of the response of the cell with no filter to that with a yellow filter. R.F. Bottlinger introduced two color filters, yellow and blue, and obtained color indices for hundreds of stars.

STEBBINS and his collaborators applied the two-filter method to the determination of color excesses of early type stars. The rather narrow wavelength range of the potassium hydride cathode offered a base line for color measurement less than half that of the International color index system, but the precision of the photoelectric measurement more than compensated for the compressed scale.

The development of the cesium-oxide-on-silver cathode by L.R. KOLLER in 1929 opened the way to photoelectric measurements of starlight over a considerably wider range of wavelengths. The new cathode (Sect. 5) was sensitive over the range from 3000 to 12000 Å, but as a consequence of the low work function of the surface, the thermionic emission at room temperature was much larger than the photoelectric emission to be expected from starlight. J.S. HALL overcame this difficulty by refrigerating the photoelectric cell with solid carbon dioxide. Under these conditions the dark current became negligible. The sensitivity of the new cathode to red and infrared light gave a great advantage in measuring the light of K and M stars.

HALL used the new cathode in the first spectrophotometric studies by the photoelectric method. A moving exit slit selected a band of wavelengths 480 Å wide in the spectrum of a star formed by an objective grating. The band centers ranged from 4550 to 9830 Å. STEBBINS and WHITFORD used the same type of photoelectric cell for spectrophotometric studies with a series of six color filters with band centers ranging from 3530 to 10300 Å.

The antimony-cesium cathode, developed by P. GÖRLICH in 1938, increased the efficiency of photoelectric cells by at least an order of magnitude (Sect. 5). In the following decade its use became nearly universal in astronomical photoelectric photometry, except for problems involving the red and infrared portions of the spectrum.

Improvements in the associated current-measuring devices have contributed the major share of the gain of the order of 10^4 between the faintest sources attempted with the early photoelectric photometer and those possible in more recent times. The Wulf string electrometer and the Lindemann electrometer, although convenient for use on a moving telescope, were recognized as being far short of the ultimate in measuring extremely small photoelectric currents. In 1926 E. STEINKE demonstrated in the laboratory that the limit set by the statistical fluctuations of the photoelectric emission could be reached through straightforward methods of charge collection. He used the Hoffmann vacuum duant electrometer. By careful elimination of disturbing influences of mechanical, thermal, and electrical origin, he was able to measure currents as small as 30 electrons per second (5×10^{-18} amperes). In 1932 SINCLAIR SMITH used such an instrument at the fixed coudé focus of the 60-inch telescope of the Mount Wilson Observatory and measured equally small currents.

The amplification of very small photocurrents by means of an ordinary thermionic vacuum tube was not successful until a tube with extremely low grid current was developed. METCALF and THOMPSON developed an electrometer tube with a grid current of only 10^{-15} ampere in 1930. A.E. WHITFORD used it in a practical astronomical photoelectric photometer in 1932. As with the Hoffmann electrometer, it was found advantageous to encase the photoelectric cell and amplifier tube in an evacuated container to avoid disturbances from cosmic ionization. Currents as small as 10^{-16} ampere could be measured with an accuracy of a few per cent. This method of current measurement was widely adopted in the following decade.

An alternative method of amplification became feasible when it was found in 1935 that sensitive photocathode materials also gave a good secondary emission yield; i.e. at bombardment potentials of the order of 100 volts, several secondary electrons were emitted for each primary striking the multiplying surface. At first only one stage of multiplication was used, but Zworykin, Morton and Malter introduced the multistage principle, which made possible overall multiplication factors of the order of 10^6. Their magnetically focused tube was used by Whitford and Kron[1] in an automatic guiding device for telescopes. The electrostatic focusing principle, and the introduction of the antimony-cesium cathode material as a multiplying surface brought further improvements, reported by Rajchman and Snyder[2] in 1940.

Secondary emission multiplication not only provided a method of achieving stable overall amplification of 10^6 or more in a simple, compact phototube, but also offered a close approach to a perfect amplifier; i.e. the amplification did not add appreciably to the fluctuations or noise present in the cathode emission, and the fundamental limit set by the unavoidable randomness of the photoelectric process was now attainable as a routine matter. G. E. Kron[3] was the first to exploit the advantages offered by multiplier phototubes in astronomical photometry. Their use soon became nearly universal, and measurements of the faintest stars photographed with large reflectors became possible. Since each photoelectron produced an easily detected burst of 10^6 electrons at the output of the multiplier phototube, registration and counting of individual photoelectric events became straightforward. Pulse counting methods for the measurement of starlight were developed by G. G. Yates at Cambridge (England), W. Blitzstein at the University of Pennsylvania, and W. A. Baum at the Mount Wilson and Palomar Observatories.

The latter stages of this historical development will be the subjects of more detailed discussion in other sections of this article.

II. Photoelectric detectors of radiation.

4. Radiation from astronomical bodies. In order to predict or assess the performace of radiation detectors on a telescope, it is necessary to establish a connection between the stellar magnitude system—long the despair of physicists and engineers—and a more fundamental system of physical units. The Sun provides a convenient reference point for comparing stellar magnitudes, photometric units used in the laboratory, and energy units. From the fact that the visual stellar magnitude[4] of the Sun is $m_v = -26.73$ and the illumination produced by the Sun on a normal surface above the atmosphere is 1.35×10^5 meter-candles[5], it may be calculated that the illumination from a star of magnitude $m_v = 0$ is 2.74×10^{-10} lumens cm^{-2} at the top of the atmosphere. The luminous flux L in lumens entering a telescope from a star of visual magnitude m_v is then found to be (atmospheric absorption neglected)

$$^{10}\log L = 2\,{}^{10}\log D - 0.4 m_v - 9.66 \tag{4.1}$$

where D is the diameter of the telescope in centimeters. If D is in inches the numerical constant is -8.85. The clear-weather atmospheric absorption at the mean visual wavelength will reduce this flux by 15 to 20% for a star in the zenith; loss in the telescope optics will reduce it further.

[1] A. E. Whitford and G. E. Kron: Rev. Sci. Instrum. **8**, 78 (1937).
[2] J. Rajchman and R. L. Snyder: Electronics, Dec. issue, p. 21 (1940).
[3] G. E. Kron: Astrophys. J. **103**, 326 (1946).
[4] J. Stebbins and G. E. Kron: Astrophys. J. **126**, 266 (1957).
[5] H. N. Russell: Astrophys. J. **43**, 103 (1916).

The laboratory test rating of the widely used antimony-cesium cathode ($S4$, $S11$) is based on a standard tungsten test source at a color temperature of $2870°$ K, and cannot be directly applied to calculating the current from a given stellar source. The yield in microamperes per lumen from the stars will be higher than the laboratory rating because the stars are hotter than the tungsten source and correspondingly richer in the blue region of the spectrum where the cathode has its peak response. Rather than to attempt to allow for this by integration of source and response functions over the wavelength range, it is preferable to make the calculation in energy units, at the effective wavelength of each filter band.

For purposes of calculation of this sort, the Sun may be approximated by a black body at $6000°$ K, producing an integrated flux[1] of 2.00 cal cm^{-2} min$^{-1}=$ 0.139 watts cm^{-2} at the top of the Earth's atmosphere.

The ratio of the monochromatic radiation J_λ given by PLANCK's equation,

$$J_\lambda = c_1 \lambda^{-5} (e^{c_2/\lambda T} - 1)^{-1},$$

to the total radiation J_t given by the Stefan-Boltzmann equation, $J_t = \sigma T^4$ may be shown to be

$$\frac{J_\lambda \, d\lambda}{J_t} = \frac{15 \, x^4}{\pi^4 (e^x - 1)} \cdot \frac{d\lambda}{\lambda} \tag{4.2}$$

where

$$c_1 = 2\pi h c^2,$$

$c_2 = hc/k = 14380$ micron deg, $\sigma = 2\pi^5 k^4/15 c^2 h^3 = \pi^4 c_1/15 c_2^4$ and $x = c_2/\lambda T$.

At $0.55 \, \mu$, the approximate effective wavelength for visual magnitudes, this equation says that the solar energy per micron is 1.31 of the total or 0.182 watts cm$^{-2} \, \mu^{-1}$ at the top of the atmosphere. The Smithsonian measurements[2] give 0.192 watts cm$^{-2} \, \mu^{-1}$ at $0.55 \, \mu$. Adopting the latter figure we find, by the same procedure used for lumens from the Sun, that the energy in watts cm^{-2} received by a telescope of diameter D (cm) from a star of visual magnitude m_v is

$$^{10}\log W(0.55 \, \mu) = 2 \, ^{10}\log D - 0.4 \, m_v + \, ^{10}\log \Delta\lambda - 11.51 \tag{4.3}$$

where $\Delta\lambda$ is the width of the observed band near $0.55 \, \mu$, expressed in microns. Atmospheric absorption is again neglected.

For other temperatures and wavelengths the monochromatic energy may be derived with sufficient accuracy by using WIEN's approximation to the Planck radiation law for the hemispherical radiation per unit area per unit wavelength interval

$$J(\lambda, T) = c_1 \lambda^{-5} e^{-c_2/\lambda T}. \tag{4.4}$$

In order to make the comparison between stars of the same visual magnitude, that is between stars of the same observed energy per unit wavelength interval at $0.55 \, \mu$, it is necessary to normalize the Planck function to constant energy at that wavelength. This may be done by dividing $J(\lambda, T)$ by the factor that the hemispherical radiation per unit area changes at $0.55 \, \mu$ in going to a temperature other than $6000°$ K. If λ_0 and T_0 are the reference wavelength and temperature, the Wien approximation gives a normalization factor

$$J(\lambda_0, T)/J(\lambda_0, T_0) = e^{-c_2/\lambda_0 T}/e^{-c_2/\lambda_0 T_0}. \tag{4.5}$$

[1] Smithsonian Physical Tables, 9th ed., p. 719. Washington 1954.
[2] Smithsonian Physical Tables, 9th ed., p. 722. Washington 1954.

The energy in a wavelength band $\Delta\lambda$ centered at wavelength λ, for a star of visual magnitude m_v and temperature T, is then found to be

$$W(\lambda,\,T) = W(\lambda_0,\,T_0) \cdot [J(\lambda,\,T)/J(\lambda_0,\,T_0)] \cdot [J(\lambda_0,\,T_0)/J(\lambda_0,\,T)].$$

In logarithmic form, using Eq. (4.2) for $W(\lambda_0,\,T_0) = W(0.55\,\mu,\,6000°)$

$$\begin{aligned}
{}^{10}\log W(\lambda,\,T) = 2\,{}^{10}\log D &- 0.4\,m_v + {}^{10}\log \Delta\lambda - \\
&- 11.51 - 5\,{}^{10}\log \frac{\lambda}{0.55} - \frac{6240}{T}\left(\frac{1}{\lambda} - 1.82\right)
\end{aligned} \tag{4.6}$$

where the constant 6240 micron degrees in the last term is $c_2\,{}^{10}\log e$. As an example, suppose a 9th magnitude star of temperature $3300°$ K is to be measured in a band $0.2\,\mu$ wide centered at $2.1\,\mu$. A telescope of 200 cm aperture is available. Eq. (4.6) then gives the result

$$\begin{aligned}
{}^{10}\log W(2.1\,\mu,\,3300°) \\
= 4.60 - 3.60 - 0.70 - 11.51 - \\
- 2.91 + 2.53 = -11.59
\end{aligned}$$

or

$$W = 2.5 \times 10^{-12}\ \text{watts}.$$

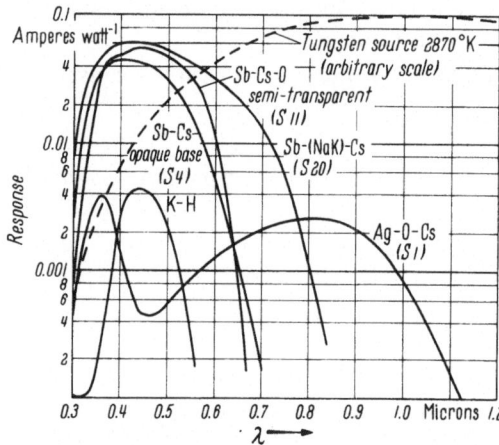

Fig. 1. Response of various cathodes as a function of the wavelength of the incident radiation.

Reference to the properties of lead sulfide photoconductors shows that this is a measurable quantity. Losses in the atmosphere and optics must still be allowed for.

As a second example we consider a high-resolution scanning spectrograph which admits $0.1\ \text{Å} = 10^{-5}\,\mu$ to the photoelectric receiver. Hot stars at $25\,000°$ K are being examined at $0.4\,\mu$ with a telescope of 300 cm aperture. It is estimated that only 25% of the incident light gets through the atmosphere and the optical system. The cathode yield is 0.04 amperes per watt and a current of 10^{-17} ampere or 60 electrons per second will give a satisfactory measurement. What is the limiting visual magnitude? From the two latter figures the required luminous flux at the top of the atmosphere is found to be 1.0×10^{-15} watt. Then,

$$-15.00 = 4.96 - 0.4\,m_v - 5.00 - 11.51 + 0.69 - 0.17$$

or

$$m_v = 9.9.$$

When dealing with a high temperature source observed in the deep red or infrared the Wien approximation breaks down and the full Planck equation must be used.

5. The photoelectric cathode. The present state of knowledge regarding the basic physics of photoelectric emission from solids is reviewed in another article[1] in this Encyclopedia. For purposes of astronomical photometry, only complex surfaces have a sufficiently good quantum efficiency to be of interest. This section will not be concerned with understanding the physical processes in these cathode materials, or with methods of preparation, but rather with a description of their practical properties as they affect their usefulness in photometry. The

---

[1] G. L. WEISSLER: Vol. XXI, p. 304 of this Encyclopedia. Berlin-Göttingen-Heidelberg: Springer 1956.

yield of several cathodes is shown graphically in two forms: (a) Fig. 1 shows the yield in amperes per watt of incident radiant energy at each wavelength, and (b) Fig. 2 shows the quantum efficiency as a function of wavelength. The quantum efficiency is defined as the ratio of the number of emitted electrons to the number of incident photons at each wavelength. The relation between the quantum efficiency q and the current yield Y is given by

$$q = \frac{1.240}{\lambda \,(\text{microns})} \, Y \text{ (amperes watt}^{-1}). \tag{5.1}$$

The potassium hydride cathode is included for historical interest, because a considerable amount of astronomical photometry has been done with it. The yield data are from SUHRMANN[1]. Next in historical order is the cesium-oxide-on-silver cathode, developed in 1930 by KOLLER[2]. It has been given the technical designation $S1$ by manufacturers of electron tubes in the United States. It is remarkable for its response in the deep red and infrared, and its sensitivity to a wide range of wavelengths is useful in astronomical spectro-photometry. The relative height of the two peaks and the infrared cutoff can be influenced by the processing technique, as shown by PRESCOTT and KELLY[3].

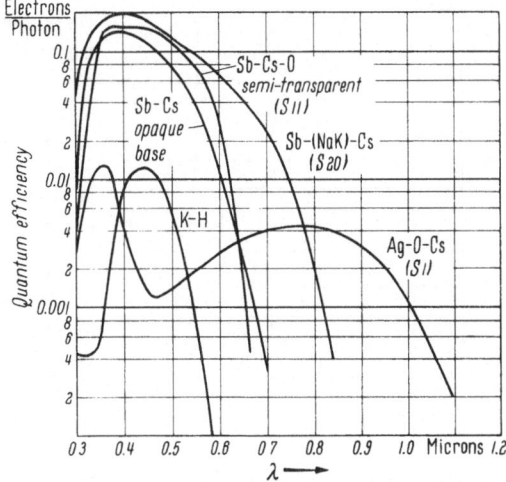

Fig. 2. Quantum efficiency of various cathodes as a function of wavelength.

The antimony-cesium cathode was developed by GÖRLICH[4] in the course of attempts to develop a semi-transparent cathode, i.e., a layer from which the photoelectrons could be drawn off on the side of the layer opposite to the incident light. The excellent quantum yield has led to its adoption in all applications where response to the red or infrared is not required. On an opaque backing it has the technical designation $S4$; as a semi-transparent cathode it is called $S11$. In passing through the reddish brown antimony base film the incident light suffers some absorption in the violet and the longer wavelengths are favored. Both response curves are shown. The peak response is little affected by the absorption.

The tri-alkali cathode was announced by SOMMER[5] in 1955. The three alkali metals sodium, potassium, and cesium are combined with antimony to form a semi-transparent cathode which has the highest quantum yield of any surface yet produced. Its technical designation is $S20$. The room temperature thermionic emission is below that of antimony-cesium, in spite of the response extending well beyond 7000 Å.

For purposes of engineering computations where the source is a tungsten filament lamp, photoelectric cathodes are rated according to their response in microamperes per lumen. The source is taken to be a tungsten filament at a color

---

[1] R. SUHRMANN: Phys. Z. **29**, 811 (1928).

[2] L. R. KOLLER: Phys. Rev. **36**, 1639 (1930).

[3] C. H. PRESCOTT and M. J. KELLY: Bell Syst. Techn. J. **11**, 334 (1932).

[4] P. GÖRLICH: Z. Physik **101**, 335 (1936).

[5] A. H. SOMMER: Rev. Sci. Instrum. **26**, 725 (1955).

temperature of 2870° K. The spectral energy curve of such a source is shown in Fig. 1. It is obvious that except for the $S1$ cathode the rating is quite sensitive to the shape of the red tail of the response curve of a blue-sensitive cathode. Typical ratings are as follows:

$$\begin{array}{ll}
\text{Cesium oxide on silver } (S1) & \text{15 to } 30 \,\mu\text{A/lumen} \\
\text{Antimony-cesium } (S4, S11) & \text{40 to } 70 \\
\text{Tri-alkali } (S20) & \text{100 to } 150
\end{array}$$

The preparation of photoelectric cathodes is still very much an art and considerable individual variation may be expected in tubes from the same manufacturer, both in absolute yield and spectral response. Exceptional tubes, having characteristics well above the curves shown here, may occasionally be found, as well as tubes below average.

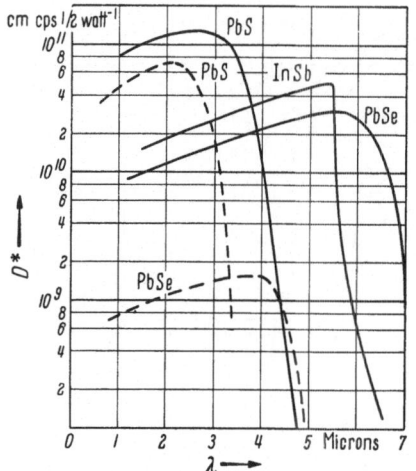

Fig. 3. Detectivity for various photoconductive cells as a function of the wavelength of the incident radiation. The solid lines are for detector surfaces at −195 °C; the dashed lines for surfaces at 25 °C.

A more extended discussion of photoelectric cathodes may be found in the treatises by Zworykin and Ramberg [4], and by Simon and Suhrmann [3].

6. Photoconductive cells. Although the photoconductivity of selenium was discovered in 1873, it was not until the time of World War II that it was discovered that thallium sulfide and lead sulfide could be made highly sensitive detectors for the near infrared. With the increased knowledge of the solid state that developed during the postwar years, progress was rapid, and a number of detectors that approach the theoretical limits set by the photon statistics have been developed. The basic physics of the fundamental processes of photoconductivity is fairly well understood. Both the principles and the practical performance have been reviewed by Smith, Jones, and Chasmar [1] and by Petritz[1].

The signal from photoconductive cells is an increase in the normal current flowing in the unilluminated condition. The signal voltage to the amplifier is proportional to the normal current, and so, in general, is the noise. For a cell whose room temperature resistance is 10^6 ohms, in series with a load resistor of equal value, and supplied with a voltage that gives a normal current of 10^{-4} ampere, the response may be of the order of 10^4 volts per watt of incident energy. But since there is no general way of specifying the output for a given amount of radiant energy, the spectral response curves are usually on a relative rather than an absolute scale.

Since the ultimate sensitivity of photoconductive cells is determined by their internal noise, and not by other parts of the apparatus, it has become customary to plot the spectral response with *detectivity* as ordinate. This quantity as defined by Jones[2] is the reciprocal of the noise equivalent input power, expressed in radiant watts, for 1 cm² of sensitive area and 1 cycle sec⁻¹ bandwidth; for the standard area and bandwidth it is called D^* (D-star).

[1] R.L. Petritz: Proc. Inst. Radio Engrs. **47**, 1458 (1959).
[2] R. Clark Jones: Proc. Inst. Radio Engrs. **47**, 1495 (1959).

The detectivity for this class of detectors varies as the square root of both of these parameters (Sect. 7). The chopping frequency and the temperature of the cell are also parameters that must be known.

Fig. 3 shows the detectivity for two lead-salt film detectors[1], both cooled and at room temperature, and for cooled indium antimonide[2], a single-crystal intrinsic semiconductor.

7. Sources of noise. If the observer were in possession of an ideal detector which could record the arrival of each photon in the incoming beam of radiation, the limits of error and the time required for each measurement could be calculated from the theoretical fluctuations in a stream of radiation. Since photons obey Bose-Einstein statistics, there is a degree of coherence or clumping that increases the mean square of fluctuations over those predicted by the classical statistics of randomly spaced events by a factor $1 + (e^x - 1)^{-1}$, where $x = c_2/\lambda T$ as in Eq. (4.2)[3]. Since this correction factor is not appreciably larger than unity for most situations in which astronomical radiation is measured by photoelectric methods, classical statistics may be adopted and the events assumed to have a Poisson distribution. If, in a specified spectral range, the average rate of arrival of photons in the signal beam is n_s, the mean square difference between the actually observed number $n_0 t$ arriving in a particular time interval t and the average number $n_s t$ derived from a large number of such intervals is

$$\overline{(n_0 t - n_s t)^2} = n_s t \qquad (n_0 t \gg 1). \tag{7.1}$$

The root-mean-square fluctuation in the number observed in successive time intervals represents an irreducible uncertainty equal to the square root of the number of observed events. The precision p of the observation, defined as the ratio of signal S to root-mean-square noise N is

$$p = S/N = \sqrt{n_s t}. \tag{7.2}$$

Practical detectors of astronomical radiation fall short of this ideal performance for a variety of reasons; among them are the effects of the Earth's atmosphere, the unavoidable additional noise introduced by the associated electrical circuits, and the inefficient utilization of the incident photons. Nevertheless, under the best conditions the combination of multiplier phototubes and a pulse counter is a sufficiently close approach to the ideal detector to make the latter a useful concept against which to compare the performance of all radiation detectors. The factors which influence the actual performance will be taken up in turn.

α) Quantum efficiency. The quantum efficiency of the most commonly used photoemissive cathodes has been given as a function of wavelength in Sect. 5. If the number of observed photoevents is only the fraction q of the number of incident photons, the precision given by Eq. (7.2) will be smaller in the ratio $\sqrt{q}$, and the time to attain a specified precision will be increased by a factor q.

The quantum efficiency of photoconductive cells[4], measured in terms of the effectiveness of incident photons in lifting charge carriers to the conduction band, is very nearly unity. Under ordinary circumstances, however, the noise from fluctuations in the rate of spontaneous generation and recombination of charge carriers, current noise, and noise from fluctuations in the rate of exchange

---

[1] R. J. Cashman: Proc. Inst. Radio Engrs. **47**, 1471 (1959); also footnote 2 on p. 248.
[2] F. F. Rieke, L. H. DeVaux and A. J. Tuzzolino: Proc. Inst. Radio Engrs. **47**, 1471 (1959); also footnote 2 on p. 248.
[3] Ref. [1], p. 209.
[4] H. E. Spencer: Phys. Rev. **109**, 1974 (1958).

of quanta between the detector surface and its container add up to a total far greater than the noise arising from fluctuations in the rate of arrival of signal quanta. These questions are considered in greater detail in later paragraphs.

Since photographic plates have a nonlinear response to radiation, the quantum efficiency is not a unique quantity. The recordable event in the emulsion is a developed plate grain. In preference to counting grains, however, there are reasons for basing the operational determination of quantum efficiency on fluctuations in the photographic density, a procedure parallel to that used in detecting and measuring sources of radiation. This "equivalent" quantum efficiency, first proposed by Rose[1], has been measured by Fellgett[2] and Jones[3] for certain fast emulsions made by the Eastman Kodak Company and found to have a maximum value of 0.9%. Baum[4] found that a quantum efficiency of 0.1% was consistent with the recognition of threshold images on direct photographs with large telescopes. Hiltner[5] found a similar efficiency from calculation of the number of received quanta necessary to achieve a given accuracy in evaluating the intensity of one picture element on a stellar spectrogram. The laboratory measurements showing the higher efficiency were for short exposures on emulsion materials having a considerable reciprocity failure. The efficiency was highest for a rather low-density, low-contrast part of the characteristic curve, and fell rapidly with increasing density. The estimates showing lower efficiency under conditions of actual astronomical use were for longer exposure times and were for the "full-exposure" density found to give optimum visual contrast.

β) *Cathode dark emission.* Antimony-cesium ($S4$, $S11$) cathodes commonly have a thermionic emission of a few thousand electrons per second per cm^2 at room temperature. By processing such a cathode so as to reduce its red response the dark emission at 20° C can be reduced a hundred times or more with a reduction of no more than half in the peak response in the blue[6]. For infrared-sensitive cesium-oxide-on-silver cathodes ($S1$), which must necessarily have a lower work function, the room-temperature thermionic emission is 10^5 to 10^6 electrons sec^{-1} per cm^2. Currents of the latter magnitude are much larger than those encountered in astronomical photometry, and would be the dominant source of noise if the cathode were operated in the unrefrigerated condition. In any attempt to reach the faintest sources cathodes of all types are of course refrigerated. A dark emission of the order of 1 electron $sec^{-1} cm^{-2}$ can be attained at a temperature of $-78°$ C, a value so much lower than the photoelectric emission induced by the background light of the night sky unavoidably included in the focal plane diaphragm as to be a negligible factor in the total noise.

γ) *Background radiation in the signal beam.* The background light of the night sky which comes through the focal-plane diaphragm along with the star (Sect. 14) may increase considerably the total radiation falling on the cathode, and increase the fluctuation noise over that expected for the star alone.

The visual brightness of the overhead sky is about magnitude 21.5 per square second[7]; in photographic light a brightness of magnitude 22.0 per square second is representative. Considerably higher values are found in the Milky Way, near

[1] A. Rose: J. Soc. Motion Picture Engrs. **47**, 273 (1946).
[2] P. Fellgett: Monthly Notices Roy. Astronom. Soc. London **118**, 224 (1958).
[3] R. Clark Jones: Adv. Electronics **11**, 140 (1959).
[4] W. A. Baum: Trans. Internat. Astronom. Union **9**, 684 (1957).
[5] W. A. Hiltner: Trans. Internat. Astronom. Union **9**, 688 (1957).
[6] J. Sharpe: EMI Electronics Ltd., Ref. Doc. CP 154.
[7] C. W. Allen: Astrophysical Quantities, p. 125. London 1955.

the horizon (airglow), or near the ecliptic (zodiacal light). The airglow[1] shows intense OH emission bands in the infrared, particularly at 1.04 μ. The full moon increases the average visual brightness of the night sky about 4 magnitudes; more efficient atmospheric scattering in the blue and ultraviolet makes a larger increase at these wavelengths.

The minimum diameter of the diaphragm that may be used, assuming that the clock drive and the guiding arrangements (Sect. 14) do not impose any limitation, is determined by the distribution of light in the star image formed by the telescope optics and by the size of the seeing tremor disk. The latter may lie between 1″ and 2″ in reasonably good seeing. The star image has a low-intensity fringe, not ordinarily seen or photographed, that carries an appreciable fraction of the total light. In order to avoid apparent intensity fluctuations caused by wandering and seeing "blowups", it is necessary to include most of this fringe. The smallest diaphragm usable with safety lies between 5″ and 10″; the latter with an area of 78 square seconds is the normal working minimum. The sky light admitted by such a diaphragm is equivalent to a star of visual magnitude 16.8. For stars fainter than this, background illumination will be dominant in setting the noise level, independent of the aperture of the telescope.

If n_b is the number of background photons coming through the diaphragm in unit time, and n_d is the number of electrons emitted by the totally dark cathode per unit time, then the signal-to-noise ratio or precision p will be

$$p = S/N = \frac{\frac{1}{2} q \, n_s \, t}{[(n_d + \frac{1}{2} q \, n_s + q \, n_b) \, t \,]^{\frac{1}{2}}}. \tag{7.3}$$

Since half of the observing time must be devoted to comparison exposures on the sky, signal photons will be received for only half of the time. When the background becomes negligible, this expression reduces to Eq. (7.2) if the dark emission $n_d \ll q \, n_s$. Ordinarily attention must be given to the latter requirement only if measurements are being made in very narrow filter bands or in a dispersed beam which will greatly reduce the sky background. Usually, however, the background photons are the major source of noise for faint stars. If

$$n_b \gg n_s \gg n_d/q,$$

then

$$p = \frac{q^{\frac{1}{2}} n_s \, t^{\frac{1}{2}}}{2 n_b^{\frac{1}{2}}}. \tag{7.4}$$

The dependency factors for background-limited photometry may be read from this equation. If a *fixed time* is devoted to a measurement, then the precision with a constant background varies *directly* as the source intensity, rather than as the square root as is the case for background-free measurements. The precision varies inversely as the square root of background intensity; increased background from artificial sky illumination or moonlight, if constant in intensity, builds up error slowly. Any instrumental change which affects the number of signal and background electrons in the same ratio, such as quantum efficiency, width of spectral band admitted, or area of the telescope objective, affects the precision as the square root of the factor of change.

If, on the other hand, a *fixed precision* is demanded, the time required varies inversely as the square of the source intensity, and directly as the first power of the total background radiation. This means the time required varies as the

---

[1] J. W. CHAMBERLAIN: Physics of the Aurora and Airglow, p. 363. New York and London: Academic Press 1961.

square of the diameter of the focal plane diaphragm. Instrumental changes which affect signal and background in the same ratio, such as quantum efficiency, width of spectral band admitted, or area of the telescope objective, affect the time required in inverse ratio to the factor of change.

As an example suppose a star of visual magnitude 21.0 is to be measured with a telescope of 300 cm aperture, using a cathode whose quantum efficiency is 0.07 at the visual wavelength 0.55 μ, with a filter band 0.10 μ wide. A diaphragm of 10 seconds diameter is to be used. A loss of 0.4 magnitude in the atmosphere and telescope optics must be allowed, making the star equivalent to one of $m_v =$ 21.4 at the top of the atmosphere. Reference to Eq. (4.3), shows that 7.8×10^{-17} watt are available from the star, while the sky will be 4.2 magnitudes or 47 times brighter than the star. There are 2.74×10^{18} quanta $sec^{-1} watt^{-1}$ at 0.55 μ. There will be 210 quanta or 15 electrons sec^{-1} from the star, and 700 electrons sec^{-1} from the sky. If a precision of 20 (0.05 mag. mean error) is demanded, then Eq. (7.4) says that the total exposure time must be 5100 sec or 85 min, somewhat longer than the exposure time for a photovisual plate-filter combination. Against a totally dark background, the same precision could be realized in only 27 sec, as a consequence of the higher quantum efficiency of the cathode relative to the photographic plate. The need for a diaphragm large enough to give a margin of safety around the fringes of the star image includes enough sky background, however, to increase the exposure by a factor of 47. Another factor of 4 comes from the need for alternation between sky and star. If observing conditions permit reducing the diaphragm to 5 seconds the exposure time is cut by a factor of 4. For still fainter stars the exposure time increases by a factor of $2.51^2 = 6.3$ for each magnitude. The justification for these long exposure times is the linearity of the method. The uncertainties of calibrating threshold photographic images are eliminated.

Because photoelectric measurements are based on a linear, non-saturable process, there is no "limiting magnitude" with a given telescope. In principle extending the observing time will always extend the range of the telescope to fainter objects. If, however, a maximum allowable observing time is established, analogous to the full-exposure time of the photographic plate, there is a minimum source that will give a preassigned acceptable precision. The dependence of this minimum source on the aperture of the telescope may be seen by introducing an area factor in Eq. (7.4). If A is the area of the telescope, u_s is the number of signal photons per unit area per unit time, and u_b is the number of background photons per unit area per unit time from the area on the sky defined by a focal plane diaphragm of constant angular size, then $n_s = u_s A$ and $n_b = u_b A$. Substituting

$$u_s = \frac{2p\, u_b^{\frac{1}{2}}}{q^{\frac{1}{2}}\, t^{\frac{1}{2}}} \cdot A^{-\frac{1}{2}}. \tag{7.5}$$

Thus the brightness of the minimum source measurable to a given precision in a given time varies inversely as the square root of the area of the telescope, and not inversely as the area itself.

δ) *Ambient background radiation.* For radiation detectors sensitive to infrared radiation beyond 1 μ, the concept of totally dark conditions can no longer be maintained. As soon as an appreciable fraction of the energy involved in the normal exchange of radiation between the sensitive surface and the container walls is carried by wavelengths to which the detector responds, then statistical fluctuations in the rate of emission and absorption of quanta can compete with similar fluctuations in the signal beam as a source of noise.

For thermal detectors, ambient radiation sets an inherent lower limit for the minimum detectable energy. This fact was recognized by MILATZ and VAN DER VELDEN[1] in 1943, on the basis of the temperature fluctuations of the receiver predicted by general considerations of statistical mechanics. The same result was derived by W.B. LEWIS[2] in 1947 by considering statistical fluctuations in streams of radiation. For a black body of area A the root-mean-square variation in the radiative energy exchanged with a surrounding cavity at temperature T is

$$\Delta W_T = (16\sigma\, kT^5\, A\, \Delta f)^{\frac{1}{2}} \qquad (7.6)$$

where σ is the Stefan radiation constant, k is the Boltzmann constant and Δf is the bandwidth in cycles per second. For the equations of the previous paragraphs, based on count integration times, the bandwidth is equal to the reciprocal of twice the integration time in seconds[3]. For a receiver in an enclosure at 300° K

$$\Delta W_T = 5.5 \times 10^{-11} \text{ watt cm}^{-1} \text{ sec}^{-\frac{1}{2}}. \qquad (7.7)$$

This quantity is obviously reduced by using small receivers, and, as for all statistical fluctuations, by extending the averaging time through a narrower bandwidth. Low emissivity on one or both sides of the receiver will also reduce it; various configurations have been considered by SMITH, JONES, and CHASMAR[4].

For quantum detectors, only the wavelengths shorter than the cutoff wavelength for release of electrons are effective in the exchange of quanta with the ambient radiation field. Amid surroundings at 300° K, where the maximum of the Planck radiation function comes at 10 μ, only an infinitesimal fraction of the photons come at wavelengths shorter than 1.0 μ to which photoemissive cathodes respond. Such cathodes are therefore blind to virtually all of the photons which carry on the energy exchange with the container. This fact, which permits such cathodes to have a noise equivalent input far below that of thermal detectors, was pointed out by FELLGETT[5] in 1949. Detailed calculations have been given by SMITH, JONES, and CHASMAR[6].

Photoconductive cells have a cutoff wavelength which includes an appreciable fraction of the quanta under the energy curve of a black body at room temperature. The number of effective photons may be determined by integration of the Planck function to the wavelength in question, and the fluctuations which determine the noise level then computed. PETRITZ[7] has given a plot from which may be read the noise equivalent input of an ideal detector surrounded by a hemispherical black body at 300° K, for various cutoff wavelengths. Calculations by SMITH, JONES, and CHASMAR[8] which take account of the gradual cutoff lead to very nearly the same result. Two examples may be quoted. For a PbS photoconductive cell at 300° K, the cutoff is at about 2.9 μ. For detecting radiation at 2 μ, the background-limited minimum detectable power is

$$W_{\min}(\text{PbS, 2 μ, 300° K}) = 1.9 \times 10^{-12} \text{ watts cm}^{-1} \text{ sec}^{-\frac{1}{2}}. \qquad (7.8)$$

As for the thermal case a receiver of 1 cm² area and a bandwidth of 1 cycle sec⁻¹ are assumed. For a PbSe cell with a cutoff at 4.7 μ the equivalent figure for detect-

[1] J.O.W. MILATZ and H.A. VAN DER VELDEN: Physica, Haag **10**, 309 (1943).
[2] W.B. LEWIS: Proc. Phys. Soc. Lond. **59**, 34 (1947).
[3] Ref. [1], p. 244.
[4] Ref. [1], p. 207.
[5] P.B. FELLGETT: J. Opt. Soc. Amer. **39**, 70 (1949).
[6] Ref. [1], p. 283.
[7] R.L. PETRITZ: Proc. Inst. Radio Engrs. **47**, 1458 (1959).
[8] Ref. [1], p. 295.

ing radiation at $4\,\mu$ is

$$W_{\min}(\text{PbSe},\ 4\,\mu,\ 300^\circ\,\text{K}) = 8\times 10^{-12}\ \text{watts cm}^{-1}\ \text{sec}^{-\frac{1}{2}}. \qquad (7.9)$$

In these two examples the noise equivalent power is respectively $\frac{1}{30}$ and $\frac{1}{7}$ of the black body case.

When a cooled detector faces cooled surroundings the energy exchange and the noise arising from it are greatly reduced[1]. The optical system for admitting the signal beam may introduce background radiation from a higher temperature over an appreciable solid angle, however, and become the principal source of photon noise.

ε) *Electrical noise.* In addition to the photon noise considered in the previous paragraphs, the total noise in the output may contain inherent electrical noise arising in the detector itself or in the associated amplifier or indicating instrument. In many cases the additional noise can be kept small compared to the photon noise; in others the external noise dominates. The detailed situation for each type of detector is set forth in the sections describing amplifying methods.

ζ) *Scintillation.* For bright stars the calculated relative errors from photon noise and electrical noise are very small and if the source were of constant intensity a very short exposure would give adequate accuracy. Scintillation noise then becomes the principal contributor to the uncertainty. Photoelectric studies of the frequencies and amplitudes involved have been made by Whitford and Stebbins[2], Nettelblad[3], and Mikesell[4]. The amplitudes vary by a factor of 5 or more with differing meteorological conditions. The noise expressed as a percentage of the signal is smaller as the aperture of the telescope increases, owing to the averaging effects over the larger area; the variation is roughly in inverse ratio to the diameter.

The frequency spectrum of the scintillation noise, according to Mikesell, is nearly flat for frequencies between 0.1 and 10 cycles per second. Reducing the bandwidth therefore has the same effect as for "white" photon noise, and the accuracy increases as the square root of the observing time. At higher frequencies, the scintillation noise per cycle decreases, virtually disappearing at 500 cycles. Hall[5] has pointed out that if a rapidly varying phenomenon which precluded a narrow bandwidth were being followed, scintillation noise could be made negligible by using an a.c. method and a chopper frequency at least as high as 500 cycles per second.

Hall[6] has analyzed the effect of scintillation noise on the overall accuracy of stellar photometry with a 40-inch telescope. Scintillation noise exceeded the inherent shot noise of the cathode photo-emission when the emission was greater than about 700 electrons per second.

η) *Variations of sky brightness and transparency.* In practical astronomical photometry, particularly in cases where the telescope must be moved between measurements of one star and the next, the accuracy is usually not appreciably better than 1% even though the predicted noise is well below this level. Variable transparancy of the atmosphere is the cause. Differential methods giving simul-

---

[1] See F. J. Low: J. Opt. Soc. Amer. **51**, 1300 (1961) for a report on a low-temperature germanium bolometer with very low noise.

[2] A. E. Whitford and J. Stebbins: Publ. Amer. Astronom. Soc. **8**, 228 (1936). — J. Stebbins: Sky and Telescope **3**, No. 4, 5 (1944).

[3] R. Nettelblad: Medd. Lund Obs., Ser. II, No. 130 (1953).

[4] A. H. Mikesell: Publ. U.S. Naval Obs. **17**, Part 4 (1955).

[5] Ref. [5], p. 61.

[6] Ref. [5], p. 47.

taneous or quickly alternating measurements are frequently used to make the precision less dependent on sky quality. Examples are explained in Sect. 15.

Measurements of faint objects under sky-limited noise conditions (Sect. 7γ) often require exposures of 2 minutes or more to give sufficient integration time. Even though the shift to the sky area loses little time, the sky, which may be 10 to 50 times brighter than the star, can vary enough to affect the accuracy. The airglow component of the light of the night sky is known to vary through the night[1]. Simultaneous exposures of sky and star in a dual channel photometer (Sect. 15) offer a remedy.

III. Amplifying and indicating instruments.

8. Direct amplification of photoelectric currents. The initial cathode emission obtained in the ordinary course of astronomical photoelectric photometry ranges from 10^{-12} ampere down to 10^{-17} ampere. Current measurement in this range was for many years carried out exclusively by electrometers[2]. Development of techniques for amplification of these small currents has been so successful, however, that electrometer methods have largely become obsolete. This is especially true where the photoelectric apparatus must be mounted on a moving telescope. This section considers amplification methods in which the current is taken directly from the anode of a simple photoelectric cell as the input to an external amplifier. Direct current methods

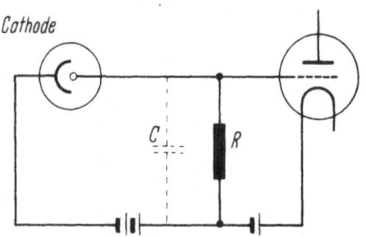

Fig. 4. Input circuit of d.c. amplifier for photocurrents.

are almost universally used, and indeed at the lower limit are unavoidable, since stray capacitance introduces time constants too long for any practical chopper frequency.

The input circuit of a typical amplifier is shown in Fig. 4. Since the output impedance of photoelectric cells is extremely high, they are essentially current generators, and the load resistor R may, as far as the cell characteristics are concerned, be as large as it is practical to make it. The larger the signal voltage applied to the grid of the first amplifying tube, the less attention need be paid to flicker noise in that tube and to stability of supply voltages. The load resistor is shunted by the leakage resistance of the photocell, and of all other supports and tube seals connected with the grid lead. By choice of suitable insulating materials (teflon, silicone-coated glass) these conductances can be reduced to a negligible fraction of that of the resistor itself.

For the input tube it is necessary to use a so-called electrometer tube, a type designed to have the highest possible input impedance at the grid terminal. By reducing all sources of currents to or from the control grid, a grid current of 10^{-15} ampere or less can be achieved. The dynamic input impedance, defined by the slope de_g/di_g, is greater than 10^{16} ohms. Such tubes have a low temperature filament, very few grid turns, and an amplification factor of about 1.0. The plate potential is about 4 volts, and the plate current only 10 to 15 microamperes. The Victoreen Type 5800[3] is a subminiature tube meeting these specifications.

---

[1] J. W. CHAMBERLAIN: Physics of the Aurora and Airglow, p. 503. New York and London: Academic Press 1961.

[2] For a review of instruments and techniques, see H. V. NEHER, Chap. 6 of Procedures in Experimental Physics, ed. by J. STRONG. New York: Prentice-Hall 1938; also Ref. [3], p. 377.

[3] Obtainable from The Victoreen Instrument Co., Cleveland 3, Ohio, U.S.A.

Type 5886[1], a similar screen-grid type with considerable voltage gain, has a grid current characteristic that is only slightly poorer. Either tube has an input impedance considerably greater than the value of the highest practical load resistor and may therefore be put in parallel with negligible shunting effect.

The electrical noise in the input circuit also influences the choice of the input resistor. The shot effect in the cathode current I_c produces a mean square fluctuation in the current through the resistor

$$\overline{i_s^2} = 2e\,I_c\,\Delta f \tag{8.1}$$

where e is the electronic charge and Δf the bandwidth in cycles per second. Added to this irreducible noise, which is identical with the noise previously considered in Sect. 7, there is the shot effect of the grid current I_g and the thermal noise (Johnson-Nyquist noise) in the resistor itself. In terms of voltages

$$\overline{v_n^2} = \overline{v_s^2} + \overline{v_t^2} \tag{8.2}$$

$$= 2e\,(|I_c| + |I_g|)\,R^2\,\Delta f + 4kT\,R\,\Delta f. \tag{8.3}$$

The mean-square shot noise from the two components of current through the input resistor adds without regard to sign. At $300°$ K this reduces to

$$\overline{v_n^2} = 1.66 \times 10^{-20}\,R\,(1 + 19.3\,I_g\,R + 19.3\,I_c\,R)\,\Delta f. \tag{8.4}$$

Each shot noise term will equal the thermal noise if the IR drop in the input resistor is 0.052 volt. For $I_g = 10^{-15}$ ampere, the resistor will then be $R = 5 \times 10^{13}$ ohms, and I_c must exceed 2×10^{-15} amperes to avoid amplifier noise being larger than the photon noise in the output of the photocell. As will be seen later $R = 5 \times 10^{13}$ ohms is at the upper limit of practical resistance values. Since $I_c = 10^{-15}$ ampere is not a particularly small signal current, it must be concluded that circuit noise will set the limit for the minimum incident radiation that may be measured by direct amplification of photocurrents from a vacuum photocell. If $I_c > 10^{-14}$ ampere, however, thermal noise does not prevent accurate measurements by straightforward techniques.

H.L. Johnson[2] has shown that when allowance is made for gaseous amplification in the original phototube a somewhat better limiting noise level is possible. The highest multiplication factor consistent with stability is about $G = 10$ at room temperature. Kron[3] has found that at $195°$ K, the temperature at which the cathode is usually operated in order to reduce thermionic emission, a stable gain $G = 25$ is possible. Multiplication by gaseous ionization does not add appreciably to the statistical fluctuations already present in the cathode emission; Jones[4] has reviewed the literature on this point. The unit of charge in the shot effect of the photocurrent may then be taken to be Ge, rather than e. If all parts of the input circuit are at $195°$ K, as was found practical by Kron, and if it be assumed that $I_g = 10^{-16}$ ampere under these conditions, as suggested by Kron's results, then for $R = 5 \times 10^{13}$ ohms, $G = 25$, and $\Delta f = 0.02$ cycle sec^{-1}

$$\overline{v_n^2} = (4 \times 10^8)\,I_c + 0.16 \times 10^{-8} + 1.08 \times 10^{-8}. \tag{8.5}$$

The first term is the shot noise of the multiplied cathode current, the second the shot noise of the grid current, and the third the thermal noise in the resistor.

[1] Obtainable from Raytheon Manufacturing Co., Waltham, Mass., U.S.A. Philips Type 4068 is equivalent.
[2] H.L. Johnson: Astrophys. J. **107**, 34 (1948).
[3] G.E. Kron: Astrophys. J. **115**, 1 (1952).
[4] Ref. [2], p. 61.

The root-mean-square value of the latter is about 100 microvolts. It is apparent that for cathode currents larger than 2.5×10^{-17} amperes, or 160 electrons per second, photon noise will prevail. For a cathode of reasonably good quantum efficiency, the light from the sky background will produce a current of this order or larger, as shown in Sect. 7.

With a load resistance as high as 5×10^{13} ohms, however, the time constant of the input circuit becomes prohibitively long. The combined capacity of all parts attached to the grid lead, designated by the dotted-line capacitor C in Fig. 4, can hardly be less than 10^{-11} farad, giving a time constant of 500 seconds. The response time may be greatly reduced by use of the feedback circuit of Fig. 5, suggested by H. L. JOHNSON[1]. With a loop gain of the order of

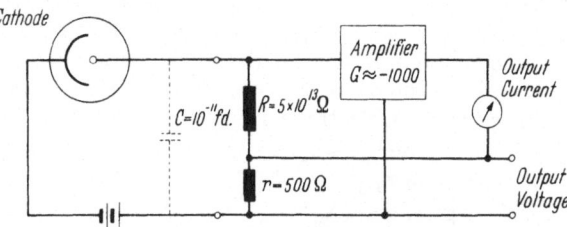

Fig. 5. Circuit for reduction of input time constant by negative feedback. With the resistances R and r shown the current amplification is 10^{11}.

$G = -1000$ (the negative sign indicates phase reversal), the input grid of the amplifier undergoes a potential change with respect to ground that is only 0.1 % of the voltage drop produced by the photocurrent in the load resistor R, and the effective input resistance across the terminals of C is reduced a thousandfold. The time constant is not reduced proportionately because the distributed capacity across R alone is not affected by feedback. Nevertheless, an effective time constant of the order of 2 seconds is possible, as shown by KRON[2].

An amplifier of this general configuration may be achieved by adding an electrometer tube preamplifier to a direct-current amplifier of the type described in Sect. 10. The refrigerated preamplifier unit containing phototube, load resistor, and electrometer tube may be separated by some distance from the main am-

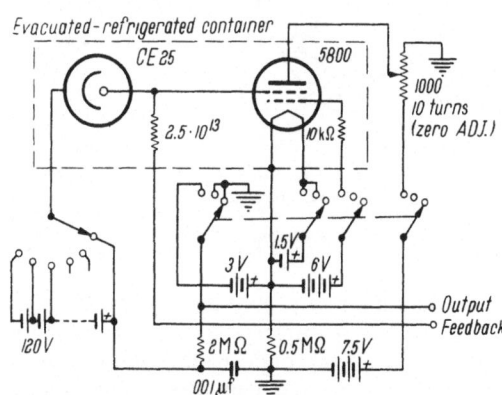

Fig. 6. Preamplifier circuit using an electrometer tetrode in a cathode follower circuit. The switching arrangement insures that the filament is heated before other voltages are applied.

plifier and power supplies. KRON's[3] battery-operated cathode follower circuit, shown in Fig. 6, provides an input stage without phase reversal, and the feedback line can be transferred from the normal point in the main amplifier to the appropriate point in the preamplifier circuit. If the circuit of the main amplifier can be modified to give a feedback voltage of phase opposite to normal, then a plate-loaded preamplifier stage with the filament at a fixed potential can be used. LITTAUER[4] has shown how an electrometer tube may be operated from a potential divider fed by a stabilized supply, rather than from batteries.

[1] H. L. JOHNSON: Publ. Astronom. Soc. Pacific **60**, 303 (1948).
[2] See footnote 3, p. 256.
[3] See footnote 3, p. 256.
[4] R. LITTAUER: Rev. Sci. Instrum. **25**, 148 (1954).

Electrometer tube circuits must be encased in an evacuated container in order to eliminate the noise arising from cosmic ray ions settling on the sensitive grid lead[1]. Since the refrigerated input circuit must in any case be in a sealed enclosure to avoid moisture condensation, a vacuum is not a serious additional requirement.

9. Multiplier phototubes. Multiplication of the photoelectrons by secondary emission from targets in the same envelope with the cathode offers an alternative method of amplifying the small initial currents encountered in the photometry of faint sources. Because of its simplicity and its wide-band, nearly noise-free amplification, the newer method has almost completely displaced thermionic vacuum tube amplifiers for the crucial first stages.

The basic phenomenon involved is the emission of a number of secondary electrons when an electron with an energy of the order of 100 volts strikes a suitably prepared target. The results of research in this field of the physics of solids have been reviewed in another volume of this Encyclopedia by KOLLATH[2]. Other reviews emphasizing aspects relevant to the use of secondary emission in current multipliers have been given by ECKART[3], by BRUINING[4], and by RODDA[5].

The most important factor in devising an amplifying system based on cascaded multiplying surfaces is the secondary yield at a given target. The two most widely used cathode materials, antimony-cesium compounds and cesium-oxide-on-silver, give a satisfactory yield. Another useful multiplying surface is a silver-magnesium alloy, modified in processing by treatment with oxygen and cesium. The yield on any of these surfaces may be 8 or more for 400 volt primary electrons. In order to avoid too high overall voltages in a cascaded system, it is usual to operate with a gain of about 4 per stage and a potential between 75 and 150 volts per stage. A total multiplication of 10^6 can then be realized with 10 stages. The intermediate multiplying electrodes are called dynodes.

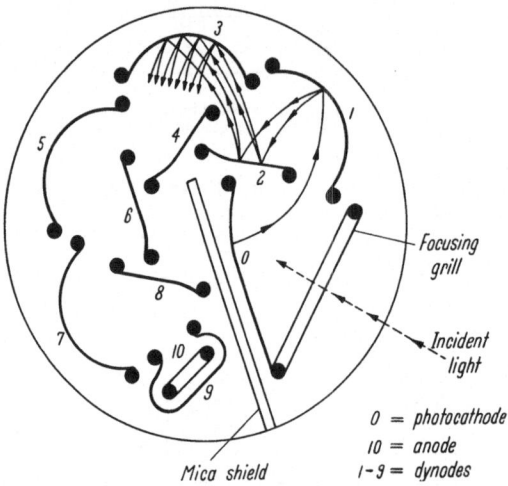

0 = photocathode
10 = anode
1—9 = dynodes

Focusing grill

Incident light

Mica shield

Fig. 7. Focusing type dynode structure for multiplier phototube. The curved electrodes produce an electrostatic field which insures impact of all secondaries on the next more positive surface. The mica shield is a barrier to prevent ion feedback between anode and cathode.

Three forms of electrode geometry have been adopted for multiplier phototubes in regular commercial production. In the type shown in Fig. 7 the electrodes are shaped to produce a strong focusing action in the electrostatic field between one dynode and the next[6]. This results in high collection efficiency, but makes the tube more susceptible to external magnetic fields. A further consequence is the sharp variation of sensitivity across the inclined cathode[7]. Both

[1] A. E. WHITFORD: Astrophys. J. **76**, 213 (1932).

[2] R. KOLLATH: This Encyclopedia, Vol. XXI, p. 232.   Berlin-Göttingen-Heidelberg: Springer 1956.

[3] Ref. [3], p. 326.

[4] H. BRUINING: Physics and Applications of Secondary Electron Emission. London: Pergamon 1954.

[5] S. RODDA: Photo-electric Multipliers. London: McDonald 1953.

[6] J. A. RAJCHMAN and R. L. SNYDER: Electronics, Dec. issue, p. 21 (1940).

[7] RODDA, l. c., p. 69.

the inclined cathode and the wire grill in front of it are unfavorable to imaging the telescope objective on the cathode (Sect. 14). The RCA 1P 21 multiplier phototube, used widely in astronomical photometry because of its compactness and its low dark current when cooled, is based on this configuration. In tubes with a semitransparent photocathode on the end of the envelope, the same multiplying structure is used, but the former inclined cathode becomes the first dynode. The RCA 6199 with an S11 cathode and the RCA 7102 with an S1 cathode are of this type.

The Venetian blind dynode structure, described by SOMMER and TURK[1], is used in many of the EMI multiplier phototubes[2]. It is shown in Fig. 8. There is a relatively weak field urging secondaries to the next dynode, and no focusing action. The efficiency is lower, and the volts per stage necessary to get a gain per stage of 4 somewhat higher than in other types. The dynode structure is quite opaque to the ion feedback discussed in a later paragraph.

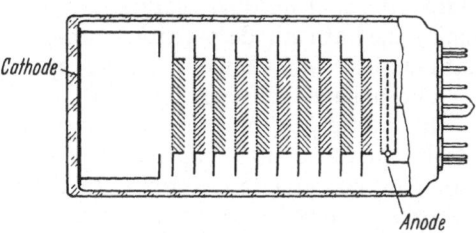

Fig. 8. Multiplier phototube employing Venetian-blind-type dynodes. A fine-mesh grid at the entrance of each structure shields secondary electrons from the field of the previous dynode, and prevents them from falling back on the surface from which they were released.

The third type of dynode geometry, the box-and-grid structure, s shown in Fig. 9. It has a weak iextraction field, is unfocused, and has a high efficiency and high gain at a given voltage.

The voltage supply to the various electrodes of a multiplier phototube is furnished by a voltage divider, usually made up out of small composition resistors soldered to the socket at the base of the multiplier

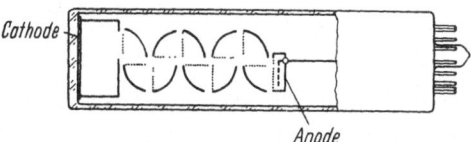

Fig. 9. Box-and-grid dynode structure for multiplier phototubes.

or in some cases to the base pins themselves. The maximum anode current in astronomical photometry is rarely allowed to exceed 1 microampere, in order to avoid fatigue. Therefore a divider current of 100 to 200 microamperes is sufficient.

Since the gain at one dynode varies as $V^{0.6}$ to $V^{0.7}$, the overall dependence of gain on voltage is as high as the 7th power of the overall supply voltage. For stable gain a carefully regulated supply is therefore required. Suitable circuits for power-line operated supplies may be found in the literature. Commercial units of adequate stability and freedom from hum are available. Hearing-aid batteries are suitable for portable or field use.

Ideally the anode current from a multiplier in the dark should arise entirely from amplified thermionic emission at the cathode. As shown by ENGSTROM[3] there are, however, leakage currents, varying roughly as the first power of the voltage rather than the 7th power. These currents can cause noise. If the leakage is not the result of contaminated insulating surfaces inside the envelope, it can be reduced by attention to the socket, the base and the glass between seal-in wires. On the 1P 21 where the anode pin is not provided with guard pins, it is often helpful to unsolder the anode wire from its base pin and to remove the pin from the bakelite base. If suitable insulation and shielding are provided for the anode wire, the leakage is reduced to that in the glass envelope alone.

[1] A. SOMMER and W.E. TURK: J. Sci. Instrum. 27, 113 (1950).
[2] Obtainable from EMI Electronics Ltd., Hayes, Middlesex, England.
[3] R.W. ENGSTROM: J. Opt. Sci. Amer. 37, 420 (1947).

A second source of noise comes from ion feedback, a term that is used to describe the bursts of charge that result when an ion formed by collision of one of the multiplied electrons with the residual gas strikes the most negative electrode, the cathode. Instability sets in when multiplication gain is high enough so that an electron leaving the cathode is almost certain to produce a feedback ion. The overall voltage at which a given multiplier breaks over into a condition of erratic, pulsating noise depends on the residual gas pressure and perhaps other factors, such as field emission. Enough external amplifier gain must be available so the multiplier may be operated well below the threshold of instability.

Fluorescence of the glass envelope under electron bombardment is another possible source of feedback noise. This can be eliminated, along with any electrostatic effects of residual charges on the glass, if the outer surface of the tube is maintained at cathode potential through a coating of aquadag[1] connected to the cathode base pin. A metal foil wrapping is an alternative.

The inherent noise in the anode current, assuming that sources extraneous to the multiplying process at the dynodes have been eliminated, is determined by the statistical spread in the number of electrons arriving at the anode for different initial cathode electrons. In the perfect amplifier, where every cathode electron produced exactly $G=m^s$ electrons at the anode, G being the overall gain and m the multiplying factor at each of the s dynodes, the fractional root-mean-square fluctuations in the anode current would be the same as those in the cathode current

$$(\overline{i_c^2})^{\frac{1}{2}}/I_c = (\overline{i_a^2})^{\frac{1}{2}}/I_a \qquad (9.1)$$

or

$$(2e\,\Delta f/I_c)^{\frac{1}{2}} = (2G\,e\,\Delta f/I_a)^{\frac{1}{2}}. \qquad (9.2)$$

In other words the shot noise in the anode current would be calculated with a unit charge G times the electronic charge. On the assumption of a Poisson distribution of the actual multiplying factor about the mean value m at each dynode, Shockley and Pierce[2] showed that the noise enhancement factor a giving the fractional increase of the anode noise over the cathode noise is

$$a = \left(\frac{m}{m-1}\right)^{\frac{1}{2}}. \qquad (9.3)$$

In a more detailed discussion which allowed for collection efficiency and differing gains at the various dynodes, Morton[3] and Sharpe[4] showed the importance of a higher than average gain at the first dynode. For this reason manufacturers' specifications often recommend a double voltage between the cathode and the first dynode.

For an average dynode multiplying factor of 4, Eq. (9.1) gives a noise enhancement factor $a=1.15$. This value is so near unity that it is ordinarily considered sufficiently accurate to assume a perfect amplifier and use cathode statistics in estimating fluctuations in counts or anode currents obtained with a multiplier phototube. The reason for this superiority of the secondary emission process as a method of amplification lies in the elimination of the load resistor in the cathode circuit. As seen in the previous section this resistor generates noise and limits the bandwidth. It is necessary in thermionic vacuum tube amplifiers, which

---

[1] A suspension of colloidal graphite obtainable from Acheson Colloids Corp., Port Huron, Michigan, U.S.A.

[2] W. Shockley and J. R. Pierce: Proc. Inst. Radio Engrs. **26**, 321 (1938).

[3] G. A. Morton: R.C.A.-Review **10**, No. 4 (1949).

[4] J. Sharpe: Nuclear Radiation Detectors. London: Methuen 1955.

require a voltage for an input signal. In a system of current amplification, however, voltage signals are not needed.

A tabulation of the characteristics of multiplier phototubes furnished by manufacturers in several different countries is given by ECKART[1].

10. Amplifiers following multiplier phototubes. The output current of a multiplier phototube as used in astronomical photometry usually lies in the range from 10^{-6} to 10^{-11} ampere. The amplification that follows such a tube will depend on the requirements of the final indicating instrument. Almost universally this is a strip-chart recording meter in which a servo-driven pen writes a complete record of all measurements. The advantages of a permanent, impersonal record are obvious. One type of recording meter, used in many observatories, is the so-called Brown recorder[2]; it is a self-balancing potentiometer that derives its error voltage from vibrator contacts at the input. The error tolerance and reading accuracy is 0.1% of full scale; the instrumental error does not exceed 0.25% of full scale. It is commonly used with an input resistor (say 10 ohms for a unit requiring 10 millivolts full scale) that converts the unit to a current meter reading 1 milliampere full scale. Another type of strip-chart recorder is the Kipp micrograph[3], which uses a d'Arsonval moving coil system as the sensing element. Currents in the microampere range are adequate for a full scale reading.

Direct-current amplifiers based on the 100% feedback principle shown in Fig. 5 have been found quite satisfactory for furnishing the output current needed to operate a recording meter. The output voltage of the amplifier is to a high approximation the same as the input voltage, and the current gain is to the same approximation simply the ratio of the input resistor to the feedback resistor. Since the gain is almost independent of supply voltage changes or vacuum tube characteristics a very stable amplification factor is obtained. The input (and output) voltage for full scale may conveniently be taken to be 0.1 volt; the drift rate can be kept low for this range. Then the multiplier output currents given in the previous paragraph require input resistors of 10^5 to 10^{10} ohms. The feedback resistor for 1 milliampere amplifier output current is 100 ohms.

The stability and noise level of such an amplifier-recorder combination is entirely determined by the multiplier phototube which feeds into it. The anode current of the multiplier has a "grain size" a millionfold greater than that of the electronic charge. When this larger unit charge is substituted in Eq. (8.4) it is seen that the shot noise will dominate if the potential drop in the input resistor of the amplifier exceeds 0.05 microvolt, a value thousands of times smaller than any encountered in practice.

A typical amplifier circuit due to WEITBRECHT[4] is shown in Fig. 10. The selector switch for the input resistor gives steps of a factor of 10 or 2.5 magnitudes. For the finer steps, the feedback resistor is changed in steps of 0.5 magnitudes or a factor of 1.585 in resistance ratio. When this attenuator reads 2.5 magnitudes, the input resistor becomes 1000 ohms, and the input voltage for full scale is 1.0 volt. The essential parts of this amplifier, including the monitoring meter, the two gain controls, and the zero adjustment, can be mounted in a small box next to the eye end of the telescope. The connecting cable to the multiplier housing can then be

[1] Ref. [3], p. 363.

[2] Minneapolis-Honeywell Regulator Co., Brown Instruments Division, Philadelphia 44, Pa., U.S.A. Similar instruments are made by Leeds and Northrup Co., also Philadelphia 44, Pa., and by other manufacturers.

[3] P. J. Kipp Zonen, Delft, Holland.

[4] R. H. WEITBRECHT: Rev. Sci. Instrum. **28**, 883 (1957); A similar amplifier was described by J. BORGMAN: Bull. Astronom. Inst. Netherl. **15**, 251 (1960).

quite short. The cables bringing in the stabilized supply voltages and carrying the amplified output signal to the recording meter have no high impedance circuits in them and can be as long as needed. It is customary, however, to provide shielding and to run the recorder wires in a separately shielded cable.

The stability and drift characteristics of the amplifier are determined by the electrometer tube in the input. Flicker noise is negligible. In order to provide first-order compensation of supply voltage changes, LITTAUER[1] used balanced electrometer tubes in the first stage. A single tube seems to give equally good results if the filament current comes from a well stabilized source, or from a battery. Since the electrometer tube requies a filament current of only 10 milliamperes, a small battery is sufficient.

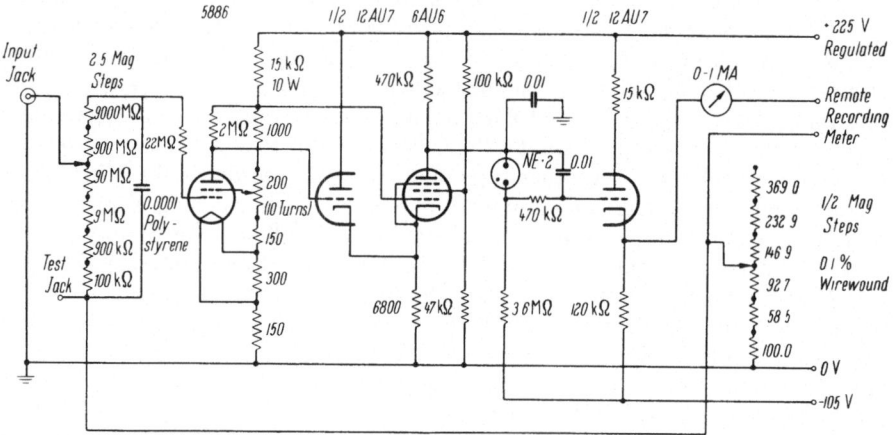

Fig. 10. D.c. amplifier to follow multiplier phototube. Full scale deflection may be obtained for input currents ranging from 10^{-5} to 10^{-11} ampere.

The residual drift appears to arise from slow changes in the properties of the emitting layer on the filament[2]. After warmup this drift does not exceed two millivolts per hour.

Commercial amplifiers covering the desired range are available[3]; they are designed to work into recording potentiometers. The gain steps are, however, considerably coarser than the half-magnitude steps found convenient by astronomers. The necessary modification to provide the desired steps in the feedback resistor is easily accomplished in these units. Wire-wound resistors having a low temperature coefficient should be used. Another modification that is necessary is the insertion of a time-constant network in the input. The capacitors used at this point must be of the low-leakage, low-soak polystyrene type[4]. The wiring and insulation in the input grid circuit must be of a high standard. Ceramic-insulated selector switches of the type used in the electronic industry have adequate insulation if kept clean and dry. Fortunately the natural power dissipation in the amplifier provides enough warmth to insure against the collection of moisture.

Since commercially-built amplifiers contain their own stabilized power supply the entire unit may be found inconveniently large to mount at the eye end of a telescope. The multiplier output may be fed through a shielded cable for a distance of 25 feet or so to the amplifier mounted at a more convenient place.

[1] R. LITTAUER: Rev. Sci. Instrum. 25, 148 (1954).

[2] J.W. GRAY: p. 730 in Vacuum Tube Amplifiers ed. by G.E. VALLEY and H. WALLMAN. New York-Toronto-London: McGraw-Hill 1948.

[3] General Radio Company, West Concord, Mass., U.S.A.; Keithley Instruments, Cleveland 6, Ohio, U.S.A.

[4] John E. Fast Co., Chicago 18, Ill., U.S.A.

Polyethylene insulated cables have adequate insulation and teflon insulated connectors are both satisfactory and convenient. Flexure and swaying of the interconnecting cable can produce "cable noise". This can be considerably reduced by a special conducting coating between the outer surface of the polyethylene and the braided shield.

For precision work the gain steps in such an amplifier must be periodically calibrated. The half-magnitude steps of the feedback chain may be tested by injecting precisely known voltages in the input in place of the photoelectric currents. Amplifiers with the feedback and grounding connection as shown in the Weitbrecht circuit of Fig. 10 work only as current meters, since the feedback is not powerful enough to overcome the very low impedance of a battery connected to the input. Consequently the test voltages are connected directly across the high resistance alone. The calibration voltages may be derived from a laboratory-type potentiometer. If a series of voltages having the ratio 0.100, 0.1585, 0.2512, etc. are injected, the half-magnitude feedback steps may be changed in a reverse fashion to give a constant indication on the pen chart recorder, preferably near the top of the scale. Any small differences remaining can be read with an accuracy of approximately 0.1%.

The ratios of the input resistors may be checked by balancing a change from one to the next against a compensating change of 2.5 magnitudes on the previously calibrated feedback selector switch. This may be done either with multiplier anode currents or with a calibration unit consisting of a high resistance and a switch in a shielded box. The resistance is put in series with a battery across the amplifier input. If the calibrator unit is connected to the amplifier test terminals that respond to voltage, the voltage drop in the external series resistor must then be allowed for. The series resistor should be 100 times larger than the input resistor being calibrated. If, however, the battery return is connected to ground and the amplifier is operated as a current meter, the feedback cancels this voltage drop and it may be neglected. Resistors comparable in value with those being calibrated are then safe to use.

11. Charge integration and pulse counting. Direct amplification of photocurrents gives an immediately visible indication of what is being measured and a continuous check on sky transparency variations and guiding. For this reason it is preferred by most observers when measuring relatively bright stars. When working near the limit of a telescope, however, the unavoidable background noise from the sky illumination may become as large or larger than the deflection from the star itself. A calculation of the fluctuations in the currents encountered in the example given in Sect. 7γ will illustrate this.

The bandwidth of the RC smoothing circuit at the input of the amplifier may be found by calculating the mean square voltage developed across the parallel RC combination by the mean square shot fluctuation current. The impedance is $Z(f) = R(1 + 4\pi^2 f^2 R^2 C^2)^{-\frac{1}{2}}$. The contributions from frequencies above $f = 1/(2\pi RC)$ decline rapidly, and integration over all frequencies yields the result

$$\Delta f = \frac{1}{4RC}. \tag{11.1}$$

For small currents the time constant $t = RC$ will be made long and 5 seconds may be assumed. Then, according to Eq. (8.1) the root-mean-square noise current for a sky current of 700 electrons sec^{-1} = 1.1×10^{-16} amperes amounts to 1.3×10^{-18} amperes or 1.2% of the total amount. The star signal of 15 electrons sec^{-1} amounts to 2.4×10^{-18} amperes or 1.8 times the root-mean-square noise fluctuations. This is about half the peak-to-peak noise. As would be expected from the calculations in the previous section, repeated intercomparisons of the star area and sky area are needed to obtain a measurement of acceptable accuracy.

The accuracy achieved in a given time is not as high as calculated from the equations of Sect. 7γ, which assumed perfect integration, or counting of photo-electrons. If the observation time for the direct-current amplifier be taken as $t = 4RC = 1/\Delta f$, then the precision or signal-to-noise ratio for the amplifier is $I_c/(2e\,I_c\,\Delta f)^{\frac{1}{2}} = (I_c\,t/2e)^{\frac{1}{2}}$. This is poorer by a factor $2^{\frac{1}{2}}$ than the ratio $(q\,n\,t)^{\frac{1}{2}} = (I_c\,t/e)^{\frac{1}{2}}$ given by the simple formula for statistical fluctuations. The reason for the difference is the fading memory of the RC circuit for events in the early parts of the exposure interval. Many observers consider that a longer exposure of, say, $16RC$ gives not just 4, but something like 8 independent judgments of the deflection, and the factor of $2^{\frac{1}{2}}$ is thereby overcome. There is still an element of judgment that enters into drawing an average line through a ragged pattern of fluctuations, and in the end the measurement is not entirely impersonal.

Charge integration and pulse counting offer two practical methods of realizing the full statistical accuracy calculated in Sect. 7. With either system the end result is an instrumentally produced number which leaves no room for judgment. The integrator units are simpler electronically and have more flexibility and a wider range of linear operation. The method blindly integrates "bad" (i.e. non-statistical) noise such as ion bursts. These events are subconsciously disregarded in the steady deflection method and have low weight in a pulse counter. The pulse counter discriminates against leakage noise and other sources of output current not originating in the cathode emission.

The principle of the electronic integrator is a simple one. A low-leakage condenser is substituted for the input resistor in a feedback amplifier of the type shown in Fig. 10. Whatever the charge that flows into the condenser from the anode of the multiplier, the 100% feedback in the amplifier will insure that the potential of the grid of the first amplifier tube will remain unchanged. The terminal of the condenser connected to the feedback line will therefore change in proportion to the accumulated charge, and the current through the feedback resistor will be a measure of this charge. An amplifier of this type may be made into a combination current-meter and integrator by providing an extra terminal for a capacitor on the selector switch for the input resistors. A suitable discharge connection must be included. When the amplifier is integrating, the recording pen draws a slanting line on the chart, the slope of which is a measure of the input current.

Weitbrecht[1] has given full details of an integrator unit of this type for use in routine photometry at the telescope. By choice of input capacitor, feedback resistor, and integration time, a very large range of currents may be measured with precision. The system is well adapted to multiplex photometry where the charge accumulated simultaneously on several condensers during a given exposure may be quickly read off and recorded on the strip chart potentiometer through a programming unit that samples and discharges each condenser in turn. In this case the indication on the chart is a level line.

The technique of pulse counting has been highly developed because of its widespread application in nuclear physics. Pulse amplifiers and counting scalers have therefore become standard commercial instruments. These instruments are designed to accept the pulses from the multiplier phototubes that are coupled to the scintillation crystals used as detectors of nuclear radiations, and may therefore be applied to astronomical photometry with no modification. Since the

---

[1] R.H. Weitbrecht: Rev. Sci. Instrum. **28**, 883 (1957).

electronic details have been fully described in the literature[1], only a resumé of the basic principles will be given here.

Each cathode photoelectron produces a charge of approximately 10^6 electrons or 1.6×10^{-13} coulombs at the anode of the multiplier, the time spread of the pulse being less than 10^{-8} sec. There is no strong reason to operate at an average pulse rate higher than 10^4 counts per second, since a current of 1.6×10^{-9} amperes is easily and precisely measurable with a standard direct-current amplifier. The electromechanical impulse counter that follows the scaler unit should not be called upon to register more than 10 counts per second. This means that the division ratio in the scaler should be at least 1000.

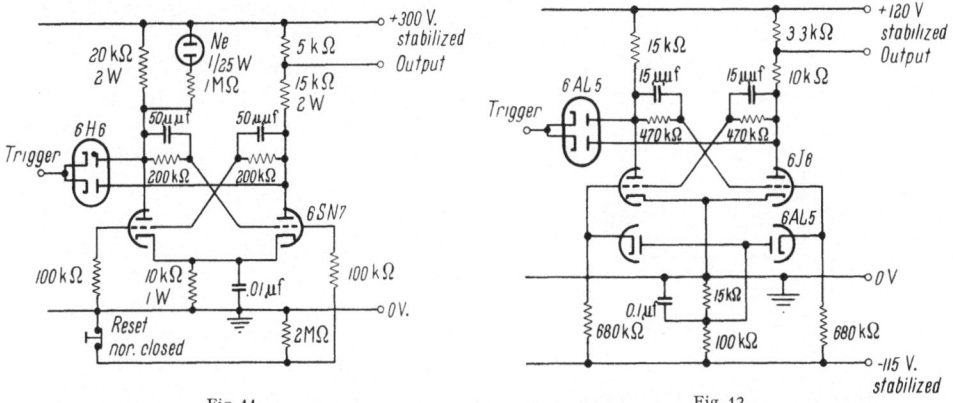

Fig. 11.   Fig. 12.

Fig. 11. Standard scale-of-two circuit, or "flip-flop". The circuit will remain in a stable condition, with either the right or left triode conducting, and the other cut off. A negative trigger transfers the circuit to the opposite stable condition. The neon glow lamp indicates that the left triode is conducting.

Fig. 12. Fast scale-of-two circuit. A diode catches the grid of the triode going into the non-conducting state at a potential just below cutoff, thereby reducing the recovery time.

The time resolution of the scaler has an effect on the linearity of the response at high counting rates, since pulses arriving during the "dead time" will be missed. If τ is the dead time, defined as the minimum time between resolved pulses, and n is the average number of randomly spaced pulses per second, then a total counting time $n\,\tau$ will be lost in each second and the observed counting rate n_1 will be

$$n_1 = n\,(1 - n\,\tau) \qquad (n\,\tau \ll 1). \tag{11.2}$$

To keep the fractional counting loss below 1%, then, the time resolution of the scaler should be 1 microsecond for a maximum counting rate of 10^4 per second. The standard scale-of-two circuit[2] used in all scaler circuits, is a bistable multivibrator, sometimes called the Eccles-Jordan circuit. With the circuit values shown in Fig. 11, it has a resolving time of 5 microseconds. Corrections for counting losses not exceeding 5% are easily introduced, and will in fact rarely be needed. It is not difficult, however, as shown by WOODBURY and HOLDAM[3], to reduce the dead time to less than 1 microsecond. A simple modification of the circuit of Fig. 11, given by YATES[4], is shown in Fig. 12. The resolving time is 0.6 microsecond. Beyond the first two stages of the scaler a lower time resolution is of course acceptable.

---

[1] Ref. [6].

[2] Ref. [6], p. 209.

[3] R.B. WOODBURY and J.V. HOLDAM: Chap. 17 of Waveforms ed. by B. CHANCE et al. New York-Toronto-London: McGraw-Hill 1949.

[4] G.G. YATES: Ref. [5], p. 99.

The scale-of-two dividers may be cascaded to give a total division factor of at least $2^{10} = 1024$, or they may be arranged in three subunits, each of which becomes a scale-of-ten divider. The feedback connections to accomplish this have been given by ELMORE and SANDS[1]. The same authors have also described the output circuits needed to drive various electromechanical registers.

Since all scaling circuits require a pulse height of at least several volts to trigger a count, a pulse amplifier following the multiplier is needed. A typical section of such an amplifier is shown in Fig. 13. This section, called the preamplifier, would be mounted very close to the anode of the multiplier; if a connecting cable is needed it should be less than a foot long. If it be assumed that the total distributed capacity of the input circuit is $30\ \mu\mu F$ then the average pulse of 10^6 electrons will produce a negative step-function of 50 millivolts on the first grid. At some stage in the amplifier chain a short time constant which differentiates the step function must be introduced in order to make the output to the scaler a pulse that quickly returns to the base line. As shown by ELMORE and SANDS[2] it is preferable to have only one such differentiation in order to avoid an undershoot which effectively reduces the output pulse height of closely following pulses. The other time constants in coupling circuits are kept long in comparison.

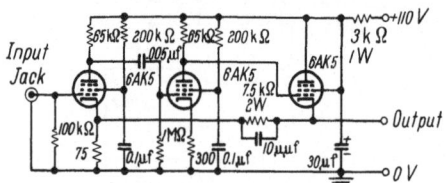

Fig. 13. Preampliifer for pulse counter. Negative feedback to the first cathode gives a stable voltage gain of 100.

Since the input step function from the multiplier is well above amplifier noise, the differentiation can come directly at the input, and is so shown in Fig. 13. A decay constant of 3 microseconds is satisfactory for the resolving time of the scaler of Fig. 11. The gain of the preamplifier is very nearly equal to the feedback ratio, and would be 100 for the resistors shown. Since it may be preferable from the point of view of multiplier noise to use less multiplication, two such amplifier sections in cascade would be used to obtain the output pulse to the scaler, which should have an available maximum value of the order of 50 volts.

The pulse amplifier must not overload on any pulse or succession of pulses from the anode. If the grid of one of the stages goes positive, the charge thereby deposited on the coupling condenser must leak off to restore normal response. This is less of a problem in astronomical photometry, since the pulses vary in height only by the normal statistical spread in the multiplication factor, and the very large bursts produced by scintillation crystals do not occur. A non-overloading pulse amplifier of advanced design has been described by FAIRSTEIN[3].

The standard scale-of-two circuit of Fig. 11 will operate on a negative input pulse of a few volts, the exact value depending on the ratio of the cross-coupling condenser to the distributed capacity in the grid circuit of each tube. A potentiometer following the pulse amplifier may be adjusted to give reliable counting while still avoiding noise triggering. A much more precise and stable choice of countable pulses is preferable, however. A discriminator stage which gives a standard output pulse for all inputs exceeding an accurately predetermined height is described by ELMORE and SANDS[4]. It would be inserted between the pulse amplifier and the scaler.

[1] Ref. [6], p. 211.
[2] Ref. [6], p. 130.
[3] E. FAIRSTEIN: Rev. Sci. Instrum. 27, 475 (1956).
[4] Ref. [6], p. 203.

The proper setting of the discrimination level is a matter of opinion. If it is set too low some spurious pulses originating in the first dynode, or in multiplier noise will be counted. If set too high to avoid this danger, some true cathode pulses on the lower side of the statistical size distribution will be rejected and the effective quantum efficiency goes down. The usual procedure is to compare the counts as a function of discriminator setting, first with the cathode dark, and then illuminated to give an easily handled counting rate. The excess of the latter over the dark count usually shows a plateau as the setting is advanced to admit more pulses, and a setting near the lower edge of this plateau gives a good operating point, insensitive to fluctuations in the multiplier supply voltage.

YATES[1] and BLITZSTEIN[2] have given full descriptions of pulse counting photometers.

12. Amplifiers for photoconductive cells. The chopping frequency with photoconductive cells is usually of the order of 1000 cycles per second, and is imposed by a multi-vane toothed wheel which interrupts the signal beam at a point near where it comes to a focus. If the width of the beam is an appreciable fraction of the tooth spacing, rounding of the wave and a reduced modulation factor will result. The frequency is not very critical, but low frequencies, such as 30 cycles per second, are unfavorable because of increasing current noise in

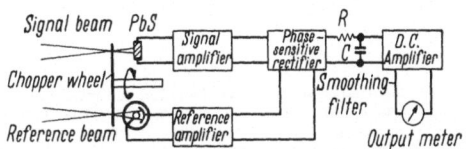

Fig. 14. Block diagram of chopper and phase-sensitive demodulator system for use with photoconductive cells.

the semiconductor. Cooled photoconductive cells have a longer internal time constant[3], and must be used at a lower chopping frequency. The resistance of a cell increases considerably when cooled, and distributed capacity in the input circuit may lower the maximum frequency which can be amplified without a large phase shift. The feedback circuit of Fig. 5 can be used to improve the frequency response under these circumstances. Chopper frequencies in the range 100 to 200 cycles are satisfactory for these conditions where the cell current is low.

The chopper frequency is a carrier frequency, modulated in amplitude by the intensity variations introduced by the sequence of shutter and filter changes needed to make a measurement. In astronomical applications these will rarely require more than 1 cycle per second of bandwidth. The amplifier must bring the chopper signal up from an amplitude that may be less than a microvolt to a suitable level for demodulation and transfer to a recording meter. The amplifiers described by KUIPER, WILSON, and CASHMAN[4] and by BROWN[5] used inductance-capacity tuning to confine the amplification to a narrow band centered on the carrier. The demodulation was by a diode rectifier, which is not strictly linear; the Wilson amplifier was found by MOHLER et al.[6] to depart quite appreciably from linearity near zero signal.

A phase-sensitive synchronous rectifier, called a homodyne in England, offers a preferable demodulation scheme. A block diagram[7] is shown in Fig. 14. The rectifier acts as a double-pole reversing switch which turns over the negative

---

[1] G. G. YATES: Monthly Notices Roy. Astronom. Soc. London **108**, 476 (1948).

[2] W. BLITZSTEIN: Ref. [5], p. 64.

[3] Ref. [2], p. 69.

[4] G. P. KUIPER, W. WILSON and R. J. CASHMAN: Astrophys. J. **106**, 243 (1947).

[5] D. A. H. BROWN: J. Sci. Instrum. **29**, 292 (1952).

[6] Ref. [5], p. 280.

[7] Ref. [1], p. 431; see also A. E. WHITFORD: Astrophys. J. **107**, 102 (1948). — R. H. DICKE: Rev. Sci. Instrum. **17**, 268 (1946).

swings of the chopper signal, so that the output to the meter is a d.c. signal. Such a system is quite linear near zero and preserves the sign of the chopper modulation. The latter is important in astronomical work, where the chopper vane

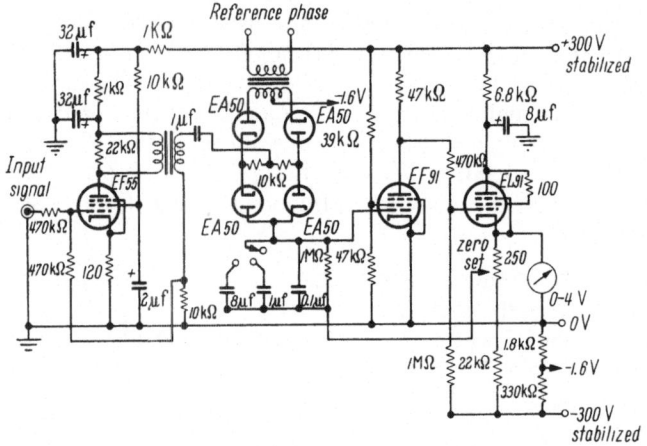

Fig. 15. Phase-sensitive (homodyne) rectifier and smoothing circuit. During one half of the square-topped reference wave, the upper diodes are conducting, clamping the right plate of the coupling condenser to the bias source; the lower diodes are non-conducting. On the other half of the chopper cycle the signal wave, either positive or negative, is impossed on the RC smoothing circuit at the input of the d.c. amplifier. Current feedback in the a.c. input stage prevents overloading on noise peaks.

may be warmer than the sky; in turning from sky to star, the sign of the modulation can reverse.

It can be shown that the bandwidth for noise is entirely determined by the time constant inserted between the synchronous rectifier and the indicating instrument. This may take the form of a 1000 μF electrolytic condenser in a relatively low impedance meter circuit. A more flexible scheme uses a high impedance RC combination, with choice of condenser. Polystyrene low-soak capacitors are advisable. To drive a recording meter from this high impedance circuit a 100% feedback d.c. amplifier of the type described in Sect. 10 would be inserted. GRAY[1] has described a convenient circuit.

CHANCE[2] has given a general discussion of phase-sensitive demodulators. A design, due to D.A.H. BROWN[3], which includes the smoothing circuit and meter-matching amplifier is shown in Fig. 15. The signal for switching the phase comes from a generator opposite the signal beam on the chopper wheel. This may be an ordinary phototube or possibly a germanium photojunction. The latter is capable of acting as the switch without further amplification.

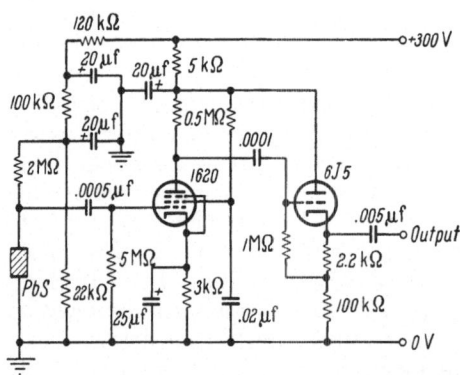

Fig. 16. Preamplifier for lead sulfide photoconductive cell. The voltage gain is about 300, and the cathode-follower output tube is capable of driving a long connecting cable which transmits the signal to the main amplifier.

[1] J.W. GRAY: p. 480 in Vacuum Tube Amplifiers ed. by G.E. VALLEY and H. WALLMAN. New York-Toronto-London: McGraw-Hill 1948.

[2] B. CHANCE: Waveforms, p. 511. New York-Toronto-London: McGraw-Hill 1949.

[3] Ref. [1], p. 433.

The amplifier which precedes such a demodulator need not be tuned and may be of the standard resistance-capacity coupled type. To avoid noise overloading in the later stages, shunt and interstage capacities may be chosen to confine the pass band somewhat without introducing serious phase shifts. Loss of the odd harmonics in the square wave does not affect the performance, since these frequencies contain noise also. A gain control before the final stage is helpful in preventing overloading.

The only critical part of the amplifier is the first stage or preamplifier section. The circuit used by KUIPER, WILSON and CASHMAN[1] is shown in Fig. 16. The photoconductive cell and a load resistor of approximately equal size are fed by a very well filtered supply. The resistor is wirewound, or of a low-noise film type, to avoid the current noise in carbon composition resistors. The tube noise in "quiet" pentodes is equivalent to the Johnson-Nyquist noise in a resistance of 50000 ohms or less and will therefore be negligible in comparison with the same type of noise in the photoconductor circuit. Eq. (8.3) gives a thermal noise of about 0.1 microvolt cycle$^{-\frac{1}{2}}$ sec$^{\frac{1}{2}}$ for a 0.5 megohm input circuit. Cell noise should of course exceed this; a properly functioning amplifier will be quieter with the cell supply voltage turned off.

Because the photoconductive cell and preamplifier are close to the chopper and chopper motor, microphonics in the preamplifier tube may be troublesome. Cushioning of the motor and of sensitive subunits may be necessary and the wiring in the preamplifier should be short, well-supported, and non-resonant to vibration.

IV. The photometer on the telescope.

13. Refrigeration of phototubes. Since the dark emission of electrons from photocathodes can be reduced to an almost negligible level by lowering the temperature to the vicinity of $-80°$ C, solid CO_2 (dry ice) is the refrigerant generally used. Its heat of sublimation is 123 cal g^{-1} or 188 cal cm^{-3}, much higher than the equivalent figures for the heat of vaporization of liquid nitrogen: 48 cal g^{-1} and 38 cal cm^{-3}. Dry ice is therefore an excellent refrigerant from the viewpoint of space and weight requirements.

The container which houses the phototube and the refrigerant must be designed with due regard to the following principles:

1. The thermal transfer between the cathode and the refrigerant must be efficient enough so that the cathode reaches a temperature very close to $-78°$ C. The phototube can be nearly but not quite surrounded by a metal sheath externally in contact with dry ice. The warming paths via conduction, convection, and radiation, particularly along the tube which admits light, should be minimized. The wires for the electrical connections must be kept thin.

2. The phototube and all parts of the electrical insulation that become cold must be in an airtight container, in order to avoid the drawing in and condensation of moisture.

3. The window which admits light must be kept warm enough so that it does not collect dew or frost. The insulation should be adequate so that the external walls do not get cool enough to sweat under ordinary conditions.

---

[1] Ref. [4], p. 267.

The thermal conductivity of some of the better insulating materials may be listed for reference:

Material	Thermal conductivity k in units of 10^{-5} cal cm^{-1} sec^{-1} deg^{-1}
Balsa wood	8.4
Hair felt	8.6
Polystyrene foam . .	8.6
Silica aerogel[1]	4.4
Polyether plastic foam[2]	4.6

The external wall of the insulated container will be cooled below ambient until the transfer by convection and radiation supplies the heat conducted inward. An approximate prediction can be made from the rule that this transfer will be[3] 1.0 to 1.5×10^{-3} cal cm^{-2} ΔT^{-1} (°C). Setting the conduction loss equal to the heat transfer at $\Delta T = 5°$ C, corresponding to the dew point depression at a relative

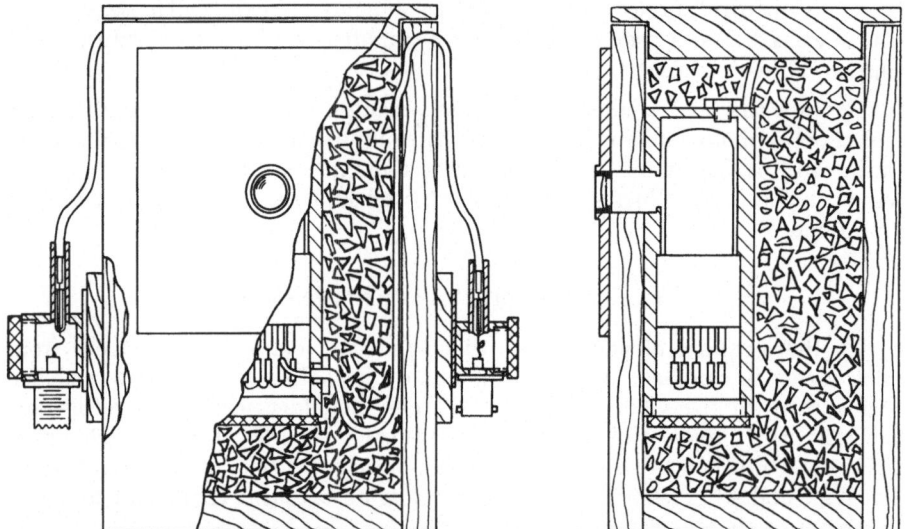

Fig. 17. Refrigerated housing for 1 P21-type multiplier phototube. The connector for the high voltage supply is on the left, and the resistors forming the voltage divider for the dynodes are soldered to the basepins of the multiplier. The connector for the anode cable is on the right.

humidity of 70%, we find that for a temperature differential of 100° C the thickness l of the insulating layer must be $2 \times 10^4 k$ (cm). Thicknesses from 1 to 2 cm are adequate.

The tube admitting light to the cathode is best made of metal in order to maintain an airtight seal. The nickel-chromium alloy inconel[4] has the very low conductivity of 0.036 cal cm^{-1} sec^{-1} deg^{-1} and excellent mechanical properties. By making the walls of the tube only 0.005 inches thick the thermal loss along the tube may be made a small fraction of the loss through the insulated walls of the refrigerant chamber.

A simple but effective mounting for photomultiplier tubes of the 1 P 21 type is shown in Fig. 17. It was developed by KRON and has been extensively used at Lick Observatory. The insulating walls are made of balsa wood $\frac{9}{16}$ inch thick,

---

[1] Santocel A, Monsanto Chemical Co., St. Louis, Mo., U.S.A.
[2] American Latex Products, Hawthorne, Calif., U.S.A.
[3] Int. Critical Tables, Vol. 5, p. 234. New York-Toronto-London: McGraw-Hill 1929.
[4] International Nickel Co., New York, N.Y., U.S.A.

treated with paraffin to make them impervious to moisture. The multiplier is sealed in a cavity in an aluminum block. Long, thin metal tubes running through the dry ice chamber bring in the high voltage to the potential divider mounted on the base of the tube, and carry away the anode current to a connector on the outer wall. The wires inside the tubes are teflon-insulated. The energy dissipated in the divider can be made only a few hundredths of a calorie sec^{-1}, a negligible factor in the temperature of the cathode. The airtight seal of these metal connections to the aluminum multiplier block is made with an epoxy cement on a threaded bushing or plug. The seal at the external housing for the electrical connectors is maintained by means of a wire sealed in a glass tube, this tube being waxed to the surrounding metal.

The light is admitted through a thin-walled inconel tube chosen to be of such a length that the Fabry lens which forms the light window will produce a spot 3 to 5 millimeters in diameter on the cathode. Epoxy cement forms the seal between the various parts of the light tube: the aluminum block, the inconel tube, the brass lens mount and the lens itself. The lens mount is sufficiently cooled by conduction so that condensation must be prevented by warming it through contact with its surroundings. Close thermal contact between the lens mount and the heavy metal parts of the photometer base is usually adequate, but sometimes a small amount of electrical heat must be added to the adjacent metal parts.

A somewhat more elaborate refrigerated container based on all-soldered construction and a liquid-tight refrigeration space is shown in

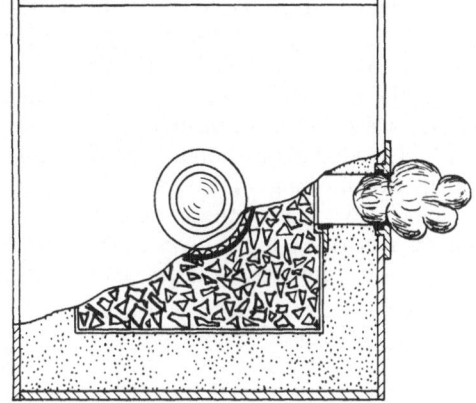

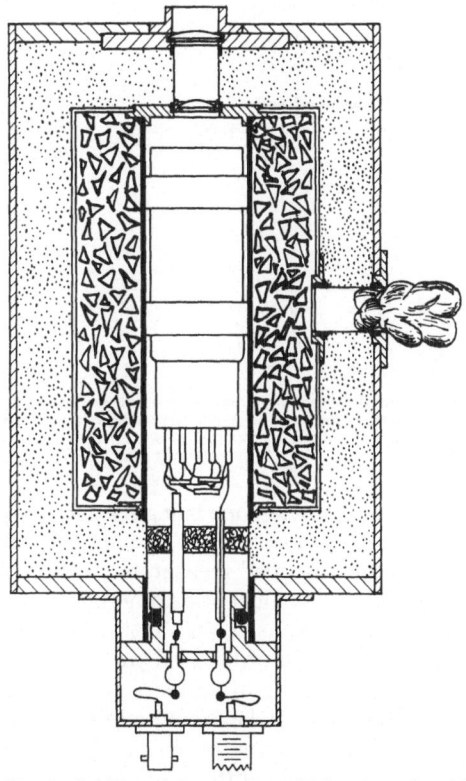

Fig. 18. Refrigerated housing for multiplier phototube with semitransparent cathode. Both the inconel tube around the multiplier and the surrounding copper box are made leak-proof by hard soldering with a brazing alloy. The inner chamber is supported on three very thin inconel tube sections which pass through the insulating layer. A heavy felt washer breaks up convection around the electrical wires. The voltage supply and the anode wires are led out of the airtight space on glass bead hermetic seals. A thin rubber gasket under the outer lens and an O-ring gasket in the electrical plug are also a part of the sealing system.

Fig. 18. It is designed for multiplier phototubes with semi-transparent cathodes. Phototubes may be quickly changed without elaborate disassembly and unsealing, a feature which does make an additional thermal leak. The liquid-tight refrigerant space makes it possible to use liquid nitrogen and a narrow filling neck is therefore necessary. Pulverized dry ice can be introduced through this neck by means of a funnel. Acetone or isopropyl alcohol added to the pulverized dry ice gives excellent thermal contact with the walls around the phototube and counteracts the tendency of dry ice to form an insulating cushion of gas between itself and the container wall. The electrical terminals are brought out through glass bead seals[1] soldered to the exit bushing. The seals to the light window and to the exit bushing never become cold and may therefore be made with rubber gaskets. The two-component Fabry lens at the input gives better imaging over a wide field, and the cold inner component provides a baffle against radiation and convection along the thin-walled tube which leads to the warm outer wall around the outer component. Electrical heating is provided around the outer lens and around the bead seals.

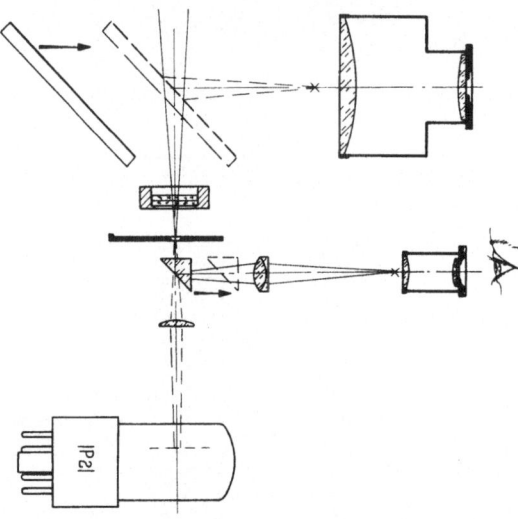

For structures which require gasketing that must withstand the cycling between room temperature and $-78°$ C soft metal gaskets are satisfactory. Rubber gaskets of the "$\bigcirc$-ring" type are less desirable because of the loss of resiliency of the rubber at dry ice temperature. Gasket manufacturers are able to furnish silicone-type products which remain resilient at this temperature.

Fig. 19. Optical schematic of viewing arrangements for an $f/15$ beam from telescope objective. The upper large-field eyepiece is used for identification and coarse centering, the lower for precise centering of the star in the diaphragm.

14. Optical arrangements.

The optical arrangements at the focal plane of a telescope being used for photoelectric photometry must be designed to fit the focal ratio of the telescope. The complexity of the arrangements depends on whether the intended measurements push the technique to the point where seeing-limited operation must be attained. If this be the case, then the smallest possible diaphragm must be used, the optical and mechanical performance of the telescope must be of a high order, and there must be provision for controlling the guiding of the telescope by an eyepiece centered on an auxiliary star. These refinements will in general not be worth while on a telescope of moderate size because the greater light-gathering power of a larger telescope would make the same measurements a routine operation. If poor mechanical performance of the telescope or the chromatic aberrations of a refractor require the use of a diaphragm of the order of 1 minute of arc or larger, then the photometry will become sky-limited at a much higher level than would be the case with seeing-limited diaphragms.

When the photometry is to take place at the cassegrain focus, the focal ratio of the telescope is usually in the range between $f/11$ and $f/18$. With the relatively large scale of the star field and the relatively slow convergence of the star beam, the optical and mechanical arrangements do not have to meet the exacting tolerances that would be the case in an $f/5$ beam. Fig. 19 shows an optical schematic of the various parts under these conditions. The diaphragm in the focal plane

---

[1] Hermetic Seal Corp., Newark 7, N.J., U.S.A.

is a carefully made hole in an interchangeable slide. By suitable champfering the edges of the hole may be made nearly a knife edge in the actual focal plane. The steps between the various hole sizes are usually in the ratio 1.4 to 1.6. Alternatively there may be a single long slide with a number of holes each brought on center by detent action. The various holes must be concentric with each other to a small fraction of the diameter of the smallest. By partial withdrawal of the diaphragm a field of the order of 5 minutes of arc should be available for inspection.

A sliding right-angle prism directly behind the diaphragm reflects the beam to one side where a relay lens re-images the diaphragm in the focal plane of the eyepiece. Illuminated cross hairs in the focal plane of the eyepiece may be adjusted for optical coincidence with the center of the diaphragm, thus providing exact centering. Because of the danger of slight shifts in the prism, however, it is preferable to see the edges of the diaphragm. Illumination by scattered light on the objective is often used though the brightening of the field makes it difficult to see faint stars. Tangential illumination on the back of the diaphragm from a small red bulb makes the edges visible and leaves the star aperture dark. The overall power of the telescope through this optical system should be adequate to make a 1 second seeing disk easily visible as an extended object.

When the sliding prism is removed from the beam, the light of the star goes on to the phototube housing. At the entrance of the phototube housing is a Fabry lens which images the objective of the telescope on the sensitive cathode. In general this illuminated spot should be as large as the most sensitive area of the cathode will permit, in order to make the output insensitive to small shifts due to flexure and changing alignment.

In an $f/15$ beam a simple fused quartz lens is entirely adequate for the quality of imaging that is required. Provided that the lens is large enough to contain the beam for the largest diaphragm opening, the illuminated area on the cathode is independent of the size of the diaphragm or the position of the star within it.

A radioactively-excited test source is mounted in an extra socket on the filter slide in order to provide a continuously available check on the overall response of the photometer that is independent of the vagaries of the sky. The color and brightness of the source can be regulated by the choice of the materials in the mixture of phosphor and exciting agent. A surface brightness of the order of 1 microlambert, or ten times that of the night sky, is convenient. Excitation by β-particles as from the weakly radioactive C^{14}, produces a glow that is free from the scintillations that accompany α-excitation. Suppliers of radioactive materials can furnish an encapsulated button which prevents escape of radioactive particles. The size of the star diaphragm regulates the amount of light reaching the cathode. In a more sophisticated design, an auxiliary lens makes the luminous surface appear to coincide with the telescope objective, and to have the same angular extent. Thus the test source illuminates exactly the same cathode area as does a star.

If the setting circles and the finder of the telescope are inadequate to identify the field being observed and to place the desired star within the field of the viewing prism, a removable mirror some distance ahead of the focal plane may be used to reflect the field into a low-power eyepiece, thus giving the full aperture of the main telescope for inspection and preliminary centering. Once the star is in the center of the diaphragm the driving mechanism of the telescope must be trusted to keep it there. Inspection at time intervals of the order of 2 minutes may be necessary for stars which require small diaphragms. It is preferable to use a guide star picked up by a maneuverable prism and eyepiece just ahead of the focal plane. If the mechanical rigidity of this mechanism when clamped is adequate, the star may be observed for long periods without interruption for checking or centering.

In order to make possible rapid and precise offset to a sky area and return to the star, the whole diaphragm and phototube assembly is mounted on a cross slide. A crank handle with a detent for whole revolutions is a convenient adjunct.

When the photometry is to take place at the prime or Newtonian focus of a large reflector, the rapid convergence of an $f/4$ or $f/5$ beam requires some modification of the optical arrangements. The Fabry lens system must be placed as close to the focal plane as is practical, owing to the rapid divergence of the beam beyond the focus. A 2-component lens with a weak first component permits a thicker layer of insulation on the side of the phototube from which the light enters; an example is shown in Fig. 18.

With such an arrangement, either a hinge or a slide mechanism must be provided to get the refrigeration box to one side for inspection of the centering of the star in the diaphragm. The inspection eyepiece is mounted on an arm which swings aside when the photometer box is again to be put on axis. By suitable steps and flanges, or by the use of black velvet gaskets, the light path through the Fabry lens system can be made sufficiently leak proof.

Since this arrangement puts the filter slide in a strongly convergent star beam, it is highly desirable that all filters be of the same thickness in order to preserve a common focus. A clear piece of fused quartz or glass may be cemented to the thinner ones in order to bring them up to equality with the thickest.

Fig. 20 shows the coordinate offset mechanisms needed when seeing-limited photometry is to be attempted. The whole base for the photometer is best mounted on a graduated ring, so that the x- and y-coordinates may be set to any position angle with respect to the celestial coordinates. With such a ring, the offset may be made in polar coordinates, and only one coordinate of transverse motion on the cross slide bearing the diaphragm and phototube is necessary. The desired sky comparison area is selected from photographs of the area in question and the cross motion set up to provide the desired offset in millimeters. Adjustable mechanical stops to limit the motion to this distance are a convenience and remove the need for any light to read scales during this shift.

The auxiliary eyepiece for the guide star is mounted on transverse ways giving both x- and y-motion. Differential motions up to at least 100 seconds of arc must be readable to a few tenths of a second of arc. It is desirable to have a pickup prism which can reach into the part of the star field near the axis where the coma is small.

Since the photometer will, under seeing-limited conditions, often be called upon to measure stars that are too faint to be seen with the eye, offset coordinates determined from a previous photograph must be used to center the star in the diaphragm. Unless the photograph of the region includes a star trail to establish the orientation of the coordinates, it will be necessary to measure the x- and y-coordinates of objects on it with respect to an arbitrary direction, such as the edge of the plate. The x- and y-offsets of the object to be measured are determined with respect to a nearby star, bright enough to be seen with the eye. Short offsets, usually not over 50 seconds of arc, are of course to be preferred. The coordinates of other bright stars in the area must also be measured in order to establish the orientation.

The observer at the telescope first checks that the orientation of the photometer is such that the coordinates of bright stars measured on the telescope agree with those measured on the plate. Usually a small rotation must be introduced to get the correct orientation. For this check the x- and y-coordinates which move the diaphragm are convenient; the guiding eyepiece is held fixed on any

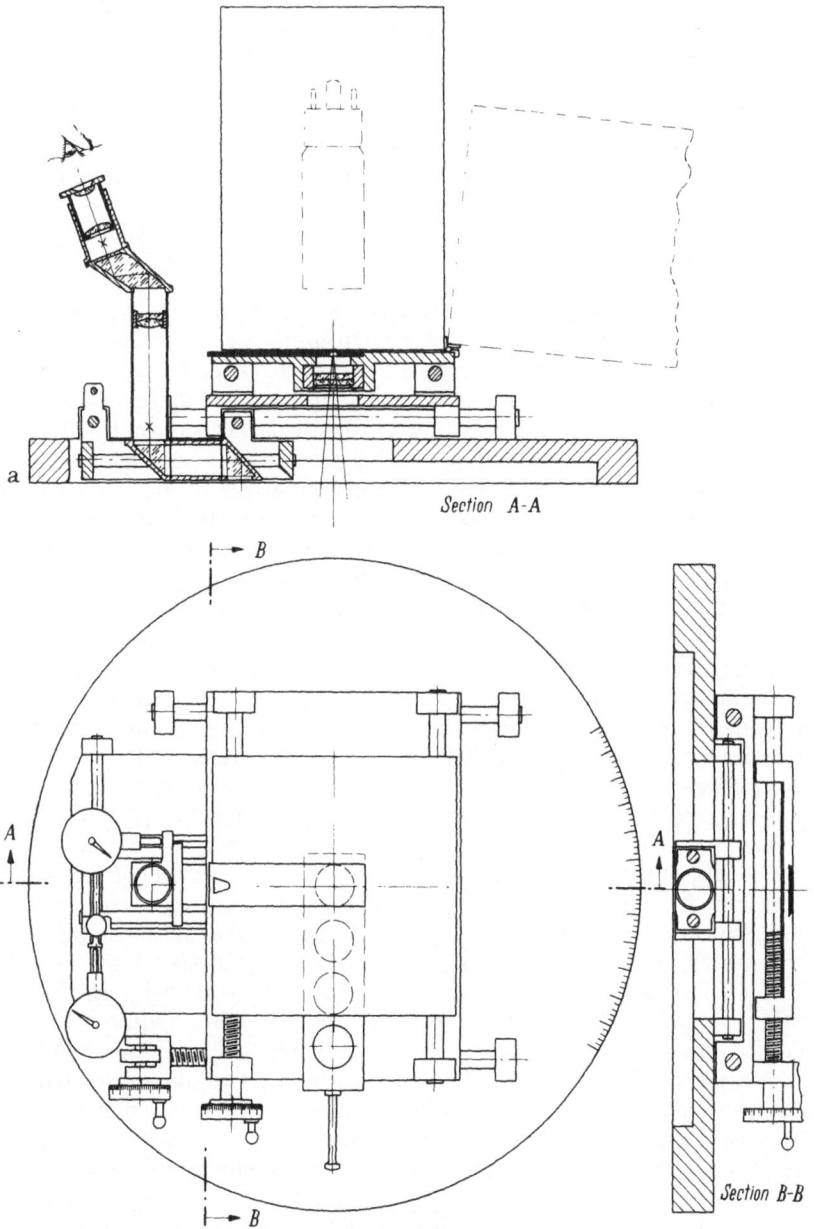

Fig. 20. Photometer base with graduated cross slides and provision for offset guiding. The offsets needed to bring an invisible object into the center of the diaphragm are read on machinist's dial gauges. The clamps and fine adjustment screws on the guiding eyepiece carriage are not shown. The refrigerated multiplier housing folds back for inspection of the field with a large eyepiece. The eyepiece and supporting bracket are not shown.

convenient star. After the orientation adjustment is satisfactory, the reference star near the object to be measured is carefully centered in a small diaphragm and a guide star of suitable brilliance is selected by scanning the available off-axis area. Next the x- and y-coordinates of the guiding eyepiece are shifted by

18*

the measured offset distances. Then when the whole telescope is moved to bring the same guide star back on the cross hairs, the object to be measured moves into the center of the diaphragm. The accuracy of the procedure may be tested by using two easily visible stars. Clamps for locking the guiding eyepiece in its final position are a good safety measure if the program of exposures is to be a long one.

Offset mechanisms which must work on a moving telescope require more care in their construction than is necessary with similar mechanisms that work with a constant direction of gravity. If machinist's dial gauges are used to measure the actual displacements, the ways and screws need serve only as guides and actuators. Then lost motion and variable loading of the screws does not require such careful attention.

15. Differential and multiplex methods. The sequence of exposures with a photoelectric photometer is always chosen to minimize the variation of factors that affect all measurements of light alike, that is, to make the measurements as differential as can be. In a series of measurements through several color filters on a bright star a very small percentage of them will be spent on the dark or sky and the maximum on the uninterrupted sequence of star exposures. The sequence of filters is passed in front of the aperture first in direct and then in inverse sequence in order to provide first-order cancellation of any drift in sky transparency, amplifier gain, or cathode fatigue. If, on the other hand, the sky background is a considerable fraction of the total measured, the star brightness will be obtained by moving the diaphragm back and forth several times between the star and sky.

In spite of these precautions, however, variations in sky brightness and sky transparency are a limiting factor in achieving high accuracy or in measuring a very faint star against a very large sky background. In similar situations in other fields of physical measurements, such as radio astronomy, the standard practice has been to go to a.c. methods. A small differential on the top of a slowly varying background can thereby be sorted out and accurately measured. These schemes have found only limited application in astronomical photometry.

WALRAVEN[1] has described a comparison photometer for variable stars in which the light of the two stars passes alternately to the cathode of a single phototube through an oscillating diaphragm mechanism in front of a large Fabry lens. After amplification of the signal, rectification by means of a contact on the magnetic vibrator which shifts the diaphragm holes converts the signal to a d.c. signal which is used to drive a servomotor attached to a neutral wedge in the path of the brighter star. The number of rotations of this motor required to produce balance is transferred to a moving pen on a strip-chart recorder to give a logarithmic indication of the difference of brightness in the two stars. Oscillation frequencies between 20 and 50 cycles are used. The accuracy was shown to be consistent with the expected photon noise associated with the bandwidth employed and approximately 2.5 times better than results on the same stars by the previous method of shifting the telescope alternately from one star to the other.

Alternating current methods are also well suited to the measurement of polarization of starlight. HALL and MIKESELL[2] have described an a.c. amplifier and synchronous rectifier system operating on a carrier frequency of 30 cycles per second. The alternating current of the signal is introduced by rotating a polarizing prism in the star beam, and the synchronizing signal for the rectifier is derived by a commutator on a shaft rotating at the same speed as the prism. The details

[1] Ref. [5], p. 114.
[2] J. S. HALL and A. H. MIKESELL: Publ. U.S. Naval Obs. **17**, Part 1 (1950).

of the optical and electrical arrangements are shown in Fig. 21. The rectification of the signal is done electronically by the suppressor grids of a push-pull pair of pentodes in the final step of the amplifier.

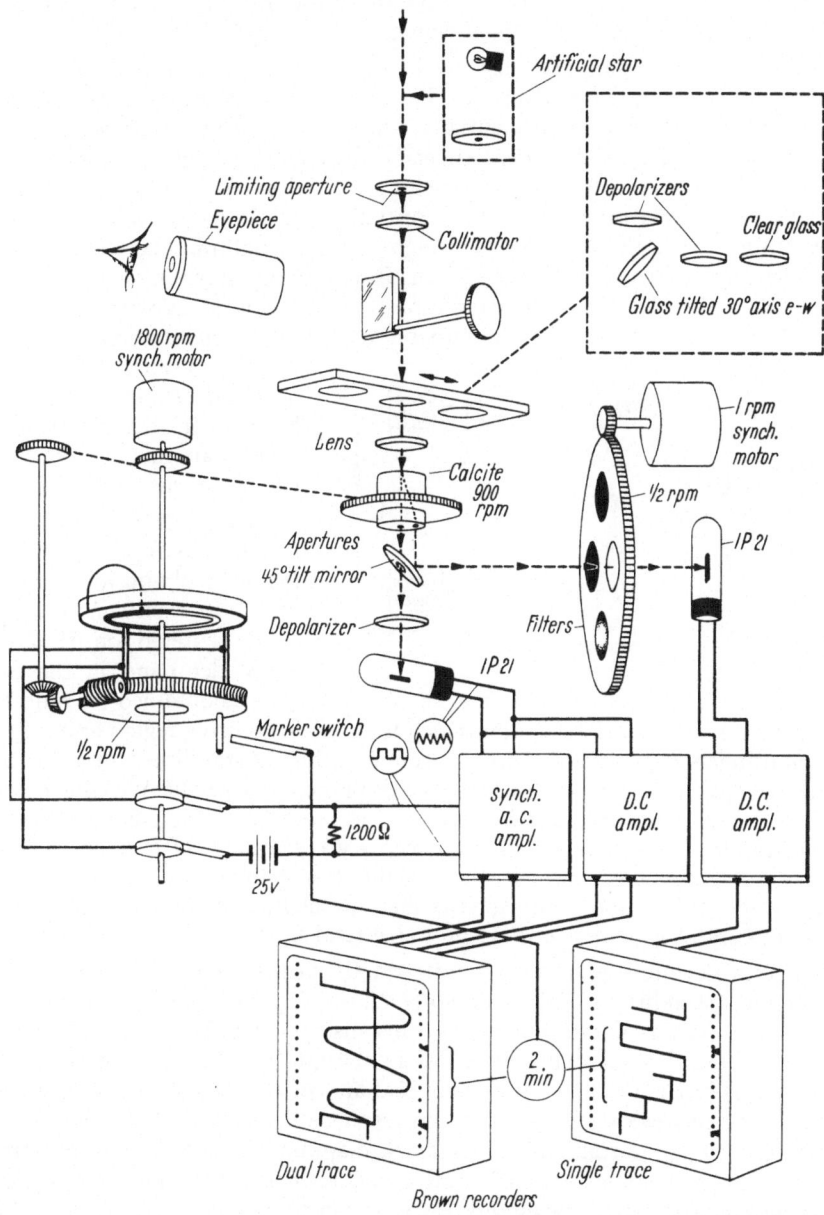

Fig. 21. Photometer for measuring polarization of starlight by an a.c. method. The phase of the reference signal for the synchronous rectifier is advanced by one cycle in a two-minute period; the degree and position angle of the polarization are then indicated by the amplitude and phase of a sine wave on the recorder. A parallel d.c. amplifier records the average intensity in light of the same color used for the polarization measurement. The extraordinary ray from the rotating calcite prism is used for simultaneous measurement of color through standard filters and a second d.c. amplifier.

Schemes such as the foregoing have the advantage that there is only one cathode and one amplifier, so that fatigue of the cathode, variations in supply voltage and amplifier gain affect the two fields being compared alike. The two

optical paths to the cathode that are employed in a two-star or a star-sky photometer must be frequently checked for constancy of alignment and constant transmission factors. The electronic parts of the differential amplifier and synchronous
rectifier may be made sufficiently stable and noise-free so as not to add to the
photon noise present in the original signal.

If the problem of gain stability can be satisfactorily solved, multiple channel
photometry has an obvious advantage, because it registers two or more quantities
simultaneously rather than sequentially, thus making a more efficient use of a
given amount of telescope time. An early demonstration of the possibility of
this method was given by Hiltner[1] who achieved extremely high accuracy in
a simultaneous measurement of two mutually perpendicular polarized beams
derived from the light of a star. Through a differential amplifier following the
multiplier phototubes two polarized beams from a bright star were compared with
a probable error of ± 0.00016 magnitude, a value consistent with the photon
noise. Such accuracy would be impossible in sequential measurements because
of the residual scintillation
noise. Since, however, the two
beams were derived from the
same star, this component of
the noise was cancelled.

Dual-channel filter photometry was used by Craw-
ford[2] to obtain a comparison
of the intensity in a narrow
filter centered on H β with
a broader band in the same

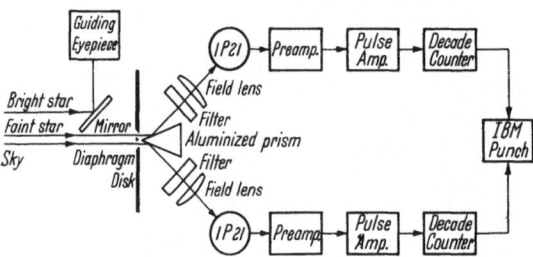

Fig. 22. Block diagram of Johnson's two-channel pulse-counting
photometer.

part of the spectrum. In one scheme, the broad band was obtained by reflection
from the back of a multilayer filter which isolated the narrow region around H β.
Both multipliers were operated from a common, highly stabilized voltage supply
and the output was read on an integrating photometer of the Weitbrecht type
which had high gain stability because of the large feedback factor. As in all
photometry the consistency of the apparatus was checked by frequent recourse
to standard stars; here, however, the variable transparency of the sky was cancelled by the differential nature of the measurement, and the check mainly involved the stability of the optical and electrical systems.

For a two-channel photometer designed to measure very faint stars under
seeing-limited conditions H.L. Johnson[3] chose a pulse-counting technique to
measure the output of the multipliers. Because the number of pulses counted
is not very sensitive to small variations in the multiplier supply voltage, there
was less concern about the relative gain stability of the two channels than there
would be for two integrators, where the response depends on both the number
and the size of the output pulses. A block diagram of Johnson's apparatus is
given in Fig. 22. In order to facilitate the machine computation of data, the output of the counters was registered directly on punch cards.

The advantages of simultaneous registration of star and sky are apparent
under the conditions illustrated in the example of Sect. 7γ, where the light of the star
amounted to only 2% of the light of the sky. If the sky were to vary by as much
as 1% during the typical alternation period of two minutes between star and sky,

[1] W.A. Hiltner: Observatory **71**, 234 (1951).
[2] D. L. Crawford: Astrophys. J. **128**, 185 (1958).
[3] H.L. Johnson: Sky and Telescope **17**, 558 (1958).

a fluctuating component equal to 50% of the light of the star would be introduced. The signal-to-noise ratio is also increased by a factor of $\sqrt{2}$ in a two-channel system because twice as many electrons are counted in the same observing period.

V. Photoelectric spectrophotometry.

16. Multicolor filter photometry. Determination of colors of stars by photo-electric methods has an advantage beyond that of the linearity and precision that go with all photoelectric methods. The two filter measurements are made on the same detector in quick succession and do not depend upon differences between separately determined magnitude scales made at different times. A probable error of the order of 0.005 to 0.007 magnitude is common for a single determination and is somewhat better than the magnitude accuracy which depends upon intercomparison of stars some distances apart in the sky.

Although a star color based upon two suitably chosen color filters is adequate uniquely to determine the color temperature of a black or grey body (assuming that an absolute determination has been made for one or more standard objects), it is of interest to consider what information can be derived by additional filter bands that further subdivide the total range of the spectral response of a cathode. Particularly with cesium-oxide cathodes ($S1$) a range far wider than is ordinarily covered by photography can be surveyed at a single setting of the telescope on a given star.

In the 6-color photometry of STEBBINS and WHITFORD[1] the filter bands were centered from 3530 to 10300 Å. This extended range is particularly valuable in studying the features of the energy distribution of objects which cannot be associated with the intrinsic properties of any one star. For example, interstellar absorption, over a short range, is indistinguishable from temperature reddening. Over the extended 6-color range the differential effects of interstellar absorption are shown to be quite different in the green-red-infrared and in the violet-ultraviolet. Similarly composite objects made up of a mixture of hot and cool stars, such as galaxies and globular clusters, can be matched to a single star over a short range, but over the extended range show either excess infrared or ultraviolet, and thus give clues to the makeup of the stellar population. Compositeness in unresolved pairs of stars[2] of differing temperature may likewise be gauged.

If the intrinsic energy curves of the stars were smooth and contained no irregularities of a scale smaller than the width of the filter bands, then multicolor comparison of any star with one which has been calibrated on an absolute basis would provide a rapid and reliable method of determining energy distribution. The method could be extended to very faint objects because the filter bands are wide enough to admit considerable light. Actually, as is well known, the energy spectra are distorted by band absorptions, by strong line absorption, by blanketing by numerous faint lines, and by absorption edges such as the Balmer discontinuity. When these features are kept in mind, comparative multicolor measures convey a great deal of information. Temperature effects usually override the irregularities. But as demonstrated by CODE[3] neglect of the near-discontinuity in the deep violet part of the spectrum of K stars, resulting from a concentration of metallic lines, can be quite misleading in deriving the absolute energy curve of galaxies. This group of lines is sensitive to the metal content of the stars, and a correction derived for one source may be applied to another only with caution.

---

[1] J. STEBBINS and A. E. WHITFORD: Astrophys. J. **98**, 20 (1943); **102**, 318 (1945); **108**, 413 (1948).

[2] J. D. R. BAHNG: Astrophys. J. **128**, 572 (1958).

[3] A. D. CODE: Publ. Astronom. Soc. Pacific **71**, 118 (1959).

Photoelectrically determined energy curves are best undertaken by the scanning method of the next section.

Multicolor filter combinations have had their greatest astronomical usefulness when they have been designed to exploit the absorption features and to derive information from them. The 3-color UBV system of JOHNSON and MORGAN[1] uses the Balmer discontinuity as such a feature, and also the previously mentioned smaller slope of the interstellar reddening curve in the ultraviolet, compared to the visible. Thus main sequence stars, giants, and supergiants may be distinguished, and so may reddened and unreddened stars.

STRÖMGREN[2] has shown that interference filters 30 to 100 Å wide can, if located over strategic features of the spectrum, gauge the absolute magnitude of a star from the strength of the H β absorption, measure the Balmer discontinuity, and determine a metallic index from the blanketing in the deep violet. Other narrow-band filters at selected parts of the continuum give a temperature index and provide symmetrical points about each "sensitive" filter that make the measured quantity nearly independent of space reddening.

A very efficient method of isolating the spectral bands was used by T. and J.H. WALRAVEN[3]. Successive orders of interference maxima in a polarizing monochromator were made to coincide with astrophysically important bands in early type stars. These bands were then separated by two prisms in series and fed simultaneously to five multiplier phototubes followed by five Weitbrecht integrators. The parallel channels not only saved time but resulted in higher accuracy because of cancellation of sky fluctuations that might have affected the channels differently had the observations been taken sequentially. Not only was the peak transmission higher than that ordinarily obtained with interference filters, but the effective bandwidth was two to three times greater. This permitted reaching fainter stars with ease, but did not seriously affect the astrophysical effectiveness of the bands in deriving the information about the luminosity, temperature, and space reddening of the stars.

17. Scanning of spectra. In the infrared beyond 1.1 μ, where photographic plates are extremely slow and there is no other high-resolution image forming detector, the delineation of spectrum detail has depended on sequential scanning in a system where only one resolved element of the spectrum is registered at a time. The amount of energy available from the stars and the noise equivalent input of the available detectors has not permitted examination of the spectra of the stars with any but the lowest resolution. KUIPER, WILSON, and CASHMAN[4] scanned the spectra of planets and bright stars out to 2.5 μ using a lead sulfide cell as the detector. The resolving power was approximately 80. The results on the stars were of value in showing the energy distribution curve but only the broadest spectral lines could be observed with the low resolving power. MOHLER, PIERCE, McMATH, and GOLDBERG[5] scanned the spectrum of the Sun from 0.84 to 2.52 μ with a resolving power of 30000, using photoconductive cells. In order to achieve this resolution it was necessary to devote considerable attention to a grating rotation mechanism that could move smoothly and uniformly at the rate of only a few seconds of arc per minute.

[1] H.L. JOHNSON and W.W. MORGAN: Astrophys. J. **117**, 313 (1953).
[2] B. STRÖMGREN: Stellar Populations, p. 385. New York: Interscience Publishers 1958.
[3] T. and J.H. WALRAVEN: Bull. Astronom. Inst. Netherl. **15**, 67 (1960).
[4] G.P. KUIPER, W. WILSON and R. J. CASHMAN: Astrophys. J. **106**, 243 (1947).
[5] O.C. MOHLER, A.K. PIERCE, R.R. McMATH and L. GOLDBERG: Photometric Atlas of the Near Infra-red Solar Spectrum. Ann Arbor: Univ. of Michigan 1950.

In most parts of the spectrum, where photography is a successful method for simultaneously recording many thousands of details, photoelectric recording of the entire length of the spectrum is so wasteful of telescope time that it has not been attempted. At the rather moderate resolving powers from 300 to 400 it does have advantages, however, in portraying the energy curve of stars and galaxies. The observing times are not prohibitive. With an antimony-cesium cathode the region from 3400 to 6000 Å may be scanned with a resolving power of 10 Å in about 15 minutes with a dwell-time of 3 or 4 seconds on each element of the spectrum. Although the weaker spectrum lines are unresolved, the broader bands and stronger lines are clearly shown upon the trace and form a much better basis for the study of the relative and absolute energy curve of the stars than the heterochromatic intensity measurements derived by multicolor measurements through filters. CODE[1], and OKE and BONSACK[2] have published examples of scans of stars showing resolution of about this order.

The design of scanning spectrographs is affected by the degree of resolution that can be obtained if the entire seeing disk of the star is admitted through a broad entrance slit, as was done in the examples just cited. An examination of the optical parameters shows that the critical factors are the diameter of the collimator and the angular dispersion produced by the grating or prism.

The angular width β of the slit or diaphragm at the entrance of the spectrograph, seen at the collimator, is $\alpha F/f$, where α is the angular width of the same slit seen from the objective of the telescope, and F and f are the focal lengths of objective and collimator. Since the focal lengths are in the same ratio as the diameters D and d of the objective and collimator, $\beta = \alpha D/d$. Then if $d\beta/d\lambda$ is the angular dispersion of the dispersing element (grating or prism), and $\varDelta\lambda$ is the desired resolution in the camera section of the spectrograph, the maximum angular diameter on the sky that can be admitted at the entrance slit is

$$\alpha_{\max} = \frac{d}{D}\beta = \frac{d}{D}\frac{d\beta}{d\lambda}\cdot\varDelta\lambda. \tag{17.1}$$

Since the exit slit can be adjusted in proportion to the focal length of the camera, the design of this part of the scanning spectrograph is simply a matter of choosing a convenient scale. If the spectrograph, considered as a monochromator, has equal entrance and exit slits, expressed in wavelength units, there will be a small loss of resolving power. If the star image is appreciably smaller in wavelength range than the exit slit, the latter will control the resolving power; if it is larger the exit slit can be widened to collect more light without loss of resolution.

For a diffraction grating of grating space ε as the dispersing element, β becomes a small increment in the angle of incidence i, measured relative to the grating normal, and the angular dispersion in the n th order on the collimator side is

$$\frac{d\beta}{d\lambda} = \frac{di}{d\lambda} = \frac{n}{\varepsilon\cos i}. \tag{17.2}$$

Substituting in (17.1),

$$\alpha_{\max} = \frac{n\,d}{\varepsilon\,D\cos i}\varDelta\lambda. \tag{17.3}$$

In the second order of a grating of 600 lines per mm, with a collimator such that $d/D = \frac{1}{30}$, we have, for $i = 30°$ and $\varDelta\lambda = 10$ Å $= 10^{-7}$ cm,

$$\alpha_{\max} = 4.6 \times 10^{-5} \text{ radians} = 9.''5.$$

---

[1] A.D. CODE: Proc. Conf. on Stellar Atmospheres. Indiana Univ., Bloomington, 1954.
[2] J.B. OKE and S.J. BONSACK: Astrophys. J. 132, 417 (1960).

Since this angle is larger than the seeing disk of the star in any but the poorest seeing, a circular aperture of 9″.5 in the focal plane of the telescope will not damage the resolution as determined by the exit slit, and will provide an adequate reserve for fringe illumination around the core of the image, as mentioned in Sect. 7γ.

The usual method of scanning is rotation of the grating mount with fixed entrance and exit apertures. A tangent arm[1] pushed by the tip of a micrometer screw can be made to give smooth and precise rotation and is simpler than a worm wheel. If the tangent arm is perpendicular to the screw axis for the grating position at which the zero order image is in the exit slit, the wavelengths in the exit slit will be exactly proportional to the number of rotations of the tangent screw, and direct-reading wavelength dials may be devised.

Even with the narrow entrance slit needed to give the high resolving power of a large coudé spectrograph, enough of the star image comes through for detailed scanning of the spectrum of bright stars[2]. To keep the observing time within reasonable limits only isolated single features would be covered. The problem of the smooth rotation of the grating at extremely low angular rates can therefore be avoided by moving the photoelectric receiver and exit slit together on guides in the plane of the plateholder.

A much more serious problem comes from the highly variable fraction of the star's light coming through the entrance slit as a result of seeing variations and imperfect guiding. A second monitoring channel taking its light from a fairly broad part of the spectrum near the feature being scanned should in principle provide the information for compensating out the fluctuations. Hiltner and Code[3] have described a method for taking the ratio automatically. The output of the amplifier following the monitoring multiplier phototube is brought to a level where it can substitute for the battery furnishing the slide-wire current of the recording potentiometer which receives the scanner output as its input signal. Rogerson, Spitzer, and Bahng[4] used parallel-channel pulse counting. Color dispersion in the seeing waves which cross the objective, motion of the star's atmospheric refraction spectrum across the slit, and uneven sensitivity over the cathodes of the two multipliers all combine to make complete compensation difficult to achieve. There is an advantage in deriving the light for the monitoring channel from two samples of the continuum symmetrically placed on either side of the spectrum feature being scanned.

High resolution scanning makes very inefficient use of the available photons. Rogerson et al. estimated that only one photon in 5000 incident on the top of the atmosphere produced a counted pulse. Ring and Woolf[5] have emphasized the value of the pressure-scanning Fabry-Pérot interferometer for high-resolution studies of single spectrum features. Because of its extremely high dispersive power, the interferometer can achieve the desired resolution with a wide entrance aperture admitting virtually the whole star image. The requirements of the rather powerful accompanying spectrograph needed to separate adjacent orders limit the aperture, and seeing compensation is still required. Woolf[6] has reported a tenfold gain in efficiency over the experience of Rogerson et al.

[1] W. G. Fastie: J. Opt. Soc. Amer. **42**, 641 (1952).
[2] T. Dunham jr.: Vistas in Astronomy, Vol. 2, p. 1223. London: Pergamon 1956.
[3] W. A. Hiltner and A. D. Code: J. Opt. Soc. Amer. **49**, 149 (1950).
[4] J. B. Rogerson, L. Spitzer and J. D. Bahng: Astrophys. J. **130**, 971 (1959).
[5] J. Ring and N. J. Woolf: J. Phys. Radium **19**, 354 (1958).
[6] N. J. Woolf, in: Annual Report Lick Obs., Astronom. J. **66**, 450 (1961).

VI. Electron image tubes.

18. Statement of the problem. The multichannel photometry discussed in Sect. 15 is the first step in overcoming the basic handicap of traditional photo-electric photometry mentioned in the introduction, namely, that each cathode and associated amplifier unit can at one time record only one optically selected element of the entire star field. Parallel channels with their obvious advantages of strict linearity and photon-limited operation could profitably be used to record simultaneously quite a number of carefully selected spectrum elements, chosen with foreknowledge of the effects to be gauged. Such a system would of course fall far short of the information capacity of a photographic plate. What is needed is a combination of the image-preserving properties of a television camera tube and the wide-range linearity and freedom from instrumental background found in a single channel multiplier phototube. A television camera tube can success-fully portray about 360000 picture elements in a given scene, and maintain proportionality over a brightness range of about 100 times. Although the high quantum efficiency of the cathode enables the tube to operate at light levels well below the minimum for cinematography with the fastest film[1], the design of the tube as ordinarily used and the inherent noise level prevent even approach-ing the scene brightness encountered in astronomical photography. The average night sky has a scene brightness of 3×10^{-8} stilb[2] or 10^{-5} foot-lambert. This is about 500 times fainter than the limit of an image orthicon operating at 30 frames per second.

A successful image tube that could be put to routine use at the tasks now assigned to astronomical photography would give two immediate advantages. The higher quantum efficiency of the photoelectric cathode (Sect. 7α) would shorten the exposure time needed to gain the same information by a factor be-tween 10 and 100. On the very long photographic exposures required for spectra of faint external galaxies the time gain would be somewhat larger because of the reciprocity failure of the plate. Thus spectra now barely possible could be ob-tained in great numbers, or much finer spectrum detail could be realized in the same exposure time. A more important result would be the ability to detect fainter star images. As pointed out by BAUM[3] the saturation properties of the photographic plate limit the number of photons that are effective in producing a threshold image; further exposure adds to the background and image in such a way that there is no improvement in contrast. The decision as to the presence of a threshold image is a statistical one. ROSE[4] has shown that the eye requires an addition of approximately 5 times the root-mean-square fluctuation in the number of sky-blackened grains normally present in the area occupied by the see-ing disk of a star in order to be certain of a threshold image. If these 100-odd background grains were to become 5000 recorded events in an image tube because of the 50-fold higher efficiency of the cathode, the level of the threshold image would sink from $5 \sqrt{100}$ additional grains, or $\frac{1}{2}$ the background light, to $\frac{1}{14}$ the background light. The gain would be $\sqrt{50}$ or 2.1 magnitudes. An equal improve-ment in the ability to detect threshold images in the same exposure time by photo-graphy alone would require a seven-fold increase in the diameter of a telescope.

A non-saturating image tube would be able to realize a further gain over a saturated exposure on the photographic plate, simply by lengthening the exposure

---

[1] A. ROSE: Adv. Electronics **1**, 131 (1948).

[2] C. W. ALLEN: Astrophysical Quantities, p. 125. London: Athlone Press 1955.

[3] W. A. BAUM: Trans. Internat. Astronom. Union **9**, 681 (1957).

[4] Ref. [1], p. 283.

time. Such a gain cannot be considered as real, however, since a similar gain is possible with photographic plates by increasing the focal length of the telescope. The greater linear diameter of the seeing disk would then include a larger number of plate grains in the decision area, and the saturation exposure would be proportionately longer.

An electron image tube to be used in astronomical applications should have the following properties:

1. Each electron leaving the cathode on which the scene is imaged must produce a recordable event. The event may be sensitization of one or more developable grains on a photographic plate, or a flash on a phosphor screen bright enough to be photographed, or a modulation in a scanning beam large enough to stand out above amplifier noise.

2. The number of spurious events must be low enough so that the background over a period of one hour or more is a small fraction of the signal produced by the light in the image on the cathode.

3. The sharpness of the resolution in the output should not be appreciably poorer than that of fast photographic plates, about 25 microns. The image on the initial cathode may be enlarged optically to compensate for poor resolution, but the available field of view will be correspondingly reduced, and the requirements on the frequency of spurious events must be tightened because of the lower surface brightness in the image.

4. The storage capacity available to each picture element must be sufficient to contain enough recorded events to give adequate statistical accuracy before saturation sets in. In order to achieve a considerable range of linear operation the storage capacity must not saturate for the highest surface brightness encountered. Multiple exposures or frequent data collection may be necessary to overcome this limitation.

At the time of writing, image tubes have not become a routine piece of astronomical equipment, although there have been some successes and some promising developments. In view of the changes that are likely to be brought about by rapid development in the years just ahead, only a limited description of the basic problems will be given[1].

19. Devices employing electronography. When an electron strikes a photographic emulsion with a velocity of some tens of kilovolts, one or more developable plate grains are inevitably sensitized. This direct process, called electronography, is used in the electron microscope. The electronic camera developed by LALLEMAND[2] and his colleagues utilizes this method in the simplest possible form. The electron emission from the cathode is imaged on a photographic plate at the other end of the same evacuated chamber by an electrostatic lens system. A diagram of the essential parts of the camera is shown in Fig. 23. The photographic plates coated with a fine-grain nuclear emulsion are loaded in a magazine in the lower section and cooled by liquid nitrogen. The cathode, previously prepared and sealed in an evacuated glass capsule, is stored in a side tube. All other parts of the tube can be baked during the evacuation process. When the exposure cycle is about to begin, the capsule is broken and the metal ring bearing the glass-supported semitransparent cathode is drawn into the upper end of the electron optical system. The cathode is kept cold by a conducting copper rod reaching into another liquid nitrogen flask. The magnification of the electron optical system is 0.7. The cathode is approximately 20 mm in diameter. A total potential of the

---

[1] See J. D. MCGEE: Reports on Progr. Phys. **24**, 167 (1961), for a detailed review of instrumental problems and techniques.

[2] A. LALLEMAND: Trans. Internat. Astronom. Union **9**, 673 (1957).

order of 25 kilovolts, furnished by an electrostatic generator, is used to accelerate the electrons through the lens system.

The resolution of the electron optical system is adequate to show 70 line pairs[1] per mm, equivalent to about 10 microns on the cathode. With Ilford G-5 plates, however, which are much faster than the maximum resolution plates needed to record such fine detail, the resolution is poorer, but nevertheless better than that of an ordinary fast photographic plate with medium grain. The plate shows no inertia and the density is proportional to the intensity. Photometry is thus straightforward and calibration spots are unnecessary.

During its use on the telescope the tube is valved off and carried away from the vapor pumps used in achieving the initial vacuum. The photographic plates, even though cooled, slowly release gases which poison the cathode and the system once prepared, cannot be maintained in an operable condition for more than a day or two. A titanium ion pump attached to the sealed-off tube is necessary for preserving the cathode for more than a few hours. The accompanying improvement in the vacuum is helpful in reducing the background, and exposures as long as 4 hours are possible[2].

The Lallemand electronic camera comes very close to meeting the specifications set

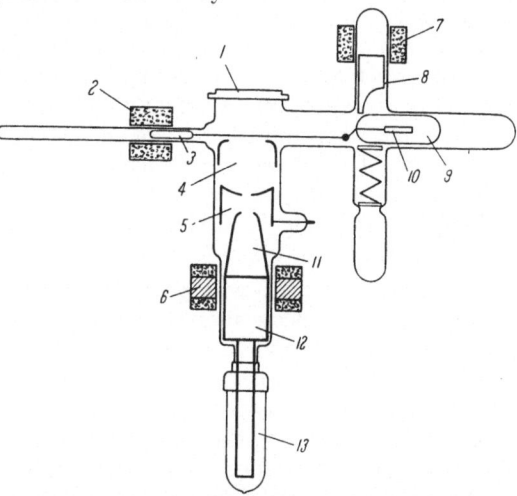

Fig. 23. Electronic camera developed by A. LALLEMAND. *1* Entrance window; *2, 3* solenoid and iron plunger for moving cathode into position; *4, 5, 11* cathode support and electrostatic lens system; *6* electro-magnet for changing plates in magazine; *7, 8* solenoid and magnetic hammer for breaking cathode ampoule; *9, 10* cathode mount and evacuated thin glass storage ampoule; *12, 13* plate magazine and liquid nitrogen flask for cooling magazine by conduction.

forth in the preceding section. It has been used successfully for recording the spectra of stars, nebulae, and galaxies at the Haute Provence Observatory[3] and at the Lick Observatory[4]. One factor which has limited more widespread use has been the extensive laboratory preparation required prior to each night's observing at the telescope. The cathode is lost when air is admitted to the tube to remove the exposed plates. The tube must be carefully cleaned, fitted with a new cathode ampoule, loaded with a new magazine of plates, and then pumped and out-gassed before the next set of exposures.

Two modifications have been attempted as a means of preserving the cathode for an indefinite number of exposures. In one[5] a thin membrane only 0.1 μ thick separates the cathode and the electrostatic lens system from the photographic plate. Electrons of sufficient energy penetrate the membrane, which, if properly prepared, is impervious to gas. In order to avoid placing full atmospheric pressure

[1] A. LALLEMAND, M. DUCHESNE and G. WLÉRICK: Adv. Electronics and Electron Physics **12**, 5 (1960).

[2] M. F. WALKER, in: Annual Report Lick Observatory. Astronom. J. **65**, 531 (1960).

[3] A. LALLEMAND, M. DUCHESNE, G. WLÉRICK, R. AUGARDE et M. DUPRÉ: Ann. d'Astrophys. **23**, 320 (1960).

[4] A. LALLEMAND, M. DUCHESNE and M. F. WALKER: Publ. Astronom. Soc. Pacific **72**, 76, 268 (1960).

[5] W. A. HILTNER and P. PESCH: Adv. Electronics and Electron Physics **12**, 17 (1960). — W. A. HILTNER and W. F. NIKLAS: Astronom. J. **66**, 286 (1960).

on the fragile membrane, a rough vacuum is maintained on the plate chamber side at all times, and the plate introduced through a vacuum lock. Membranes of aluminum on a plastic substratum, and of aluminum oxide have been used. The latter is stronger and withstands contact with the emulsion. Scattering of electrons in the membrane does not then cause an appreciable deterioration of the resolving power. The background level, originally a serious problem, can be made satisfactorily low for astronomical use. The resolution is 40 line pairs per mm.

McGEE and WHEELER[1] have developed a tube with a curved mica Lenard window which withstands atmospheric pressure. The mica is 10 μ thick and up to 1 inch in diameter. The resolution on the emulsion, which is placed in contact with the window, is 30 line pairs per mm.

A second method of protecting the cathode for many exposures has been developed by KRON and PAPIASHVILI[2]. A vacuum gate is introduced between the plate chamber and the part of the tube containing the cathode and electron lens system. It is closed during loading of the plate chamber and evacuation of that part of the tube, and opened only during exposures. The cathode is exposed to the gases evolving from the plate, as in the Lallemand camera, and similar precautions must be taken.

20. Image converters with phosphor screens. If the electrons from the cathode are focused by an electron optical system on a phosphor screen, the information is converted to a luminescent image which can be recorded by ordinary photography. This transfer is not efficient, since in the fastest lens or mirror system that may be used to image the phosphor on a photographic emulsion, only about 10% of the light leaving the phosphor screen can be captured. Single stage tubes of this sort have therefore not provided sufficient gain to be interesting.

A very direct way of improving the optical efficiency of the transfer to the photographic emulsion is to coat the phosphor on a thin mica end-window which has a photographic film in direct contact on the external surface. The first tubes of this sort were made by KRASSOVSKY[3]. They have been used successfully in a variety of investigations by the astronomers of the Sternberg Astronomical Institute of Moscow University. The resolution is of the order of 50 microns and the tube background does not come up until after an exposure of 10 hours. Semi-transparent cathodes of all three major types have been used, with a speed gain over photography of 20 times for antimony-cesium cathodes, and very large gains, amounting to several orders of magnitude, for infrared-sensitive cesium-oxide cathodes. Particularly in the region beyond 8500 Å where photographic plates are extremely slow, astronomical spectra not obtainable by other methods were recorded. Similar results have been obtained more recently at Lowell Observatory by HALL and BAUM, and by FREDERICK[4]. The resolution as observed with a microscope through the phosphor screen was 20 microns, but after diffusion in the mica window 20 microns thick the resolving power was reduced to 40 microns.

Two-stage cascaded image converter tubes capable of providing enough light output on the second phosphor screen for successful photography with a fast camera have been described by ZWORYKIN and RAMBERG[5] and by STOUDEN-HEIMER[6]. A cross section of a typical tube is shown in Fig. 24. The screen-photo-

[1] J.D. McGEE and B. WHEELER: To be published; see B. ZACHAROV and S. DOWDEN: Adv. Electronics and Electron Physics 12, 31 (1960).
[2] G.E. KRON and I.I. PAPIASHVILI: Publ. Astronom. Soc. Pacific 72, 353, 502 (1960).
[3] V.I. KRASSOVSKY: Trans. Internat. Astronom. Union 9, 693 (1957). — Draft Reports 11th Assembly Int. Astronom. Union, p. 38, 1961.
[4] Draft Reports 11th Assembly Int. Astronom. Union, p. 38, 1961.
[5] Ref. [4], Chap. 9.
[6] R.C. STOUDENHEIMER: Adv. Electronics and Electron Physics 12, 41 (1960).

cathode sandwich which transfers the picture information from the first section of the tube to the second is a membrane about 25 microns thick with a phosphor on the input side and a cathode on the output side. As a result of blurring in this interface the resolution is not as good as in a single stage image converter and a resolution of the order of 15 line pairs per mm is obtained on the output phosphor. The rather small field 10 mm in diameter on electrostatically focused tubes can be enlarged several fold by the development of magnetically focused tubes, with some improvement in resolution.

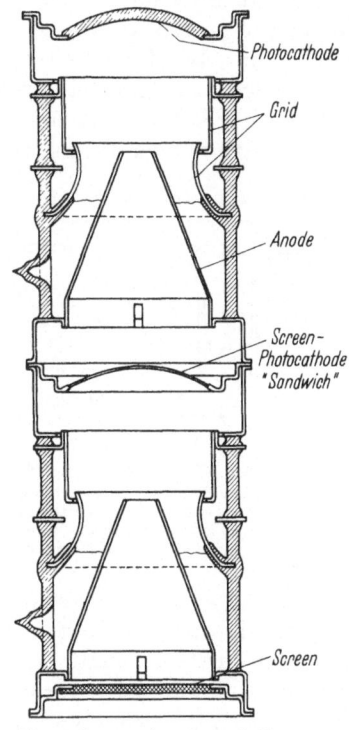

Fig. 24. Cross section of cascaded image tube.

It is possible to hold the background noise in the two-stage image converter to such a low level that it does not override the background light of the night sky focused on the initial cathode. When, however, the loss of resolution is considered and the input magnification adjusted to restore the resolution in the output, then there is little gain over the results to be obtained from ordinary photographs. This is not an inherent limitation and further development of the technique may succeed in bridging the rather narrow gap that remains in achieving photography of individual cathode events with a resolution and freedom from background ordinarily achieved by standard photographic methods.

In an alternative scheme, the brightness of the flash on the phosphor screen is increased by multiplier action following the cathode, so that a large number of electrons strike the phosphor for each photoelectron released; in simple electrostatic-lens tubes all of the energy for producing the flash comes from accelerating the photoelectron by a potential of 15 to 30 kilovolts. The multiplication takes place at a series of parallel dynode membranes. The dynodes consist of a membrane of aluminum oxide 0.1 μ thick, coated with aluminum for conductivity and a thin layer of potassium chloride as the secondary emitter. Focus through the five dynode membranes is maintained by an axial magnetic field and the resolution on the phosphor screen is of the order of 12 line pairs per mm. An overall light gain of many thousand times is possible.

Image converter tubes may be thought of as "image intensifier" devices which are inserted between the focal plane of the telescope or spectrograph and the final recording camera. They offer great attractions to the working astronomer, since the final information is recorded in a familiar way and no electronic complications are introduced. Before photometry can be attempted, however, careful mapping of the response of the cathode and the phosphor screen is necessary, in addition to the usual sensitometry of the plate.

21. Television camera tubes. The basic features of the image orthicon used in television cameras is shown in Fig. 25. The cathode emission is imaged on a storage target, where a picture element acquires a positive charge of 3 to 4 units per incident electron, owing to secondary emission multiplication. The target is scanned 30 times a second by an electron beam and the electrons that are taken from the beam to neutralize the charge in each picture element cause variations in the returning beam current which constitute the signal. Once the return beam has entered the secondary emission multiplier in the collector end of the tube, the amplification is sufficiently free from additional noise so that the information in the beam can be faithfully preserved.

Before such a device can be used to record a scene at astronomical light levels, the storage time must of course be considerably extended and the storage target scanned only as frequently as is needed to prevent saturation. The usual glass targets are made sufficiently conducting so the charge leaks away after $\frac{1}{30}$ second, in order to prevent persistence of a moving scene beyond one transit of the scanning beam. If refrigerated, these targets will store an image for much longer times, even as long as half an hour. Such refrigeration, of course, is of value in reducing the background emission of the cathode. Tubes with magnesium oxide targets have proven most successful in astronomical tests. These targets have a high resistivity and will store an image for many hours. Magnesium oxide also has a good secondary emission multiplication factor.

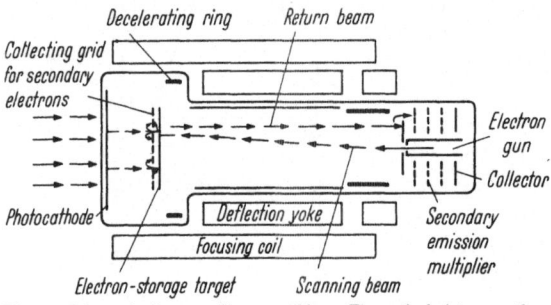

Fig. 25. Schematic diagram of image orthicon. The optical picture on the cathode becomes a charge pattern on the electron storage target. The stored information is read out periodically by means of an electron beam which scans the target line-by-line.

Image storage systems involving a scanning beam for readout have a basic limitation in the form of beam noise. If the target is to have the capacity of recording the intensity of a bright picture element, that is, say, 100 times that of the minimum detectable signal for a single scene, then a beam current large enough to neutralize this maximum stored charge will inevitably have a fluctuation level 10 times larger than that present in a beam barely adequate to record a threshold signal. For practical electron guns the clumpiness in electron beams is usually in excess of this unavoidable minimum. Consequently television type tubes have not been successful in portraying threshold images found by photography. But when the optical system can be so arranged that the surface brightnesses to be recorded are above the noise threshold set by beam noise, camera tubes have shown an enormous gain in speed over ordinary photography. The resolution capabilities are comparable with those of image converter tubes. Since the information in the electron beam is linearly related to the intensities in the elements of the cathode picture, the photometric capabilities of tubes with a storage target are superior to those of other types, where a photographic plate or a phosphor screen must be a part of the chain. The beam data may be stored on magnetic tape, rather than reassembled into a picture, and later processed by a linear analyzer.

General references.

[1] Smith, R.A., F.E. Jones and R.P. Chasmar: The Detection and Measurement of Infrared Radiation. Oxford 1957.

[2] Clark Jones, R.: Performance of Detectors for Visible and Infrared Radiation. Adv. Electronics 5, 2 (1953).

[3] Simon, H., u. R. Suhrmann, ed.: Der lichtelektrische Effekt und seine Anwendungen, 2. Aufl. Berlin-Göttingen-Heidelberg: Springer 1958.

[4] Zworykin, V.K., and E.G. Ramberg: Photoelectricity and its Application. New York: Wiley 1949.

[5] Wood, F.B., ed.: Astronomical Photoelectric Photometry. Washington: Am. Assoc. for Adv. of Science 1953.

[6] Elmore, W.C., and Matthew Sands: Electronics, Experimental Techniques. New York-Toronto-London: McGraw-Hill 1949.

Sachverzeichnis.

(Deutsch-Englisch.)

Bei gleicher Schreibweise in beiden Sprachen sind die Stichwörter nur einmal aufgeführt.

abbildende Faktoren, *image-forming factors* 144—146.

Aberration bei Antennensystemen, *aberration in aerial systems* 77.

abklingende Wellen, *evanescent waves* 65.

Abkühlung des Mondes, *cooling down of the Moon* 200, 203, 204.

Abplattung des Mars, *oblateness of Mars* 221.

Abschneidesatz für eine Antenne, *cut-off theorem for an aerial* 97.

Abschneidewellenlänge, *cut-off wavelength* 253.

Abstand, kleinster auflösbarer, *peculiar interval* 120, 121.

Abweichungen vom Reziprozitätsgesetz bei der Belichtung photographischer Platten für hohe Intensitäten, *reciprocity failure in exposure of photographic plates at high intensities* 137.

— — bei der Belichtung photographischer Platten für kleine Intensitäten, *in exposure of photographic plates at low intensities* 136, 137.

achromatischer Koronagraph, *achromatic coronagraph* 6.

Ähnlichkeitssatz, *similarity theorem* 66.

äquatoriale Aufstellung von Antennen, *equatorial mounting of aerials* 82, 88—89.

Äquatorialreflektor, *equatorial reflector* 9—11.

Äquatorialrefraktor, *equatorial refractor* 5.

Äquatorialteleskope, *equatorial telescopes* 5 bis 10.

Äquivalenzbreite, *equivalent width* 49.

Albedo einiger Asteroiden, *albedo of some asteroïds* 212.

— des Merkur, *of Mercury* 209.

Alkalimetalle, erste Verwendung in Photozellen, *alkali metals, first use in photoelectric cells* 242.

Alter der Krater, *age of craters* 195.

Ansprechwahrscheinlichkeiten verschiedener Kathoden für verschiedene Wellenlängen, *response functions of various cathodes* 246.

— für verschiedene Wellenlängen bei Farbphotometrie, *in color photometry* 153, 155, 156.

Ångströmsches Pyrheliometer, *Angström's pyrheliometer* 2—3.

Anordnungssatz, *array theorem* 66—67.

Antennen, Aufstellung und Antrieb, *aerials, mounting and driving* 82—87, 88—89.

— für Radiowellen, *for radio waves* 60—87.

—, wichtigste Typen, *main types* 68—75.

Antennenparameter, *aerial parameters* 61 bis 64.

Antennensätze, *aerial theorems* 66—67.

Antennenzuleitung, *transmission line* 75.

Antimon-Caesium-Kathode, *antimony-cesium cathode* 243, 244, 247.

Antrieb für Antennen, *driving of aerials* 86 bis 87.

Aperturausbeute, *aperture efficiency* 74, 120.

Aperturverteilung, *aperture distribution* 65.

Arsacsche Anordnung, *Arsac array* 117.

Atmosphäre der Jupitermonde, *atmosphere of the Jupiter satellites* 219, 220.

— des Mars, *of Mars* 227—228, 235.

— des Merkur, *of Mercury* 209.

Atmosphärenmodell von BEMPORAD, *Bemporad's model of atmosphere* 151.

— ebener Schichten, *plane-layer model of atmosphere* 151.

atmosphärische Extinktion, *atmospheric extinction* 149—154.

Auflösung von Sonnenspektroskopen, *resolving power of solar spectroscopes* 19.

Auflösungsvermögen photographischer Emulsionen, *resolving power of photographic emulsions* 132.

Aufstellung von Antennen, *mounting of aerials* 82—86, 88—89.

— eines Gitters, *of a grating* 20—21.

Aureolen um die großen Krater, *aureols around the big craters* 195—197.

ausgedehnte Radioquellen, *extended radio sources* 121—125.

Autokollimationsspektroskop, *autocollimating spectroscope* 21, 38.

Asteroide, Anzahl, *asteroïds, number* 211.

—, Durchmesser, *diameter* 211.

—, Helligkeitsschwankungen, *brightness variations* 211, 212.

—, Oberflächeneigenschaften, *surface properties* 211—215.

—, Photometrie, *photometry* 211.

Babcockscher Magnetograph, *Babcock's magnetograph* 32.

Subject Index.

(English-German.)

Where English and German spelling of a word is identical the German version is omitted.

Table des matières

pour la contribution écrite en français:

Audouin Dollfus: La nature de la surface des planètes et de la Lune.